RELATIVISTIC REALITY: A MODERN VIEW

SERIES ON KNOTS AND EVERYTHING

Editor-in-charge: Louis H. Kauffman

Published:

Vol. 1: Knots and Physics
L. H. Kauffman

Vol. 2: How Surfaces Intersect in Space
J. S. Carter

Vol. 3: Quantum Topology
edited by L. H. Kauffman & R. A. Baadhio

Vol. 4: Gauge Fields, Knots and Gravity
J. Baez & J. P. Muniain

Vol. 5: Gems, Computers and Attractors for 3-Manifolds
S. Lins

Vol. 6: Knots and Applications
edited by L. H. Kauffman

Vol. 7: Random Knotting and Linking
edited by K. C. Millett & D. W. Sumners

Vol. 8: Symmetric Bends: How to Join Two Lengths of Cord
R. E. Miles

Vol. 9: Combinatorial Physics
T. Bastin & C. W. Kilmister

Vol. 10: Nonstandard Logics and Nonstandard Metrics in Physics
W. M. Honig

Vol. 11: History and Science of Knots
edited by J. C. Turner & P. van de Griend

Vol. 12: Relativistic Reality: A Modern View
J. D. Edmonds, Jr.

Vol. 13: Entropic Spacetime Theory
J. Armel

Vol. 14: Diamond – A Paradox Logic
N. S. K. Hellerstein

Vol. 15: Lectures at Knots '96
edited by S. Suzuki

Vol. 16: Delta – A Paradox Logic
N. S. K. Hellerstein

K&E Series on Knots and Everything — Vol. 12

RELATIVISTIC REALITY: A MODERN VIEW

James D. Edmonds, Jr.
Department of Physics
McNeese State University
USA

World Scientific
Singapore • New Jersey • London • Hong Kong

Published by

World Scientific Publishing Co. Pte. Ltd.

P O Box 128, Farrer Road, Singapore 912805

USA office: Suite 1B, 1060 Main Street, River Edge, NJ 07661

UK office: 57 Shelton Street, Covent Garden, London WC2H 9HE

Library of Congress Cataloging-in-Publication Data

Edmonds, James D., 1939–

Relativistic reality : a modern view / James D. Edmonds, Jr.

p. cm. -- (K & E series on knots and everything ; vol. 12)

Includes bibliographical references and index.

ISBN 9810228511

1. Space and time. 2. Numbers, Complex. 3. Quaternions.

I. Title. II. Series.

QC173.59.S65E36 1996

530.1'1--dc20 96-31700

CIP

British Library Cataloguing-in-Publication Data

A catalogue record for this book is available from the British Library.

This book is printed on acid-free paper.

Printed in Singapore by Uto-Print

To Feynman

Preface

This book has its origins in my leaving applied physics in 1969 to teach at a small college and ponder the foundations of physics without pressure, even if with limited time caused by a heavy teaching load. Richard Feynman had lectured weekly during the two years I spent at Hughes Labs in Malibu, after graduating from Engineering Physics at Cornell. I had considered switching to theory at Cornell and was discouraged from doing so by visiting Hans Bethe, their ranking theorist. Feynman reawakened my love of and confidence in pursuing the foundations of physics. I felt, however unworthy it was, that he and I were kindred spirits. He searched his whole life for a new path into post QED physics and encouraged me to also do so, through his lectures.

That was a long time ago and much has happened to my career but I never gave up the quest. Now I am old and it is time to leave the insights I have gained to the next generation of mavericks to push forward. My most admired physicist is Dirac, for he seems more than anyone to have a creative spirit that pioneered so many new ideas on the foundations, even if he was a weird person. He would have loved this book I am sure, as would Feynman. I started out thinking that complex numbers fit non-relativistic quantum physics, then why not a larger number system for relativistic physics? I avoided the literature on such things fearing that I would wind up where Feynman was, and he had not found the answer. He showed us Maxwell's equations in Pauli matrices and I knew electrons obeyed laws in the extension to Dirac matrices, so I tried to generalize these number systems and see what equations I could find. I reinvented many wheels in the years that followed and wasted a lot of time doing so, but it kept me independent in outlook. That is perhaps why I have succeeded where others have failed, if I indeed have—with my weird ideas of multi-mass and extended Lorentz combined with absolute motion.

It has been a lonely vigil with no one really thinking I had the talent to do anything so basic that would be worth anything. I thank my old professor Jack Garrison and old friends Stan Klein and Larry Caroff for moral support over the years. In more recent years, I have gained some appreciation and encouragement from the physicists John A. Wheeler, Max Jammer, Egon Marx, Geoffrey Dixon, Kuni Imaeda, S. Chandrasekhar, Bill Honig, Fritz Rohrlich, and Sidney Drell. Akhlesh Lakhtakia has also been very helpful. I owe much to the pioneering mathematician/physicist Louis Kauffman for his efforts and advice in getting this book published by World Scientific. He even looked tolerably on Maheum, which I cannot leave out. The world needs that outlook a lot more than it needs a better understanding of quarks, which may follow from the rest of this book. My kids suffered a lot for this work too, as I was not there for them enough. I hope they can forgive me for that. Perhaps the world will make something wonderful of all this that will restore karma there. Both of my wives have been very understanding and they let me work hard all these past decades. Without their encouragement, I could not have finished all this. Some good better come of this work, considering the shared pain it has caused.

I wish to thank Michelle Boudreaux for her excellent typing; also Ms. H. M. Ho and her staff at World Scientific for their diligent efforts to bring this book to you, the reader. It is a bit silly in parts, but then so am I, so it is appropriate. I hope the young people can relate to it and find inspiration and guidance in it—on many fronts! Their future is rather bleak it appears to us old timers, who have gluttonously consumed all the resources of the planet. I hope there will be time and money for some of them to continue to pursue the study of the foundations of the world. It is not important, but so satisfying and awe inspiring. I look at the complicated mathematics of QED, which matches measurements so accurately, and it just overwhelms me. How can the world be so precise and mathematical? Why is it so and why does it all exist at all? These are the *important* questions but they cannot be answered. So we content ourselves with the search for more of the mathematical *details* of the design instead. It is a search for God in a sense, even though a hopeless one. We can never know the deep inner workings of the subnuclear world. But we can have some fun and wonder in the quest. To all the people who thought I was crazy to try such thinking over the years, may I say the last laugh is really great!

Lake Charles, Louisiana, February 1996

TABLE OF CONTENTS

Appendix

CHAPTER 1

Dimensions and Hypercomplex Numbers

We must remember who we are before we start to talk about space and distance. We were born, opened our eyes, and then we started our physics experiments with space. We had already done some experiments on time, in the womb. Babies gradually learn to judge distance and to stop reaching for things that look too far away. We experience space as a *three* dimensional continuum. But why only three dimensions? Well, that is just the way it is. To ask why is to ask why the universe exists at all, is it not? But we can certainly ask if there are other dimensions that we don't directly experience. That is an eternally open question, no matter how many we have found to exist at any one time, in our scientific evolution and theorizing.

There always remains the question of why the universe is stable at all, once we assume it really exists out there and generally evolves as time increases. Then there is the forward time evolution issue: "Why is *backward* in time so radically different for us, even in the womb?"

Some of these answers cannot ever be found. We must be very humble. "We don't know much and never will!" For example, suppose that what actually exists as *REAL* atomic physics is only brains in isolated capsules of real space. One brain per capsule. There is continuous and very clever creation of matter and continuous destruction of matter at the edges of the capsules. Blood and nutrients are created and pushed into the arteries. The existing blood and wastes emerge from the veins and are destroyed as the veins hit the 'edge of the world'. Such a brain really could think and could visualize space and time, though they don't really exist beyond the capsule's edges. It could even communicate with the other island brains by the rule that any output neuron signals, reaching the surface of its world, do disappear there, but they are recreated at the appropriate nerve endings of another capsule at its edge, by magic! Real magic. This way, the isolated brains could talk to each other, think they see each other, and think they are on a planet beneath them. They even would think they build and use telescopes to see the stars, etc. But nothing ever physically exists except the atoms of these island brains. The 'space' between capsules is meaningless. *This is no joke!* If I were God and *ONLY* interested in humans, e.g., in picking out the 'good' ones for 'heaven' later, and flushing the losers into 'hell', I might only make *this* much universe with exactly these rules!

A 'birth' here consists of a new capsule coming into existence with continuously created atoms forming an increasingly complex brain structure, just like a baby's brain normally develops and grows. Insidc these capsules, the physics laws would be essentially like our atomic view today of the world, with quantum physics actually dominating the atoms and their interactions. The one exception to our laws would be the continuous creation and destruction of atoms only at the boundary of these little worlds. But even that is not so strange for us. We could imagine going out into space and hitting two baseballs together, both moving at very very close to the speed of light. There is a finite chance we would produce two normal baseballs at rest and also produce (create) a new baseball and a new anti-matter baseball, all just sitting near one another in this empty space after the collision. Creation of atoms occurs here also in this example, on a massive scale! So creating atoms is no problem for us in principle. But we made anti-atoms in equal numbers here. That is the best we humanoids seem to be able to do, even in principle. God would not be limited by our humanoid restrictions!

Hawking imagines that black holes, made from ordinary atoms, will be spewing out electrons and anti-electrons until they disappear completely. The electrons and anti-electrons, so ejected, will eventually hit other electrons in the universe around the 'hole'. The anti-electrons could, in principle, hit electrons, over and over, and eventually a pair would hit slowly enough to cause them to completely annihilate each other into x-rays. Thus we could turn original atoms that made up the hole into x-rays and electrons, if this crazy idea is at all correct. That is not complete destruction of atoms into nothing, but mostly into light. Energy is conserved here.

The steady state universe theory was forced to consider continuous creation of atoms, really from nothing, to try to survive. Then, in 1965, the background, microwave, *Gamow* radiation, coming from far out in space, was discovered and the steady state theory could not cope with that, even with such a conjectured, continuous creation of atoms. The steady state theory only needed one atom of hydrogen or helium to be created in the volume of the Houston, Texas, Astrodome, every hundred years, to keep the universe the 'same', even though the clusters of galaxies separate, along with the general expansion of the universe. This expansion was discovered in the late 1920's by Hubble. The Big Bang looks to be real, but there you have a massive creation of all of the whole universe, all at once! Much from nothing is still there, one way or another; don't be smug.

No, we cannot rule out the isolated brains theory, nor could we confirm it to be really the way it is. We therefore, instead, simply GUESS that the world we experience as babies really exists around us, and that we really move through that space as time goes forward. We remember only the past, not the future, and something in the basic rules builds in this bias with respect to time. We experience a marvelous stability around us. The rules of the physics universe seem to be not changing with time at all! That means we can plan ahead and prepare ahead and thus arrange to live much longer and healthier lives. (The 'average' cave man only lived 18 years instead of 68 or more as today.) Preparing and planning have certainly benefitted our lives, at least in the physical sense. We have gathered so much astrophysical data that we can put stringent upper limits on how fast the laws of nature are changing. They have changed very little, if at all, over the past 15 billion years, since the Big Bang beginning.

So, where do we start? We GUESS that the world is out there and very stable and has *three* large dimensions of space. This we have all easily experienced, even people who are born blind. There may be many more, very small dimensions, and we keep an open mind about that. Skeptical, but open. How many dimensions for time then? Who knows? Time is too strange to really contemplate. We will never understand it. Our huge creature experience of it is one dimensional. One number is all that we need to practically deal with it. We can set up clocks all over the planet and they can all read one number at any instant. Then later they all read a larger number, *etc*. (We usually chop these numbers up with time zones, but that is not essential at all.) Time IS. It has 'moved' forward on the average from the birth of the universe until now, and will 'move' on until the death of the universe. The full span of 'existing' time appears to be a minimum of about 400 billion earth years, or so. The maximum span possible is perhaps an infinity of years. A year is of course an arbitrary unit of this time. We could use any larger or smaller increment.

Physicists have considered two dimensional time. Maybe there is one big dimension, lasting billions of years, and another little dimension like a closed loop, with total 'length' of less than, say, 10^{-15} seconds. Some physicists think 10^{-44} seconds is the smallest interval of meaningful time evolution, but that is only a guess! This is lots of fun, but totally speculative at present.

We seem to have 3 space and 1 time dimension. The two opposite directions of time are very different in a deep, physical way. The two opposite directions of space seem totally alike. Looking out in any direction, to far far away galaxies, one sees the same general pattern as looking far out in the opposite direction, or in any other direction, what-so-ever. This is a fact, and we know not why! It is nice though. No special direction in space but a bias in the direction of time. So we have at least three large dimensions of space and one large dimension of time. But why three? One dimension is sort of natural—something as opposed to nothing (like time). So why not just one space dimension also? A very good question. Any answers?

The best answer, to a lot of such basic questions, is to remember who and what we are. We are already here! We GUESSED that already. Space and time really exist, (we GUESSED also), and therefore atoms can rearrange their clusterings and they somehow make us be conscious. They created us and they sustain us, temporarily! Step on a bug and it ceases to be conscious but the atoms are all still there. Can you leave your body and look down on it? Do we need atoms to be conscious and exist? There is a can-of-worms! Most people say we don't need atoms. Then why the hell have atoms at all!?? Why not just exist as spirits from the very beginning? Atoms cause much pain and suffering in the world. People suffer disease, hunger, injury, and death.

If the atoms get slightly displaced and that consciousness cannot stay in those atoms, it ceases to exist (or instead maybe changes to some non-physics condition). The consciousness that is residing in the brain atoms of an individual exists over many decades, and our minds change in content and attitude. There is no good reason why all of these stages of our conscious being could not be **simultaneously** recreated in that spiritual state that is supposedly us after death. There is *no* conservation of souls, unless the Gods will it so by choice. You should be able, with your 40's soul, to talk to your 20's soul, in nirvana or wherever this recreated consciousness exists later. This would be just as if these two souls were two different souls, rather than different time slices of the same consciousness that evolved on.

So why not just one space dimension in our universe? Probably this is because atoms in one dimension cannot form complex enough structures to support our complex level of consciousness. In fact, two dimensions may not be enough. Even three alone may still not be enough. We know of the three large ones, remember, but maybe the little-bitsy ones are also essential! Then why not 4 big ones? Ahhh, a very good question. Why stop with just enough (if three big ones turn out to be just enough)? We can never answer that question.

There is a weird and wonderful mathematical 'thing' that may relate to the question, "Why not 4-dimensions of space?" This is the best I can do in answering this impossible question. Many theorems in math have special formulas in one and also in two dimensional applications. Yet, in three or more dimensions, only one pattern is found. A classic example is random walk. A drunk staggers randomly away from a lamp post. Each step is arbitrary in direction. As time goes to infinity, his average distance from the pole increases, but he has a finite chance of returning to the pole, over and over, if you wait long enough. Stepping randomly in three dimensions, as on the monkey bars of old playgrounds in the USA, the odds shift drastically. He 'never' returns to the center in 3, 4, 5, ···, or higher dimensions, on the average. Interesting, huh?

We have a twisted, double helix of DNA carrying the four 'letters' A,T,C,G of the code down this 'double rope', which is the length of a human being! Could this work in only two dimensions? Possibly. But the code makes long proteins along itself and these then fold up into complicated and specific three dimensional molecules called enzymes. Their specific shapes keep us alive! Maybe two dimensional enzymes just could not work well enough for complex life forms to exist. Two dimensional atoms possibly cannot sustain two dimensional conscious beings. Alright; not two; but then why three and not four? The answer may have been discovered by Hamilton in 1843.

Here is what happened, more or less. I am not an historian. Mathematicians had invented complex numbers, a + ib, with $i \equiv \sqrt{-1}$. They needed these weird new numbers to solve things like: $x^2 = -4$. The answer, x, is not on the usual number-line, so they invented i, with its own number line, and the answer is x = 2i. They had to invent negative numbers, so x + 3 = 2 has a solution. They chose (-1)(-1) ≡ +1 even though this messes up the 'balance' on the number line! Negative numbers are biased against! Today, we associate matter with positive numbers and anti-matter with negative numbers. There is a physical bias in the universe for more matter than anti-matter. At the 'beginning' of time (the Big Bang) there was almost a balance. For every 10^8 electrons there were $10^8 - 1$ anti-electrons, it seems. So all the anti-electrons got annihilated by hitting electrons and producing x-rays. The tiny percentage of electrons, left over, constitutes all the electrons in all the galaxies today! So this is a big deal choice: $(-1)(-1) = (+1)$. Why pick it? It makes other algebra properties nice, like $(ab)(c) = (a)(bc)$ and $a(b + c) = (ab) + (ac)$. These properties we use all the time in commerce calculations on our planet. Remember 'word problems'? This is 'business math' for sure, but is it physics math? It seems to be so. Nature really has some + − biases at the deepest levels!

Getting back to Hamilton: People could then describe two dimensional space locations with (x + iy) or with (x,y) in the usual grid way of graphing. But $(x + iy)(x + iy) = x^2 - y^2 + 2(xy)i$ and $(x + iy)(x - iy) = x^2 + y^2 - 0i$, so the distance to the point (x,y) from the point (0,0), could be obtained algebraically. They needed $(x + iy)^* \equiv (x - iy)$, called the complex conjugate, to easily get the 'length' needed, $x^2 + y^2$, using complex number <u>products</u>. It was also found that $[(x + iy)(z + iw)]^*[(x + iy)(z + iw)] = [(x + iy)^*(x + iy)][(z + iw)^*(z + iw)] > 0$. We say that the absolute value of a product of complex numbers is positive definite and is equal to the product of the absolute values. This is a very pretty property to our eyes. It looks like $|A \cdot B|^2 =$

$|A|^2 \cdot |B|^2 \dot{>} 0$, in compact form. So Hamilton simply wanted to generalize this, to deal with the real world, since we have three dimensions and need (x,y,z) sets of numbers to locate positions in a room (or in the galaxy). He struggled about 10 years and finally found a pretty, new pattern that worked. He carved the discovered rule into the railing of a foot bridge at the University of Dublin, Ireland: $XY = -Z = -YX$, or something like that. The new thing he had to allow was $XY = -YX$. These weird new numbers *anti*-commute! No one apparently thought of such objects before about 1843. They were not needed for algebra solutions. Hamilton was forced to them in order to keep a close parallel with the complex numbers. These same anti-commuting numbers play a central role in modern physics.

HAMILTON'S QUATERNIONS

The notation which Pauli used when he *reinvented* the quaternions in the 1920's has now become standard in physics, if not in mathematics, so we shall use it. We have a set of 8 objects, $\{\sigma_\mu, i\sigma_\mu\}$ where $\mu = 0,1,2,$ or 3 and i = $\sqrt{-1}$. The σ_0 object is the identity. The rules of multiplication are: $\sigma_1\sigma_2 \equiv +i\sigma_3 \equiv -\sigma_2\sigma_1$ and cyclic permutations. Also, $\sigma_1\sigma_1 \equiv \sigma_2\sigma_2 \equiv \sigma_3\sigma_3 \equiv \sigma_0$. Simple and elegant! We see Hamilton's original, four part quaternions as a 4-element subset, such as $\{\sigma_0, -i\sigma_k\}$, k = 1, 2, or 3. We will always include $\pm\sigma_\mu$, when we say σ_μ is in a set, and always assume $(-1)(-1) \equiv (+1)$ as usual. The $i\sigma_\mu$ terms are easily dealt with because $i\sigma_\mu \equiv \sigma_\mu i$. The i's can be 'collected' and then combined. For example: $i\sigma_1 i\sigma_2 = ii\sigma_1\sigma_2 = -\sigma_1\sigma_2 = -i\sigma_3$, *etc.* The pattern here is like that for 2×2 complex matrices; such matrices were the form that Pauli used to first describe electrons with spin, in the 1920's. They are <u>not</u> matrices, however. They are like i. We don't replace i by a matrix, although we could! Better to think of $A \equiv a^\mu\sigma_\mu + b^\mu i\sigma_\mu = (a^\mu + ib^\mu)\sigma_\mu \equiv c^\mu\sigma_\mu$ as just a 'hypercomplex' number. The repeated μ in a product means sum on μ = 0,1,2,3, so A has 8 distinct parts, where a^μ and b^μ are real numbers. For example: $A_1 = 3\sigma_0 - 2\sigma_1 + 0\sigma_2 + 0\sigma_3 + -\sqrt{2}i\sigma_0 + 0i\sigma_1 + 4i\sigma_2 - 7i\sigma_3$. We can easily describe our universe's space and time in the form

$$x \equiv ct\sigma_0 + x\sigma_1 + y\sigma_2 + z\sigma_3 \equiv x^\mu\sigma_\mu$$

or the momentum and energy of a moving mass as

$$P \equiv (E/c)\sigma_0 + P_x\sigma_1 + P_y\sigma_2 + P_z\sigma_3 \equiv P^\mu\sigma_\mu$$

There is a totally natural match between quaternions and 1+3 spacetime. (Hamilton even noticed this 1+3 match!) I would go so far as to say that our universe's spacetime is 'built' on quaternions, and that is why we have *three* large space dimensions and *one* large time dimension!! (Just a guess of course.) Four space dimensions don't fit, in a natural way, with quaternions, but see later chapters. We need to generalize the usual complex conjugation next. We have $(a + ib)^* \equiv a - ib$, or $i^* \equiv -i$. It is useful to think of this now as $(a\sigma_0 + bi\sigma_0)$ and we have $\sigma_0^* \equiv +\sigma_0$, $(i\sigma_0)^* \equiv -i\sigma_0$. Or, we could instead think of $(a\sigma_0 + bi\sigma_1)$ as representing complex numbers; therefore, we have $\sigma_0^* \equiv \sigma_0$ and $(i\sigma_1)^* \equiv -i\sigma_1$. Clearly, $\sigma_\mu^* \equiv +\sigma_\mu$ is natural. We then find that $(AB)^* = B^*A^*$, where A and B are any two complex quaternions with *eight* parts each. (Notice the order reversal on the right side.) This is beautiful and takes some patience to prove. (Grind out each side then compare, term by term.) The complex quaternions can be mathematically thought of as being 'generated' from a so-called 'direct product' of the 'pure' quaternions, $\{\sigma_0, -i\sigma_k\}$, with $\{\sigma_0, -i\sigma_1\}$. (Using $i\sigma_2$ or $i\sigma_3$ here, instead of $i\sigma_1$, would serve just as well.) But once we have the *complex* quaternions, we find another closed subset $\{\sigma_0, i\sigma_1, \sigma_2, \sigma_3\}$. Now we can form a new 'direct product' of this closed set with the 'pure' quaternion set, $\{\sigma_0, -i\sigma_1, -i\sigma_2, -i\sigma_3\}$, and get a 16 element number system $\{e_\mu, ie_\mu, f_\mu, if_\mu\}$ which has turned out to be the foundation of our relativistic quantum physics! There is obviously great wonder and 'truth' in these number systems. Generalizing them further may well lead to presently unknown future truths. That is the subject of a later chapter of this book.

Returning to the conjugation analysis, we find that there is a useful, second conjugation for the complex quaternions. It could be defined by

$$\sigma_0^{\neq} \equiv +\sigma_0, \quad \sigma_k^{\neq} \equiv -\sigma_k, \quad i^{\neq} \equiv +i$$

(It only affects σ_1, σ_2, σ_3 and $i\sigma_1$, $i\sigma_2$, $i\sigma_3$.) I call it the quaternion conjugation, but I think mathematicians call it the symplectic conjugation.

Both of these conjugations, A^* and $A^{\neq}$, are very useful in physical descriptions. We can laboriously find that $(AB)^{\neq} = B^{\neq}A^{\neq}$, in general, for complex quaternions. Again, this is very tedious to check, for A and B with 8 parts each. Grind out both sides, for practice, and compare the 64 terms which reduce to 8 terms on each side. Notice that $\{\sigma_0, i\sigma_k\}^* = \{\cdots\}^{\neq}$ and $\{i\sigma_0, \sigma_k\}^* =$

$-\{\cdots\}^{\neq}$, and we know $(\cdots)^{*}$ is antiautomorphic. This simplifies the proof that $(AB)^{\neq} = B^{\neq}A^{\neq}$; all 64 elements do not have to be dealt with. Antiautomorphic means $(AB)^{conj.} = B^{conj.}A^{conj.}$.

Since we have generalized to complex quaternions from pure quaternions, we naturally wonder if spacetime is more than 1+3 dimensional. People have examined 2+6 for example (complex spacetime). So far it has not proven fruitful, that I know of, but the complex quaternions could probably handle it.

GROUPS

The quaternions have natural groups associated with them and these are very important in the physical universe. For example, consider the pure quaternions:

$$A \equiv a^0\sigma_0 + b^k i\sigma_k$$

There are 4 arbitrary parts: $\{a^0, b^1, b^2, b^3\}$. If we choose $AA^{*} \equiv 1\sigma_0$, then we get one equation linking these four parts. Therefore, only three parts are now independent. Consider a special case

$$A = 1\sigma_0 + \epsilon^k i\sigma_k \equiv A_\epsilon$$

where $\epsilon^1\epsilon^2 \approx 0,\ \epsilon^1\epsilon^1 \approx 0,\ etc.$ In other words, the ϵ's are all very small (like 10^{-20}, so that $\epsilon\epsilon$ is like 10^{-40} and much, much smaller). Then

$$AA^{*} = (1\sigma_0 + \epsilon^k i\sigma_k)(1\sigma_0 + \epsilon^j i\sigma_j)^{*}$$

$$= \cdots = 1\sigma_0 + (\epsilon^k - \epsilon^k)i\sigma_k + \epsilon^k\epsilon^j\sigma_k\sigma_j$$

$$= 1\sigma_0 + \epsilon^k\epsilon^j\sigma_k\sigma_j \approx 1\sigma_0 + 0$$

These quaternion elements, A_ϵ, are called 'infinitesimal' members of a group. They are very 'close' to the identity element, σ_0. By multiplying them together, over and over, we can build up 'large' members of the group. The reason that we say A_ϵ is located 'close' to the identity member of the group, σ_0, is because $\sigma_0\sigma_0^{*} = \sigma_0$, so it is in the group also. A large member of the group is, for example, $A_\theta = \cos\theta\ \sigma_0 + \sin\theta\ i\sigma_1$, $A_\theta^{*} = \cos\theta\ \sigma_0 - \sin\theta\ i\sigma_1$; check that $A_\theta A_\theta^{*}$

$= \cdots = \sigma_0$, using $sin^2\theta + cos^2\theta = 1$. The infinitesimal member of this family of members is $A_\epsilon = cos\epsilon\ \sigma_0 + sin\epsilon\ i\sigma_1 \approx \sigma_0 + \epsilon i\sigma_1$, (from $cos\theta = 1 - \theta^2/2! + \theta^4/4! - \cdots$ and $sin\theta = \theta - \theta^3/3! + \theta^5/5! - \cdots$). There is an infinity of θ values, so there is an infinity of members of these groups. They are called Lie groups. We say $AA^* \equiv 1\sigma_0$ is a three parameter Lie group, because $A_\epsilon = 1\sigma_0 + \epsilon^k i\sigma_k$ has three free parameters: ϵ^1, ϵ^2, and ϵ^3.

The new group, $BB^{\neq} \equiv 1\sigma_0$, is the same group for the pure quaternions, since * and $\neq$ do the same thing there. For the complex quaternions, however, these groups are distinct. We examine their infinitesimal members to see how many free parameters they can have. Since $\epsilon\epsilon \approx 0$, we find that A^* must change the sign of each term in the infinitesimal members, (besides the σ_0 term of course). Therefore, we have

$$A_\epsilon^* = \sigma_0 - \epsilon^k i\sigma_k - \epsilon^0 i\sigma_0$$

and

$$B_\epsilon^{\neq} = \sigma_0 - \epsilon^k i\sigma_k - \delta^k \sigma_k$$

The A group obviously has 4 free parameters, if $AA^* \equiv 1\sigma_0$, and the B group has 6 free parameters, if $BB^{\neq} \equiv 1\sigma_0$. They have a common subgroup, $CC^* \equiv CC^{\neq} \equiv 1\sigma_0$, *i.e.*, $C^* = C^{\neq}$ and this is the 3 parameter group we found earlier in the pure quaternions:

$$C_\epsilon^* = (\sigma_0 + \epsilon^k i\sigma_k)^* = \sigma_0 - \epsilon^k i\sigma_k$$

Notice that these groups appear naturally in the number system, once that system is defined! The quaternion system was defined by Hamilton because nature is 3 dimensional and because complex numbers seem beautiful and powerful in applied calculus. Therefore, these groups are also natural to the real world, if the quaternions are, and they are! These groups historically have fancy names:

$$\{\epsilon^k\} \leftrightarrow SU(2)$$

$$\{\epsilon^k,\ \epsilon^0\} \leftrightarrow U(1) \otimes SU(2)$$

$$\{\epsilon^k,\ \delta^k\} \leftrightarrow SL(2,C) \approx Lorentz$$

We have not done any physics yet! We already have natural number systems and natural groups that follow from the 1+3 assumed spacetime split.

All of this algebraic stuff is a package and it goes with the 1+3 spacetime. This could have all been done by Hamilton before 1850! But Lorentz didn't find his group until 1900 and realize that it applies to the design of our real world.

GROUP REPRESENTATIONS

The group idea here is very straightforward: $AA^* \equiv 1\sigma_0$, $A = a^\mu\sigma_\mu + b^\mu i\sigma_\mu = c^\mu\sigma_\mu$ with only four of the eight numbers as independent, and they define the member A. Clearly, if $AA^* = 1\sigma_0$ and $BB^* = 1\sigma_0$, then $(AB)(AB)^* = (AB)(B^*A^*) = A(BB^*)A^* = A\sigma_0 A^* = AA^* = \sigma_0$, so the product of any two group members is also in the group here. That is why the collection is called a group. Notice that we have used **associativity** to get this result. We don't have commutivity, but we do have associativity! If nature has any *non*-associative math required, then groups will fall from grace or be generalized. See later for speculations on this.

Besides the group, we can invent things that 'change' because of a group's existence. For example: $P' \equiv A^*PA$ turns some given P into P$'$ by 'hitting' it with a group member A, $AA^* \equiv 1\sigma_0$. We can also invent $F' \equiv A^{\neq}FA$ as well, for some given F. Even though $AA^* \equiv 1\sigma_0$ here, remember $AA^{\neq} \neq \sigma_0$ in general for this group. The obvious next question is why do this complicated business? It turns out to relate to the physical world, of course, or we would not bother with it. (This is a physics book, not a math book, though you must be beginning to wonder by now.)

Notice that if $P \equiv P^*$ and $P' \equiv A^*PA$ then $P'^* = (A^*PA)^* = A^*P^*A^{**} = A^*PA = P'$, so $P' = P'^*$ and P$'$ is 'like' P, in a sense. So what? Have patience. Notice that $F^{\neq} \equiv -F$ and $F' = A^{\neq}FA$ means $F'^{\neq} = (A^{\neq}FA)^{\neq} = A^{\neq}F^{\neq}A^{\neq\neq} = A^{\neq}F^{\neq}A = A^{\neq}(-F)A = -F'$. So $F' = -F'^{\neq}$, and F$'$ is 'like' F, in a sense. All of this is true, regardless of which group A belongs to. These P's and F's are called representations of the group {A}. (Don't ask me why that name is chosen. Manifestations of the group might be a better name.) Now consider $P^{\neq}P$ and $P'^{\neq}P'$. Are they 'alike' in general? Let's see:

$$P'^{\neq}P' = (A^*PA)^{\neq}(A^*PA) = A^{\neq}P^{\neq}A^{*\neq}A^*PA = A^{\neq}P^{\neq}(AA^{\neq})^*PA,$$

so we need $AA^{\neq} \equiv 1\sigma_0$ to keep going. Then

$$P'^{\neq}P' = A^{\neq}P^{\neq}PA = ?= (P^{\neq}P)A^{\neq}A = P^{\neq}P(\sigma_0) = P^{\neq}P$$

They are 'alike' if $AA^{\neq} \equiv 1\sigma_0$ and also $P^{\neq}P$ needs to be proportional to σ_0, so it then commutes through $A^{\neq}$, on its left above. Proof that $P^{\neq}P$ commutes goes as follows: $(P^{\neq}P)^{\neq} = P^{\neq}P^{\neq\neq} = P^{\neq}P$ and since $\neq$ changes the signs of all σ_k terms, $(P^{\neq}P) = (B\sigma_0 + Ci\sigma_0)$ at most. It therefore does commute with any A.

We have found that $(P^{\neq}P)$ is an 'invariant' representation of only the group $AA^{\neq} \equiv 1\sigma_0$, in the sense that $P'^{\neq}P'$ and $P^{\neq}P$ are identical for this group. We say this $P^{\neq}P$ is an invariant of the group $AA^{\neq} = 1\sigma_0$, which is called the SL(2,C) group. We could find other representations and invariants for these groups. Another very useful representation is the spinor $\psi' \equiv A^{\neq}\psi$. We find $\psi'^*\psi' = (A^{\neq}\psi)^*(A^{\neq}\psi) = \psi^*A^{\neq *}A^{\neq}\psi = \psi^*(AA^*)^{\neq}\psi = \psi^*\psi$, if $AA^* \equiv 1\sigma_0$. So this is an invariant of another group, called SU(2)⊗U(1). But next consider

$$\psi'^*P'\psi' = (A^{\neq}\psi)^*(A^*PA)(A^{\neq}\psi) = \psi^*A^{\neq *}A^*PAA^{\neq}\psi = \psi^*(AA^{\neq})^*P(AA^{\neq})\psi$$
$$= \psi^*P\psi$$

if $AA^{\neq} \equiv 1\sigma_0$. This is an invariant of the SL(2,C) group. Clearly P, F, and ψ are all quite different. All are complex quaternions and all relate back to the same two groups, $AA^* \equiv 1\sigma_0$ and $BB^{\neq} \equiv 1\sigma_0$. They are like decorations in the picture around the central groups. The groups in turn are manifestations of the number system itself. All of this stuff 'must' be at the core of the physical world, if 1+3 dimensional spacetime is fundamental to the real universe!

OTHER CONJUGATIONS

So far we have $\{\sigma_0, i\sigma_k, i\sigma_0, \sigma_k\}$ as the basic number system. There are two basic conjugations: $(\;)^{\neq}$ and $(\;)^*$. The $\neq$ changes σ_1, σ_2, σ_3 and * changes i. These are both antiautomorphic: $(AB)^{conj.} = B^{conj.}A^{conj.}$. Are there other antiautomorphic conjugations? If so, could they have physical significance as well?

We can invent many new conjugations by the procedure:

$$A^{conj.} \equiv \sigma A^*(\sigma)^{-1}, \quad \sigma\sigma^{-1} \equiv 1\sigma_0$$

where A is any of the 8 elements and σ is any one of the 8 elements. Clearly, $\sigma \rightarrow \sigma_0$ and $\sigma \rightarrow i\sigma_0$ gives $A^{conj.} = A^*$, which we already have. For $\sigma \rightarrow \sigma_1$, we get $A^{\#1} = (\sigma_1)A^*(\sigma_1)$. Calculating $A^{\#1}$, for A equal to each basis element, is easy and good practice for you. You will find

$$\{\sigma_0, i\sigma_k, i\sigma_0, \sigma_k\}^{\#1} = \{\sigma_0, -i\sigma_1, i\sigma_2, i\sigma_3, -i\sigma_0, \sigma_1, -\sigma_2, -\sigma_3\}$$

We get the same results for $(i\sigma_1)A^*(-i\sigma_1)$. For $\sigma_2 A^* \sigma_2$ we get: $\{\sigma_0, i\sigma_1, -i\sigma_2, i\sigma_3, -i\sigma_0, -\sigma_1, \sigma_2, -\sigma_3\}$. You can now guess what $\sigma_3 A^* \sigma_3$ gives without doing the calculation.

Are these antiautomorphic conjugations? Yes! They are if A^* is antiautomorphic itself. By the way, $\sigma A^{\neq} \sigma^{-1}$ would also generate antiautomorphic conjugations then too. Are they the same or different? Find them and see for yourself. Here is the proof that $\sigma A^* \sigma^{-1}$ is antiautomorphic:

$$(\sigma A^* \sigma^{-1})(\sigma B^* \sigma^{-1}) = \sigma A^* (\sigma^{-1}\sigma) B^* \sigma^{-1} = \sigma A^* B^* \sigma^{-1}$$

and

$$\sigma (BA)^* \sigma^{-1} = \sigma A^* B^* \sigma^{-1} = (BA)^{conj.}$$

We see that $A^{conj.} B^{conj.} = (BA)^{conj.}$ in general, where A and B are any two of the 8 basis elements.

It turns out that these conjugations are apparently not so useful. Perhaps because they treat the σ_1, σ_2, σ_3 parts differently. They single out one of them and nature does not have a favorite direction in space. The * and $\neq$ conjugations don't do this. Also notice again that

$$\{\sigma_0, i\sigma_k\}^{\neq} = \{\sigma_0, i\sigma_k\}^*$$

but

$$\{i\sigma_0, \sigma_k\}^{\neq} = -\{i\sigma_0, \sigma_k\}^*$$

The $\{\sigma_0, i\sigma_k\}$ subset is closed under multiplication, whereas the other one is not. This pattern between the closed and the non-closed segments of the algebra will occur again in the larger algebras that follow in later chapters. There, we shall find more useful conjugations besides the two here.

The conjugation $\sigma_1 \rightarrow -\sigma_1$ but $\sigma_2 \rightarrow \sigma_2$ and $\sigma_3 \rightarrow \sigma_3$ is called space inversion or parity. It does play a role in the details of quantum theory. Thus we have not invented a totally useless conjugation here.

This has been a depressing start for most readers. The rest of the book is not all this tedious. But if we ask deep questions, then we must expect complicated answers about nature. That has been the pattern over time in our advancement and mastery. The deeper we go the harder it gets to still understand what is needed to be understood. At some point we will reach our limits and just not be able to go farther. How soon? No one can say at present.

CHAPTER 2

Space and Motions Therein

Of course, totally empty flat space, with time flowing forward, may have 'meaning', but it is not very interesting. Curved, empty space is interesting, especially if the curvature is changing with time—either locally, like a moving ripple, or overall. Our universe seems, as Einstein and others even before him suggested, to have an overall curvature and to be flattening everywhere as time goes psychologically forward (forward means 'away' from memories). One could imagine a curved 3-space tangling and untangling locally as time progresses, with nothing else in it but this curved and changing space. Could the tangles and their invented rules ever be complicated enough, so as to evolve self aware, conscious clumps of such tangled space? No one knows. Such things might even exist, if other universes besides our own can exist. There can be motions of ripples of curved space and so time passing can be meaningful even with nothing else present but space—if it is both curved and changing.

Our real universe used to be much more curved, over all, and is generally flattening, as time moves forward. It is either positively curved or negatively curved—meaning that parallel light beams will converge or diverge, respectively. The idea of being 'curved' has real meaning, even in an otherwise empty space! The positively curved space must close back upon itself, if uniformly curved everywhere at a fixed universal time. It has a finite volume but it has NO edge or center! Just as a balloon surface is a two dimensional closed space, as compared to a finite rubber trampoline surface. A trampoline has an edge. A negatively curved space (light beams diverge) has no natural edge, so it could be infinite! Infinite space is hard to imagine, just as infinite time is. But both are possible, we guess! We can, in principle, measure the overall curvature, but so far the measurements are inconclusive, except to say that our universe presently is only very slightly curved, at most. If positive—no problem. If negative—then potentially infinite and then devastating to our egos, as we shall see later. This is one of the most important, still unanswered questions: "Is the universe infinite, even in principle?"

We are alive and conscious and we are made of something other than complex knots in space (we hope?). The space around us is a backdrop to the action, but maybe not so passive. Little dimensions, besides the large three that we experience, may play a vital role in our being able to be conscious, for a short time in this universe. We just don't know how complicated space may really be. It 'definitely' is complicated—curved and changing overall (while remaining positive or negative) and it is definitely positively curved around objects in it.

The location of a curved wad of pure space, with a local center, could be imagined to 'move', even though it is only made of space itself. We can imagine faster moving wads and slower moving wads. We can even imagine wads approaching each other and either passing through each other, undisturbed, or producing new wad arrangements, moving out from the 'collision'. The actual collision result would depend on the rules for space wad movements and their interactions.

The next logical issue is whether there is a fastest speed for space wads. If so, then how fast and compared to what? Remember, we have no stuff here, except space itself, in this hypothetical spacetime. If there were some standard spacing of wads of space, we (ghosts) might try to measure the movement of a wad past them. But what do we do for natural time intervals? We seem to need wads of space that orbit each other at standard distances at some natural speed, some standards of length and time. This all probably seems kind of silly, but it really isn't. Simple, empty but expanding or contracting space can exist without any locally tangled wads of space in it, though these may later come to exist and even lead to self aware and conscious beings. These conscious wads can experience other wads, somehow, and make real measurements, but only if wads interact rather than pass through each other, undisturbed. A small loop of space dimension, with a much smaller wad of distorted curvature going around it, could establish a standard of time and length here. We could take this speed of rotation around the small loop space dimension as our standard of time and of speed. We might then allow other wads to move only at speeds below this loop speed, both in the tiny loop dimension or out in the other, larger space dimension loops, or even for infinite loops, should space be such. A special and constant loop, with moving kink, can give a scale to space and to motions in it. No matter how many large or small dimensions there are, (even if they are changing overall, or are only changing locally due to the moving locations of wads of curvature), we only need one loop dimension to be closed and stable. One wad moves around it at speed c, as we shall call it. This could provide a universal speed limit for motion.

Our universe seems to have such a natural speed limit, c, which is speculated to be the same value for all kinds of motions in the space—either motions of wads of space curvature or other things that can be moving through the space. So, how should things move in the space, when isolated far from other things or from local wads of space curvature? There are two logical choices, maybe: uniform motion in a straight line at the initial speed, or some decay toward zero speed in the straight line. How about uniform motion in a circle, instead? Absolutely anything is perhaps possible! Our world has its particular set of rules. It is a subset of all the possible rules that some huge beings could imagine and it may have parts that we creatures have never yet imagined. Some rules may be beyond our imaginations completely!

Quantum physics has taught us to give up trying to understand how the universe really is. We are made of zillions of lumps and it takes huge numbers of such lumps to construct any self-aware being, it seems. Therefore, we huge blobs can never experience the very small, where the action really is, perhaps. Therefore, we will get only bastardized pictures and measurements at best. Humanoid made for humanoid consumption and contemplation. Our physics, not God's physics. We have fought this creeping realization too long, but must now face it squarely and decisively. If we don't, we will never get much beyond the rudiments of subatomic physics that we now know how to deal with.

We have to stop this playing God and doing a design of the whole universe. But why stop? No one wants to quit. The philosophers do not have to stop. Nor do the poets and other artists. The physicists must stop. What distinguishes physics is that only measurements, actually made by us, are allowed as evidence in the court of ideas about TRUTH. Therefore, physicists probably miss all the best TRUTH, perhaps, but that is their game. It is very restrictive and restricted! It may even seem boring, but at least it is not up for grabs like everything else that is not physics. People argue endlessly about the REAL TRUTH and get no where. They make no progress to greater understanding. There is really no building on the past. No progress. Just wild-eyed religious speculations. Some people take these so seriously as to give their lives in defense of the particular speculations that they embrace. Belief in the REALITY of SPECULATIONS is FAITH.

Physics has its faith, of course, but it is minimal. Trust only that which anyone with ordinary faculties can see, hear, feel, and smell. Share supposed experiences with each other until you can agree on common, repetitive things existing in this limited realm of human observations. Then make models of how the world seems and predict new data. Go out and check it in the same limited and democratic way. Throw the model out or build on it, depending on the mutual observations that are found, compared to the model's predictions.

Accept any source of inspiration for a new model, but trust only in the model, if it is supported by measurements. We have gotten some good models from really bogus ideas and bogus preconceived notions, held by the model builders. It usually takes a long time to separate the two. It should not! Trust only the math, not the words! If all they have is words in their model, then be very suspicious! There is probably little there but speculation that cannot really be tested well in the lab of commonly experienced measurements. It is art then, not physics.

We—huge creatures that we are—experience 3-space, 1-time, and edges to localized objects in that space. Objects relocate as time flows forward. We don't know the real difference between forward and backward. We just experience a forward, conscious existence. But we are huge and very complex brains. Maybe both time directions are meaningful in the small world of atoms, and we will have to adjust to this in our later models that work best. We, above all, must stay humble and accept our limitations. If localized things seem to have a center and that center relocates, we have motions. If that center disappears and reappears elsewhere, or several places at once elsewhere, then we have a whole different kind of universe! Our universe seems to be such that huge things march through space, step by step. They don't disappear, jump, and reappear elsewhere. No one knows why. Yet, small things are all over the place at the same time—they even move forward and backward in time! Our size and complexity shut us out from experiencing all this wonderful complexity, somehow. So frustrating!

So, again we must keep an open mind. We don't know much and never will know much. We just know more than yesterday, and hope for more understanding tomorrow. Why should we strive to understand the universe? I guess because it is like a mountain. It is just there and our evolution/survival has programmed us to struggle. The non-strugglers existed in some tribes but didn't likely pass on their genes in the old days.

We find experimentally, without real reason that we yet know of, that big things (like us or larger) float through space in a straight line at constant speed in isolation. We experience a feeling of isolation, but we are not sure that it is real. All objects may overlap in some way—perhaps in the other small dimensions besides the three big ones that we experience directly. When blobs float into each other they grab on or they bounce off, then they go in new straight line motions afterward. That is just the way it is (or it seems to us) and we can all agree on these measurement observations, for other huge objects, like ourselves, or for bigger things.

We need clocks to measure speeds and we need rigid platforms to measure angles and directions in the sky. We actually need a fleet of platforms to really make accurate measurements, especially in curved space, and desperately so in curved and changing space-time situations!

Measuring simple motion is not simple at all. We can worry ourselves to death about how to measure it right. Calibration is a real can-of-worms, which in curved and changing space is almost hopeless to understand and attempt to do well. Even in flat and empty, static space it is not simple. But we do not have that simple universe. Ours is curved and changing radically over the long haul. It does no good to theorize about what is, in a universe that does not exist, except for engineering applications, or to get our feet wet in understanding how to calculate more realistic solutions. Simple space is OK for finding simpler solutions, but be careful with the basic concepts. For example, we should not theorize about the possible physics of an infinite, static, empty spacetime. We don't have that universe. Such is poetry, not physics. Most Relativity books are full of poetry, not physics, in the first chapter. (This one included.)

So, we must examine calibration and motion in the real universe—curved, maybe closed, and definitely expanding while getting flatter. Since the universe might be closed and this case is so much easier to grasp than infinity, we can take it as our basic model. We then explore the foundations of measurement in just such a context. This is much more realistic than the usual starting discussion in Relativity books of the past. This is truly Modern Relativity that we are about to indulge in. We still have not done any real physics so far in this book. We have just played with ideas, possibilities, fantasies, speculations, and the very important limits of real science. It is now time to get down to the real basics—how does the universe work?

A QUICK ANSWER—FOR THE YOUNG:

If you have not had any physics then this book zooms along over your head in most parts. That cannot be helped. I am trying to lay out the foundations as simply as I can.

There is the Physics-Bible already out there. This is the standard, calculus level physics book for engineers. There are several authors but their excellent books are basically about the same. They contain the basics of reality as we know them, or rather knew them before about 1930, for the most part. They may have some final chapters that deal with post 1930 physics knowledge—quarks, and many body quantum to some degree. This book does not overlap those bibles very much. You should master one of them before you really read this one with full appreciation of what you are seeing.

They start with vectors, dot and cross products, and such. They develop one dimensional kinematics of motion—a bead on a straight wire being pushed and pulled as it moves. Then they move up to ice hockey, with the puck pushed and pulled in two dimensions. Here, velocity and acceleration become complicated. My students always had the worst time of it with circular motion—a simple puck going in a circle on the end of a string. Then we hit pucks together, push them along with a stick, or attach a spring to them, etc.

We find **F** = m**a** gives the **a** acceleration response to a net force **F** acting on any mass m, changing its motion. The puck has a well defined path along the ice at all times.

Next we consider gravity deflecting the puck, from a distance, if it passes by a large metal drum filled with iron. Or we electrically charge the puck and the drum instead, and see the electrical force effects on it, as it gets deflected while passing by. We have mathematical laws for these forces and all is well, so long as we don't measure anything here to a large number of decimals, such as 15 or more digits. We can also put a magnet on the puck and see its new deflection due to a magnet located on the drum. We can attach the charged puck to the heavy drum with a spring, and let it oscillate, giving off radio waves at the same frequency.

We heat it up and define its temperature, etc. All this is done by us huge creatures, looking at effects on huge pucks (maybe 10^{24} atoms in one puck).

We could next consider red blood cells, 3 to 5 microns in diameter, moving across the super-smooth ice, bumping into each other and sticking or bouncing off. We can still see them with a flashlight and microscope and they still have a well defined path. The light bouncing off won't alter their path much at all, and we don't measure position to more than a few digits.

But we cannot keep going down in size. We cannot track individual carbon atoms moving and bumping on the ice. Our bodies are made of about 10^{27} atoms, hooked together and keeping us conscious. We cannot see one atom moving on the ice. Down there, things get strange and our idea of path gets challenged. It may disappear or become radically altered, even if it still exists. There may be a zig-zag component to motion, that we don't notice for big things, like sliding blood cells, but we could not ignore this for carbon atoms on the move, if it exists. Or, instead, all kinds of new things could be active, that we huge creatures don't experience, so did not even suspect existed. These somehow manifest themselves in only quantized jumps existing in the motions and interactions at this atom size level and smaller. Even spin rates are quantized for molecules in space. Much smaller things, like protons sliding on the ice, may involve extra dimensions to space, and what-have-you, that do not show up even at the carbon atom size level.

This book is trying to display the tools that we really can gather, to look at the physics as deeply as we can go, given our huge size limitations. We turn to math in its various forms and play with the basics there, hoping for insights that will open up the physics to us down inside protons. We need ideas and models that we can test in the lab later. We have lots of lab results and cannot explain most of them, if the scale is subatomic. At the atomic level and higher we have good models already. Thus chemistry and metallurgy are pretty well understood, in principle anyway. We cannot explain why protons are 1800 times heavier than electrons though. We know nothing about the insides of electrons and the reasons for their mass value or values.

So, this book is no substitute for reading the bible of physics. It starts where that book leaves off.

The Einstein summation convention is very compact and takes some getting used to. We say

$$\boldsymbol{P^k\sigma_k \equiv P^1\sigma_1 + P^2\sigma_2 + P^3\sigma_3 \equiv P}$$

and

$$\boldsymbol{Q^k\sigma_k \equiv Q^1\sigma_1 + Q^2\sigma_2 + Q^3\sigma_3 \equiv Q}$$

Then $PQ = (P^1\sigma_1 + \cdots)(Q^1\sigma_1 + \cdots) = (P^k\sigma_k)(Q^k\sigma_k)$, but you cannot write it this way. We need a new dummy index for Q^k. Call it Q^j. Then we have $PQ = (P^\mu\sigma_\mu)(Q^j\sigma_j) = P^\mu Q^j\sigma_\mu\sigma_j$, sum over $\mu = 1,2,3$ and $j = 1,2,3$; nine terms here. This notation is very powerful. It sometimes mesmerizes us into not realizing what a huge mess we have for a simple looking expression.

We have position $x = x^k\sigma_k$ or $x = x^k\hat{e}_k$ and we can change to spherical coordinates by defining new (r,θ,ϕ) variables, such as

$$\boldsymbol{x \equiv \rho\cos\phi,\ y \equiv \rho\sin\phi,\ z \equiv r\cos\theta,\ \rho \equiv r\sin\theta}$$

Given (r,θ,ϕ), we find (x,y,z) and ρ. We can then invent some new things, σ, or new things, ê, such that x or **x** can be written in terms of these. See the Appendix and Chapter 10 for some details. This is not a profound change. It is complicated, but not deep. It is a great aid to finding solutions in detail. It is a mathematical crutch and nothing more.

Real solutions at a deep level in nature involve partial differential equations instead of ordinary differential equations. This seems to be because we lose the concept of path for subatomic particles. The electron in hydrogen is zig-zagging like crazy, or it is instead all over the place at once, around and in the nucleus. We use $(e_\mu)\partial^\mu$ a lot in our equations. This means sum on $\mu = 0,1,2,3$ of course, but what is ∂^μ? By definition

$$\partial^0 \equiv \frac{1}{c}\frac{\partial}{\partial t},\ \partial^1 \equiv -\frac{\partial}{\partial x},\ \partial^2 \equiv -\frac{\partial}{\partial y},\ \partial^3 \equiv -\frac{\partial}{\partial z}$$

In high school calculus, students learn that

$$\frac{d}{dx}(x^n) = nx^{n-1}, \quad n = a/b \neq -1$$

for any fraction a/b. The partial derivative simply means hold other variables constant while you differentiate, when you have a function of several variables, like $\psi(x,y,z,t)$. For example,

$$\frac{\partial}{\partial x}\psi(x,y) \equiv \frac{\partial}{\partial x}(x^3y^2) = y^2\frac{d}{dx}x^3 = 3y^2x^2$$

$$\frac{\partial}{\partial x}\psi(x,y) \equiv \frac{\partial}{\partial x}(x^3 + y^2) = 3x^2 + 0$$

These partial derivatives lead to specific equations which can be solved and they predict lab results.

Moving from the hockey rink ice (flat space) to the surface of an iceberg, is a big deal change. We now have curved space. The pucks, sliding and colliding there, are in a whole other realm of complexity. The effect of these changes on $x = x^k\sigma_k$ are similar looking to the coordinate system change, but they are not alike! Curved space is different from just dealing with flat space and using a messy coördinate system. There are complicated derivative tests that one can make to tell the difference.

In this book, we generally stick to $x^\mu\sigma_\mu$ or $x^\mu(e_\mu)$ or $x^\mu i(ie_\mu)$ descriptions of flat space-time, $\{ct, x, y, z\} = \{x^\mu\}$. The physics can be nicely displayed in that form. If we want an actual solution, such as the hydrogen atom energy levels, we will need spherical coordinates to make the solution very much easier to construct. Gravity is 10^{-40} as strong as electricity, so the atom's internal space is quite flat, at least outside of the proton itself. No one knows what space is like inside the proton!

The reader need only understand the basics of derivatives and integrals for the most part, and a little vector notation, such as **F•dr**, which is a short hand for $F_x dx + F_y dy + F_z dz$ and this is the work done by the (F_x, F_y, F_z) force components on a mass m, when the mass is moved (dx, dy, dz), or $|dr|$

$= [(dx)^2 + (dy)^2 + (dz)^2]^{1/2}$. If we sum along the path steps, 1,2, etc., of the puck, we find $\mathbf{F}_1 \bullet \mathbf{dr}_1 + \mathbf{F}_2 \bullet \mathbf{dr}_2 + \cdots$. This gives the total work done by the stick on the puck as the skater pushes it along. Each $|dr|$ step here occurs in a time dt that is very small—say, 10^{-9} seconds. This means many-many steps in one meter of path.

There is much more also in this book. I hope it can be read by physics loving high school students on a superficial level, and then re-read over and over as the college years go by and readers work their way up to and through the standard bible-physics course.

At the end of the appendix, I have started the solution to the hydrogen atom energy levels. This is messy stuff, but it is the best kind of test for our theorizing! Transitions between these levels give off photons with $\Delta E = hf = hc/\lambda$ and we measure λ to 10 digits. The **good** theory matches these!

Partial differential equations are solved, if possible, by guessing that $\psi(x,y,z,t)$ can be written $\psi_1(x)\psi_2(y)\psi_3(z)\psi_4(t)$ and then substituting back into the equation. We next separate the equation by putting all parts in t on one side and in x,y,z on the other. Since these depend on independent parts, t and x,y,z, the equality requires each side to be only a constant, not a function. This constant is put down but is still unknown. It is called a quantum number. Such things happen for all partial differential equations, not just quantum physics. What is strange here is finding such partial differential equations actually apply to particles with mass and charge. We'll probably never know why this works. It just does.

The quantum numbers get fixed by the boundary conditions we guess are reasonable, from the physical context and from our extensive classical physics experience. For example, in hydrogen we guess that the wave function is less than infinite for $r \to 0$ and that it dies away like $e^{-|r|/\text{const.}}$ as $r \to \infty$. Parts that depend on angles around the nucleus, we guess, have ϕ and $\phi + 2\pi$ give the same $\psi(\phi)$ result, since we return to the same place after a 360° rotation. These guesses lock up the values of all the quantum numbers in a complicated pattern. The allowed values are related. In hydrogen, $n = 1 \Rightarrow \ell = 0$, $m_\ell = 0$ and $n = 2 \Rightarrow \ell = 1$, $m_\ell = +1, 0, -1$, or $\ell = 0$, $m_\ell = 0$. These are pretty patterns that are mind boggling. How can the real world have such things? We are talking about an electron in orbit around a proton, aren't we? I guess not. Whatever is there, inside the hydrogen atom, is something we never have and never will experience. We just do the best we can. This book has the best we know, or knew about 1930. The advances since then are not covered here. However, we shall show that perhaps we missed some important new ideas about 1930 and have been floundering for the most part ever since.

CHAPTER 3

Closed Spacetime Physics

Closed space is the simplest in a strange way. At the beginning of time, all the stuff of the universe including the space itself is confined to a very small volume. This means that everything is causally connected and on top of everything else. We can use this idea to be sure, in principle, that our cosmic clocks will all read zero together when time begins—whatever that means. At about 300,000 years after the beginning, the universe has a certain size; it is composed of thin gas (Hydrogen 80%, Helium 20% roughly); it has a nearly uniform density throughout; it has a very uniform temperature of 3,000K (red hot like a fireplace poker).

The previously ionized hydrogen and helium gas has just become neutral as the universe further expands and cools. Becoming neutral, rather than ionized, means that the light in this oven now stops running into particles very often. Meantime, the space itself continues to stretch and this somehow pulls out the wavelengths of the light in the oven. The oven cools but its light remains in a distribution of intensity vs wavelength which is typical of an oven at that new temperature (it looks just like the output of an oven operating on earth today at that same temperature). At the present time, some 15 billion years later, we see this oven still shining at us from all directions, very uniformly, and currently looking like an oven at only 2.7K. Thus 3,000K looks like 3K, by the time the light finally reaches us. It stretched along the way in between then and now. Tomorrow, when we look out, we get light from 24 light hours farther away, that left the oven there, at about the same cosmic time as the photons we captured last night. Remember, we see this same oven in all directions and it is uniform. This was first observed in 1965. Not so long ago. Before that, we were even more ignorant than today about basics. This says, by the way, that the space of the universe has stretched by $3{,}000/3 = 1{,}000\times$ since the end of the oven-light phase, called the decoupling time. The original universe had perhaps a total mass of only 10^{24} solar masses, if closed, and would then expand for only 200 billion years. From now to the end of that expansion, it will expand only by $4\times$, so any humanoids still around then will see an oven light coming in that looks like it is (3/4)K in temperature, instead of the actual 3,000K it had when the light left, some 200 billion years earlier.

The light from the distant oven would, at maximum expansion, have originated from the other side of the universe from the observer catching it. After that time, the universe contracts. The oven light (reaching anyone still alive then to measure it), would have originated from more than half way around the universe. It will look like a hotter oven also, and getting hotter and hotter every day from then on. Each day they receive radiation that started out closer to them originally, but which was originally traveling away from them back then. It has gone almost all the way around the universe to eventually reach them. As the final contraction crunch approaches, the oven light that left from very near by, but sent away, finally reaches anyone still looking out, after having traveled all the way around the universe as it stretched and then contracted over 400 billion years! What an incredible world! The old oven light would again look like an oven at 3,000K. It traveled undisturbed for 400 billion years and stretched by 4,000$\times$ in wavelength along the way; then it recontracted in wavelength back to normal before it reached a detector at the end of time. It literally went around the universe. (At the end of the world, you can almost see images of light that bounced off the back of your head at the beginning of the universe.) Of course no one lives that long but it is an interesting thought.

What a fantastic place we live in! Being in orbit about only one of these 10^{24} stars, out there right now, is intriguing. Write that number out with all 24 zeros and put in all the commas! It stops our hearts to imagine such a sea of places besides our own, some 10^{13} other galaxies with probably one or more civilization in each. Can any of the old religious myths that our planet has produced be at all reasonable in such a vast sea of places, planets, humanoids, or evolution? Our social world runs on bull.... . Is it any wonder that all religions either attack the idea of evolution or just ignore it? (But I digress foolishly.)

It is pretty obvious, in such a universe setting, that there is absolute time and absolute curvature at a given absolute time. There is also absolute motion through this space. If you are really moving, then you won't see the oven light that is coming in from all directions looking exactly the same. There will be a hotter looking direction and a cooler direction opposite to it. The difference in these measured temperatures can tell you your true speed and true direction! We now know the earth moves at (2/1000)c in the direction of the constellation Leo, specifically toward the galaxy Hydra. The really special observer platforms in space are those that see the oven light as coming in uniformly in all directions. They are the comoving observers and are really at rest with respect to the local space they are in.

It appears, experimentally, that simple movement of a clock or living being through the vacuum itself affects all the basic physics of the atoms or particles in the clock or being. The moving clock runs slower and the being thinks and ages slower. The amount of slowing appears to be given by $\sqrt{(1 - v^2/c^2)}$, where v is your absolute speed and c is the speed of light in the neighborhood.

If we could go back to the hot oven beginning Big Bang condition and track a clock that hypothetically has been ticking for 15 billion years, we would see that the comoving clocks will have the most total ticks compared to any other clocks that might have moved off from them and then came back later. All real motions slow all clocks, so moving clocks have serious problems comparing observations with each other. Even atomic clocks at different latitudes, different circle motions on the spinning earth, cannot keep in time. The ones closer to the equator fall behind due to their greater speed through space.

It also appears that clocks close to stars or planets slow down as well, even if not moving! That amount of slowing is given roughly by $\sqrt{(1 - R_s/r)}$, where $R_s \equiv 2GM/c^2$ and r is the distance from the center of the planet. M is the mass of the planet and G is a measure of how strong gravity is, as an influence on things in our universe. (The particular value of G is not yet predictable.) Again, c is the speed of light as usual. The R_s is a distance, usually well within the planet, and is called the Schwarzschild radius. It is about a centimeter for the earth, for example, and is directly proportional to the mass. The curved space around any planet is positively curved and somehow this curvature alters clocks. This is even true of subatomic muons, as clocks, and they have no known size or parts. Electron clouds, moving around nuclei in the curved space, are affected. This affects neuron signals for the mental processes of any humanoid brain.

These electron clouds, around the atomic nuclei of the atoms in a brain, somehow cause signals to move slower through the neuron network. Amazing, but apparently true! The body ages slower also. Everything slows down.

Another group of atoms, which is moving absolutely through the space and which is in orbit around the stationary planet (rather than holding steady position using a firing rocket), is additionally affected by the motion through the curved space. The orbiting object physically shrinks in the direction that it moves absolutely. This shortening is given by the same factor as the time slowing, $\sqrt{(1 - v^2/c^2)}$. An isolated and moving alarm clock then gets oval gears, an oval shape, and the escape mechanism moves back and forth more slowly. This is due to the clock moving absolutely through the space, even if it is far from all planets. The space itself apparently causes this strange change in the atoms. Space has its effects on every thing. It changes lengths and time.

Notice that $r \to R_s$ means $\Delta t \to 0$; also $v \to c$ means $\Delta \ell \to 0$ and $\Delta t \to 0$. Yet, as an object later slows back down, by retrofiring its rockets or otherwise, the atoms expand back and the alarm clock returns to being round. If we accelerate up to very high speed with the clock in our hands, and we watch it continuously, we notice no shrinking. This is because we are shrinking along with the ship's cabin and the clock, by the same amount.

It is totally unknown how much shrinking a macroscopic body can tolerate before some irreversible effects set in. For example, the brain atoms become flat disks at $v \sim c$, yet the brain remains conscious—we guess! The real experiment has never been done!! I wonder if the believers in only relative motion would volunteer to be the guinea pigs for such an experiment. How about 0.999999c, where one's 10 cm head would compress to only 1.4×10^{-2} centimeters thick, the width of a pencil line, when this speed is reached. Since big things are the sum of their little things, it may just be possible that you would live through it, if the atoms can still function ok in this flat pancake state of existence, while moving through and interacting with the space (and its sea of virtual particles).

Of course, the background radiation will be very shifted in color in the direction you are going. Movement through space makes photons appear bluer, when coming at you from the direction you are moving. It appears that the frequency change factor is $\sqrt{((c + v)/(c - v))}$, which for the speed considered here is 1,414. This means that the most intense photons in the background (Gamow) radiation, which normally have a (microwave) wavelength of about 1 millimeter, would come at you looking like photons with a wavelength of 7,071 Angstroms. Still not high enough to be a danger to you, since this is only red light. Your ship would easily deflect or absorb these photons.

Your rocket, coasting at this very very high speed, and looking like a very flat disk to the special, comoving frame, would be slowed down gradually by these high energy Gamow photons hitting it in the face. It would gradually slow down without anyone doing anything! Thus, absolute rest is really the natural state for all objects in the universe that have mass. They are naturally slowed down. (Aristotle was sort of right after all.) Photons appear to be created going only at speed c and maintain that high speed until destroyed. Neutrinos were once thought to be like this also. Now they seem to have a rest energy of maybe 10eV. Compare this to the lightest known object, the electron, with a rest energy of 500,000eV.

There is a very intense oven of neutrinos out there too, that comes at us uniformly from all directions, in principle. It has not yet been detected because neutrinos are very inert. They could pass through billions of kilometers of solid rock with only a low chance of being stopped!! You could not shield them out, when moving at .999999c, with anything. But they would be of no danger, even if coming in with great energy from the forward direction. Most would pass through the ship and you without interacting. Still, there is a finite chance of interaction, so we can always catch some neutrinos in a detector that is only meters across. To catch many, you have to be very patient, or rich (bigger detectors), and hopefully you have a high dose source of neutrinos nearby as well, like a nuclear reactor. Someday the neutrino oven will be detected; we are pretty confident that it is really there.

Little is known about neutrinos at present. They come in two varieties, at least, which interact differently with lab sized detectors. Some claim three types exist. The oven neutrinos would be left from the Big Bang, early phase of the universe, just like the photon oven, only much older because they stop running into things at a higher density. Collisions can create particles called muons, which are like heavy electrons. They decay on the average in about 10^{-6} seconds and break up into an electron and one each of the two kinds of neutrino. (There is evidence of an even heavier electron type object and so there may be three different types of neutrinos.) The muon is about 200 times heavier than the electron and its mass is totally unexplained (as are all the other masses of the particles we know of!!). We have still barely scratched the surface, even with our fairly advanced knowledge of some of the great mysteries that abound around and within us. Muons provide us with microscopic clocks whose decay changes have demonstrated the speed effects on time.

You can readily see that we are in trouble when we try to view the universe from a very rapidly moving reference frame. Our clocks are slowed and our shapes are all flattened. If we are also spinning, like a frisbee, then the flattening remains toward the direction we are moving. It changes regularly and cyclicly for the atoms as they go around. The electron clouds can apparently handle this adjustment alright. But humans currently move with the earth at (2/1000)c through absolute space. We are shortened by $\sqrt{(1 - (2/1000)^2)}$ = .999998. The motion of the earth around the sun is only at (1/10,000)c and produces a small additional flattening, in varying directions, that is pretty negligible.

The famous Michelson Morley experiment was designed to look for our absolute motion, but it was defeated by the changes in moving clocks and rods. The Gamow radiation, which was predicted by him and his colleagues in the 40's and detected in the 60's, is the best way to measure absolute motion of any object. All objects can now be referred to their absolute motion relative to THE preferred, comoving frame, after being observed from our particular moving frame.

Here is a nice thought experiment to help illustrate these ideas: Suppose we have a long and hollow rocket with flash bulbs at each end. When a bulb goes off, a beam of light flashes across the open end of this tube ship, which is at rest. Next, we build a skinny, pencil-like ship that can fit freely through the hole in the tube shaped ship, but which is 5/4 its length, so it sticks out at both ends quite a bit. At the ends of the pencil are photo cell circuits that set off a firecracker when light shines on them. The pencil floats along in empty (but slightly negatively or positively curved and expanding) real space, and goes through the tube. We can flash the end lights at the right times to successively catch the ends of the pencil at the ends of the tube. The first firecracker is triggered by the light flash and it goes off on the pencil end that is about to emerge from the tube; then, shortly thereafter, the second flash occurs, as the rear end of the pencil enters the tube and it ignites that firecracker. The delay in time will be small but finite. Now repeat the experiment over and over, each time with a greater speed for the pencil ship. The moving pencil ship really shrinks as it moves. At a speed of $v/c = 3/5$, the shortening factor is $0.8 = 4/5$:

$$\sqrt{1-\left(\frac{3/5c}{c}\right)^2} = \sqrt{1 - \frac{9}{25}} = \sqrt{\frac{25}{25} - \frac{9}{25}} = \sqrt{\frac{16}{25}} = \frac{4}{5}$$

This means that the pencil ship is the same length as the resting tube ship. We now fire both flash bulbs at the same time and they each explode the moving firecracker which is momentarily at the end of the passing pencil. Both firecrackers go off at the same time, as recorded by our flash instruments at each end of the tube ship. Seems straightforward enough, even if rather strange compared to our every day, slow speed experience.

The pancake headed people on the pencil ship will be able to also observe what is going on here and write their own account of what happened. But we don't care what they find! They are all screwed up, obviously. But let us be generous and listen to them anyway. (We will humor them.)

They swear that they see themselves as normal (although we know differently). They also swear that we have flat faces and bodies and look really ugly. They claim that our tube is even shorter than before, when they were coasting slowly through it. It was already 4/5 as long as their ship then and now seems to them to be $4/5 \times 4/5 = 3.2/5$ as long as their pencil ship. Yet the firecrackers both blew up and we swear that the two photo flashes were set off at the same time. Clearly it is impossible for them to agree with that, since the front end of their pencil ship emerged from our short tube ship well before the rear end entered the tube, as seen by them. If we guess that they really do feel normal and they do see us shrunk, then we are forced to also guess that the two flashes do NOT happen at the same time for them. The exiting front pencil flash happens first and the entering rear pencil end flash occurs later, when the skinny tube covers only the rear portion of the pencil. With this strange appendage to the guessed physics that is going on here, we are all happy! They can think we are moving and we know they are really moving. They know it too, since the Gamow radiation is not coming at them isotropically. We don't need to worry about what they see anyway, unless we are engineers, rather than physicists. We physicists are after THE rules of the game as described in the best frame, and there is a best frame, obviously!

The pencil ship people can watch a third ship, say a saucer shaped ship, go by them and by us. We disagree on the shape it has though both see it flattened to some extent. Again, their opinion does not really matter. If we send a pulse of light along our tube and therefore along the pencil ship when it catches up to us. We can tell them how fast they are really going and how shortened they measure and how slow their clocks are really running. They can use all this to figure out how long a second of time should be for them, with their weird clocks, and how long a meter stick should be on their pencil ship, with their weird measuring tapes. Using their lousy instruments, they then find that the same light beam seems to traverse their ship at the same numerical speed c. Big deal. As we said, they are all screwed up. What is more interesting is the fact, apparently, that the speed of a pulse of light that they create moves through our comoving (preferred) frame with the same speed as light we create from at-rest sources. The light moves through the space, curved and stretching, with one speed, regardless of how it got made. Is that reasonable? How should I know what is reasonable? Ask God! What is IS, and we have to make the best of it.

Speaking of what is—I am sorry but there seems to be no Warp Drive available in our universe. No worm holes either, it seems so far. The bigger the fuel supply we put on board, the faster the final speed, up to about (1/10)c. Beyond that speed, we get less and less for our money. To reach 0.999999c, as conjectured before, would take so much fuel and be so expensive that no planet would waste the money to do the experiment, to actually see if people on board survived the extreme brain shortening. This also means that the interstellar transports will move at only about (1/10)c and no more, even if we survive to live the Buck Rogers life style. (Sorry about that, I should say the Star Wars life style, since most of you younger readers never heard of Buck Rogers.) At (1/10)c it would take about 100 years of coasting to reach the nearest two stars like our sun. But you couldn't easily stop. They would have to meet you and somehow slow you to a stop, at great expense in energy consumption. Nobody is going to make the trip until the other end agrees to pay for the stopping process! They probably will not agree to the expense of sending you home again. Big time space travel is a pipe dream, it appears. Wishing it were otherwise won't change the rules of nature. You need divine intervention, rather than better engineering, to change those rules.

I have thrown at you, all at once, the many marvels of how space and time appear to us at the present epoch in our evolution. I have not diverted into how we know all this—just the facts that have been revealed over the past century or so. New physics revelations lie ahead too, though perhaps limited to the subatomic domain. We may now know all there is to know about the large blob physics phenomena at a basic level. All that is left there is engineering and bio-chemistry. The earth's magnetic field is generated somehow and the mechanism is still unknown. However, once understood, we expect the answers to be within the known physics laws. The same for weather prediction.

In the next chapter, we go into more detail about how the large blob world seems to operate. From here on I will assume some introductory calculus and introductory physics background for the reader. If you do not have such, you can skip the math and still follow the beauty, I hope! However, readers tell me that this material is too advanced and mathematical to be of interest to anyone without at least one physics degree. I have failed then, just as Feynman did in his attempt to reach the beginning students with his deep penetration into the foundations of classical physics. I am disappointed, as I am sure he must have been. In any case, read on until you burn out and get what you can. We scientists have learned to skim and to skip the math in books over our heads. Non-scientists are not used to being confronted with something that does not read so easily, so you may get easily discouraged here. Try not to, for there is a lot of the structure of nature revealed here between the equations.

The Appendix contains a diagram for communicating in a closed universe which shows how things work out if we could communicate with someone almost half way around the universe. This is very helpful since closed space is so unintuitive.

CHAPTER 4

Big Object Motions and Interactions

We are big blobs. How big? Our brain's have the essence of 'us' and they have about 10^{11} neurons. Each is connected to thousands around it. Each has millions of atoms in just the right places and of just the right types to make us be alive. But that is not enough. We need to see and touch the world and our sensory appendages obviously do that for us. They feed the brain inputs and the brain forms images and theories about the outside world. This is very very important! We are huge and we receive signals through huge numbers of atoms in the dendrites. We never can see or touch small things like those that literally make us up. There are NO little brains with self aware consciousness. How many atoms are in our brain? Well, we have a volume of about $(10\text{cm})^3$ and a density of about water, 1gm/cm^3. Thus the mass is 10^3gm or 1kg. Water has an atomic weight of 18 so 18 grams of water would have 6.2×10^{23} molecules of H_2O in it. Therefore, our consciousness is contained in a blob of atoms having about $[(1,000)/18]\ 6.2\times10^{23} = 3.4\times10^{25}$ atoms. We have calculated for a 'balloon full of water' but the numbers would be similar for an actual brain. We are 10^{25} atoms collectively thinking, yet these atoms are trying to study one atom at a time! A hopeless JOKE!

But give us 10^{24} atoms in a single blob to study, such as a baseball, and we can deal with that directly. What do we learn about baseballs in empty space? (1) Throw one and it travels a 'geodesic' in the curved and stretching but nearly flat space out there right now. This is a 'straight' line. Shine a photon in 'empty' space and it travels the same path, more or less. Make a reliable clock and use it to time the flight and find that (2) the ball travels at almost constant speed. It slows very gradually due to the Gamow radiation hitting it, as discussed in the last chapter.

Next, throw another ball at the first and measure their speeds and directions very carefully before and after they hit. We then find that (3) they 'bounce' and take up new geodesic paths with new 'constant' velocities (speeds and directions). Next throw a third ball at one of these two balls, but somehow throw it at 0.999999c. After they hit we may have NO baseballs left. Zillions of atoms, chunks, and created quanta will fly out in all directions. We no longer have a big blob to look at. End of experiment. We leave the world of big blob physics this way.

Next go back and arrange to push 'evenly' on the ball that was not 'exploded'. We find that (4) it changes speed and direction, depending on how hard and in which direction we push. When we stop pushing, it returns to geodesic motion with a new velocity (speed and direction).

But how much does the motion get changed by the pushing? Experiment shows that

$$\boldsymbol{F} = \frac{d}{dt}\left(\frac{m\,\boldsymbol{v}}{(1 - |\boldsymbol{v}|^2/c^2)^{1/2}}\right)$$

and

$$W = \int_{r_o} \boldsymbol{F} \cdot d\boldsymbol{r} = \frac{mc^2}{(1 - v^2/c^2)^{1/2}} - \frac{mc^2}{(1 - v_o^2/c^2)^{1/2}}$$

The W is work done by us (how tired we get) in pushing the ball which has mass m. Notice that, as $v \to c$, we have $W \to \infty$, so no baseball can reach the speed of light. There is no Warp Drive in our universe!!

The interesting question is, "Why should balls respond like this when pushed?" The answer is that the ball consists of about 10^{24} atoms. These atoms are little blobs and obey a whole other set of rules, to be discussed in the next chapter. When this many of them swarm together and hold on tightly to each other, the microscopic rules can be crudely approximated to describe the whole blob as one object. These microscopic laws then lead to the equations above as a crude approximation. This equation is only an engineering level, useful result. There is little deep physics in it. It says that if you tell me how you push in detail, I can tell you the path in space and time that the collective blob will take in response, and I can calculate how much energy you exhaust in the process of doing the pushing. Sounds like engineering? It is. It only works for big blobs.

There are other forms of and other consequences of these equations. These are all theorems and they tell us nothing really important, except that they too crudely mirror the deep, quantum level laws that they approximate. Besides having engineering applications, their 'mirroring' of deeper physics is their greatest usefulness. The quantum laws are very hard to find. The large blob laws give us some hints. The F equation leads directly to the W equation above, and also to momentum conservation, where

$$\boldsymbol{P} = \frac{m\,\boldsymbol{v}}{(1 - |\boldsymbol{v}|^2/c^2)^{1/2}}$$

For v/c < 1/10, we have the useful cruder approximations $F = ma$, $W = 1/2mv^2 - 1/2mv_o^2$, and $\mathbf{P} = mv$, which historically were discovered first, of course.

ELECTRICITY

Besides pushing mechanically, we can put a large electrical charge, q, of say 10^{-4} Coulombs on the ball. This corresponds to putting about 10^{15} electrons on the ball. It then responds at a distance to charged rods brought near it, and to large currents flowing in wires brought near it. We find that these pushings and pullings obey, empirically, the following force, F, rules:

$$\boldsymbol{F}_m = q\boldsymbol{E} + q\boldsymbol{V}_q \times \boldsymbol{B}$$

where E and B are the 'fields' imagined to exist at m due to the nearby big charges and currents for other big blobs. Again the real issue is, why this particular form of interaction? The answer is the same as above. The ball has 10^{24} atoms and about 10^{15} excess electrons on it. The nearby blobs that influence it are charged and either static or moving, or they may instead have maybe 10^{20} electrons/second flowing through them. These individual, atomic sized quanta interact with each other in all these big blobs and the rules down there are very complicated. They crudely approximate to the F rule above, overall. The E and B fields are not really big blobs. They consist of zillions of virtual light quanta which are individually little quantum blobs. They don't behave quite as simply as baseball blobs, when acting collectively as a 'big blob' of field.

GRAVITY

The baseball has a mass m and it is attracted to other masses M, brought near it. The force of interaction is very weak (about 10^{-40} as strong as electrical forces—no one knows why!). The nature of this pull is very complicated, even at the big blob level. Einstein guessed that M even warps the nearby space itself. That space is also a part of the general universe so it is overall warped and flattening on a huge scale. This is part of the same gravity guess, since all the mass in the universe causes curvature everywhere. Each body warps the space around it and together they warp the whole cosmos. A body moving near the body M, such as a baseball, follows a 'geodesic' in this space—a 'straight' line generalization.

Is this curved space theory also a crude approximation for quanta interacting at the little blob level? No one knows! We guess that it is, but we have NO proof yet! The pull of gravity, crudely viewed as a flat space pull, is approximately the Newton guess

$$F_m = -\left(\frac{GMm}{r^2}\right) = \frac{-mc^2R_s}{2r^2}$$

where $R_s \equiv 2GM/c^2$ is an important physical length parameter associated with the mass M. The pull is more accurately described by the precession inducing

$$F_m = \frac{-GMm}{r^2} + \frac{-6G^2M^2m/c^2}{r^3} = \frac{-mc^2R_s}{2r^2} + \frac{-mc^2R_s}{2r^2}\left(\frac{3R_s}{r}\right)$$

or even more accurately by my formula

$$F_m = \frac{-GMm}{r^2}\left(1 - \frac{R_s}{r}\right)^{-3}$$

This last form shows better that at short distances, $r \approx R_s$, strange things should happen, unless you keep the body M only at a much larger r. You may recall earlier that time slows in a gravity field by the factor $\sqrt{(1 - R_s/r)}$, so time stops at $r = R_s$, the so-called Schwarzschild radius. A black hole is a star that has radically shrunk inward until the blob is smaller than its previously theoretical Schwarzschild radius. R_s is about 4 kilometers for a crushed sun-mass star. Extremely dense, to say the least! This is a big blob but is it really a quantum blob? Most likely it is, at this density! We don't know when quantum gravity must replace the crude classical equations, for we have no quantum gravity theory yet. Light that is pulling away from a star is a quantum, tiny blob. It changes color as it pulls away, indicating a loss of energy (but not speed). This is essentially a quantum phenomenon and really needs a quantum explanation. That requires a quantum gravity theory, which does not exist yet, as we said many times.

Einstein threw out the F equations for gravity pulls, and he replaced them by curved spacetime. This mainly is important for the whole universe. We may even have a relatively small (10^{24} stars), closed universe, as described in the last chapter. His classical equations, when applied to a fairly dense, large blob universe as a whole, lead to the following equations, where 'by' means billion years and 'bℓy' means billion light years distance:

$$t = (63.4by)(\theta - \sin\theta)$$

$$R = (63.4b\ell y)(1 - \cos\theta)$$

$$d = R\theta$$

$$v = \left(\frac{\sin\theta}{63.4by(1 - \cos\theta)^2}\right)d \equiv Hd$$

Here, $\theta = \pi$(radians) gives t = 200by. This is the time of maximum expansion for this universe model. At $\theta = 2\pi$, the universe has re-collapsed to zero size, where it started at $\theta = 0$. This 'size' is given by R, such that d is the distance at t between two, say, galaxy clusters. The v is the speed away of one as 'seen' from the other if instantaneous signal movement were possible (it isn't). The universe became transparent to light at $\theta = 0.031$ radians (1.8^o) when R = 0.03bℓy, about 1000× smaller than today. This model is built on H = (1/19.7by) and a total universe mass of 10^{24} solar masses total, about 10× the mass we see now. (See Chapter 9 for more math details.)

These equations describe in detail how the positively curved and finite universe expands and recontracts, if it has a total mass of 10^{24} suns (current density 6.34×10^{-30}g/cm^3) and if the expansion rate right now is given by (1/19.7by) = (45.7km/sec/MPS). These values are only representative. If the current density is much lower than this case, then the overall curvature is instead negative and the expansion is eternal. In that case, there is no natural edge. Such a universe may be truly infinite in total mass, if negatively curved. Then this universe was already infinite from the very beginning and has had no boundary—ever. Rockets, traveling to the planets, need the right course prior to their burn out, so as to be able to reach the desired planet after years of coasting. This is an engineering problem, really. The engineers use the $F = ma$ and $F = GMm/R^2$ forms and calculate the needed heading and speed at burn out. If they wanted to hit very precisely, they would need to use the full curved space approach. The calculations would be horrendous! But accurate! Of course, if we try to get too accurate, we get back down to the quantum level. There we know nothing of gravity at present.

That's about it for the basic concepts in big blob physics. The rest is engineering applications. Hard calculations but nothing really new besides what we have covered. Electricity is really a quantum level interaction, so only crudely dealt with in the big blob interaction world. That is the world of human travel and the motion of objects our size or larger, like planets and stars.

TIME FLOW

We actually see big things move in 'flowing' time. We remember past paths for such blobs and wait to see their future paths. We cannot ever know the future paths for sure. This is because the F equation is a crude approximation. It predicts the exact future, but nature is not quite like that. For big blobs it works OK, most of the time. For little blobs it never works reliably. Therefore, it can also fail for large blobs. In other words, the future is NOT predictable, even at the big blob level! If we hit two baseballs together, head on, each moving at 0.999999c, we have quantum physics really here! There are many possible outcomes and each has only a probability. Slow collisions just boost the odds for one outcome and reduce the odds for everything else that could really happen still. Thus the outcome seems reliable. Baseball batters know the likely outcome of a careful swing.

We return now to the issue of our huge size limitations. We humans get to know about only big blob crashes and changes, ones that produce big chunks flying out from the crash. Therefore, we test our ideas only in such direct observations. We usefully theorize only about stuff that has such big blob consequences, or else we are doing art and fantasy, not physics anymore. This is a terrible perversion of reality. We hate to face it. But we must face it and deal with it. We only experience things in about millisecond time intervals: big blobs of input go to the brain and these produce outputs to the big blob motor appendages. That is us, our nature, and we want theories that work. Therefore, we cannot assume that our big blob experience is at all close to what is really going on down there in the little blob world. For example, we see things localized, such as baseballs. They have edges and march in an orderly way through space. They are never two places at once and they always move smoothly (no gaps) as time flows forward. That is our crude, direct experience of the big blob world that we can relate to.

In the little blob world, things are all over the place at once and are moving both forward and backward in time as well. It is crazy down there! I guess it is a good thing that we are large, for only there, in the large, can our consciousness make sense—orderly thoughts and memories—one after another. We are somehow more than the sum of our chaotic parts, for we are not chaotic. The microscopic chaos smooths out, somehow, to allow us to be, think, and communicate, in time.

Now to real experiments: A real, big blob experiment is very special. At a time t_1, we activate a large number of identical big blob set ups, arranged side by side. Things happen in each set up that we know not of. At a later time, t_2, we observe directly, with eyeballs or cameras, the big blob condition in each of the duplicate set ups. They may all have about the same final arrangement. In that case, we didn't need all the duplicate set ups in the first place. Such situations we call classical experiments. One set up is sufficient and the outcome is reliable and predictable. A bunch of roulette wheels is a good example. All are set up and released 'exactly' the same, at t_1, (as much the same as big blobs can control). This would result in all the balls being in the same number location on all the wheels in the end, at t_2. This is because we started everything 'exactly' the same. The other wheels tell us nothing new. A restart at t_3, with the same initial ball and wheel conditions, gives the same outcome number at t_4, and over and over again. Once in a while, we may see the ball wind up in the adjacent number slot, due to the initial conditions being at the extreme of their range that we can control. An important point here is that we must turn off the lights to do it right. We set up the t_1 conditions accurately—ball location on the lip; ball speed and direction; speed of the wheel; location of the wheel, 'exactly' when the ball is released. Then the lights go out. (The air was all pumped out in advance to avoid air currents that we cannot control.) We see and hear nothing. At t_2, the lights go on and we see the ball in one slot of the 38 possible slots. The final state is observed very carefully: The wheel's new speed; the ball's position and its movement within the slot; the wheel's location when t_2 occurred and the lights went on. Where is the physics here? The physics here is our learning the rules that lead to the t_2 observation from the t_1 observation. Suppose that the t_1 observation is a little crude, in accuracy, then the ball could land in, say, slot 13 or 14. In this case, we would need a large assembly of identical wheel set ups to get all the t_2 results we would need, in order to search for the rules that are working here. We do not know what actually happened between t_1 and t_2 in the dark. In developing a big blob theory of what happened, we would calculate what we 'think' happened here. For example, we could use the F equation and try to theoretically follow the ball in the dark as it is calculated to roll and bounce off the various projections that stick up, so as to wind up in a particular, predicted slot at time t_2. This would not mean that this actually happened. The ball may have disassembled into its 10^{23} individual atoms and they may have floated over the top of the wheel to reassemble in the final slot. They may have zipped over to other slots and even traveled backward in time for a while, when the lights were out, before settling into the final slot just as the lights were turned on at t_2. We cannot know what really happened! We theorize the simple bounce pattern as most important, compared to all the other theoretical possibilities. In this case, we are right that this gives good enough outcome probability projections. We actually just got lucky here. For little blobs, these other weird possibilities become important and cannot be neglected. So, even here in this huge blob situation, we cannot say that the ball really bounced along in a simple way to reach the final slot. That is only one of a zillion possible things that could have happened while the lights were off. The classical motion is just the very very dominate 'amplitude' among the zillion other amplitudes. This is the NEW physics of the 1920's. We don't know what we don't actually see!! Don't be a fool and trust your intuition! Fortunately for our engineering lives, we can get away with trusting our hunches as to what happens for large objects in the dark. Big blobs are pretty reliable even in the dark. We trust them to behave as expected, such as when walking up the stairs in total darkness. We trust the stairs to support us, even though they don't really have to. They could actually let us tunnel through without leaving any broken steps. The odds are very small, but we simply would find ourselves falling through space below the stairs that we were just walking on. To get such crazy results, we step on the first step and then we turn out the lights. The step atoms and our atoms are then at liberty to do all kinds of things. Most probably they will simply stay in place and we will walk normally up the stairs. But they can fly apart and zip all over the place before reassembling us at the top when the lights go on again. We cannot be sure! OK, maybe only a dozen atoms likely fly off and they are all over the place at once for a time, before returning to their proper locations on our elbow, or such. We still have to learn to think this humble way! All possibilities can occur in the dark, even ones that violate conservation of energy and momentum! With the light on, a continuous series of measurements is being made at t_1, $t_1 + \epsilon$, $t_1 + 2\epsilon$, $\cdots$, where ϵ is very very small. The possibilities are then greatly restricted.

We cannot talk about what really happened between t_1 and t_2. 'Stuff' went crazy and flew forward and backward in time 'between' t_1 and t_2. The smaller the blobs, the greater import this weird behavior has. The cooler the system, the greater import this strange behavior has, even for big blobs! Liquid helium-4 can do some very strange things in the dark due to the helium atoms being all over the vessel at once and acting like a collective thing. The liquid misses the spoon if you try to stir it—zero viscosity! It climbs the walls and pours out on the floor. The world is weird and we do not really see it well at all. We sit up here in our huge blob composite wad world 'being' one conscious thing, seeming to have a well defined boundary in 3-space dimensions, while being made up of some $10^{27}+$ 'things' that move both ways in time and are all over the place at once. They make us 'extend' our boundary inside each other in some sense, since they are not localized, either in space or in time. We are not as unique and autonomous as we let ourselves think we are. Again, we don't know very much about reality. We just muddle on doing the best we can to grasp the reality around us and inside us. We want little blob know-how, but we must wait to see big blob consequences. From those consequences we try to construct ideas about what the little blob rules are. Our big blob observing limitation is very frustrating and depressing!

If we could become spirits and leave our 10^{27} atoms behind, we could go down there and see inside protons and electrons. Just ain't possible. Too bad. Maybe God will give you that treat in the next world, if there is one. Most people could not care less what the rules are anyway. They focus happily on big blobs, and only such, all their lives. Is that alright? Who knows? It all depends on the meaning of existence if any and the purpose of existence, if any. We need some standard by which to judge what is important—which ways of spending our time and energy are foolish, if any are.

CHAPTER 5

Big Blobs Moving in Space

Moving blobs somehow interact with the curved-space-vacuum in such a way that they become shortened in the direction they are going, but not in the two perpendicular directions. This is impossible to test directly, so far. Muons, when created moving at a high speed, live longer before they explode. But they are very little blobs. We don't know what is going on inside them. Atomic clocks are large blobs and they seem to tick slower in the same way when moving. Fast motion through the vacuum slows time 'in' the moving object. The effect lasts as long as the object moves fast. This reduction in the passing of time is accumulative and 'must' be due to the vacuum that the clock is moving through. An atomic clock, that is moved slowly into a strong space-curvature region (strong gravity field), will have time passing slowly for it, all of the time that it just sits in the highly curved space. If sitting in curved space can slow time then surely moving in space, flat or curved, could also influence the passing of time. The one time effect is no more weird than the other length effect. We presently have no idea how it works. We just have empirical formulas for how big the effect is, in both of these cases. This view is heretical in the 20th century but will become orthodoxy in the 21st century. Einstein set us back in some ways, as well as advancing us a lot.

Twins that are moving out with equal speeds but in opposite directions, relative to the Newton frame (absolute rest frame), age slower biologically due to their motion in the space. If one twin retrofires her rocket and turns around to acquire enough speed to over take her sister, then she moves even faster in the space and is younger when she passes the other sister (who coasted all the time). This is reasonable.

If the twins initially started out at, say, 60° to each other in direction, then one of them can 'gently' change her direction if moving toward a very big rock in space, at rest. She whips in behind it, coming close, and out the other side moving in a new direction and back at the old speed again, when later far from the rock. While near the rock, she moved through curved space and this slowed her ageing some. She didn't stay in the curved space very long, however, and she never experienced any strong accelerations, contrary to what she would have experienced in the retrofire case above. She only felt tidal forces tending to stretch her as she flew past the big rock. So she knows it happened. She then could be headed on a course that will intersect the other twin's path. They won't both get to the crossing point at the same time unless the speed of the straight line twin is chosen correctly. The 'bending' path twin could fire up and gain some extra speed, unless she had the right amount of extra speed in the first place, or the rock was moving and she 'sling-shots' by it, gaining some speed at its expense. In either case, the bending twin will, of course, be younger (for both speed and curvature reasons) when they pass each other the second time. There is no paradox here. Both twins have pancake faces, as they travel, and what they see, looking at each other, is not important. We can predict how much older each is when they meet again the second time. That is physics—what is seen in the Newton (God's) frame. We must be able to make successful predictions in that frame, to say we know something. We still don't have the mechanism at hand. How does space do this to people?? The answer lies at the little blob level, of course, but it is still unknown to us at present.

Exactly how much does a moving clock or twin age when moving in locally empty (fairly flat and slowly stretching) space? The answer seems to be

$$\Delta t' = \int dt' = \int \frac{dt}{\sqrt{1 - v^2/c^2}}$$

where v is the absolute speed compared to c, the local light speed, and dt is the time interval, in the Newton frame, during which they are 'seen' to be moving at speed $|\mathbf{v}|$. Notice that their direction does not matter here—straight line or circle, or whatever the rocket can produce. The speed does not need to be constant. This is a rather amazing result. The accelerations experienced do not show up here, even though they could kill the twin if strong enough! In that case, the formula only applies for the 'clocks' on board that are not 'broken' by the strong accelerations. Verification of this result exists only for gentle accelerations of large clocks, like the ones carried around the world on regularly scheduled airlines. But it has been verified for muons going straight and for muons curving sharply in a strong magnetic field, using the $q\mathbf{v}\times\mathbf{B}$ (sideways) force referred to earlier. Accelerations here are in the ballpark of 10^{18} gees. The accelerating muons live the same extra time before exploding as muons going straight at the same speed. Nuclei accelerate violently back and forth about their average positions in a hot iron bar. The photons that they shoot out from some nuclei have a very accurate color that depends on the speed of the oscillating nucleus, when it shot out the photon. Analysis of the color range obtained shows again that the accelerations don't seem to influence the nuclear emissions directly. Only the speed does.

Rindler calls these independence of direction and acceleration effects The Clock Axiom. His relativity book discusses it more than any others I have seen.

It would be wonderful if we could somehow come up with a more comprehensive algorithm to tie all these empirical motion effects together. There is length contraction, time dilation (independent of accelerations), loss of simultaneousness (simultaneity), and color shift for light emissions from a moving atom, called the Doppler effect. (Doppler originally observed this effect on sound waves from moving trumpets on a rail car.) The shifts are different in detail from sound, for a moving atom in space, but source motions produce a frequency shift for all wave motions. The moving atom interacts directly with the space and gets flattened as well as slowed in 'internal' time. These somehow result in the photon, that is created and simultaneously emitted, coming out with its color depending on its direction of ejection, relative to the atom's speed direction through the vacuum. This is not surprising in principle, since all atoms are messed up by their motion, but determining just how much the color is affected is not so easy to calculate. See Chapter 6.

WHY RELATIVITY

So, now we get down to it. WHY is the vacuum able to do all these wild and wonderful things to moving big blobs and to their light emissions (such as from a moving and rotating flash light, sending beams in different directions, even with a narrow band, color filter over the lens)? Why does the world work this way? Watch out. We are treading on theology. Why did God design the universe so and so—because he/she/it felt like it—who the hell are you to question it! We cannot go very far in that direction.

The only answer I have is, again, complex QUATERNIONS. We saw earlier that they seem to be the corner stone of the design. They fit with $3+1$ space and time dimensions in a natural way. They have two natural, antiautomorphic conjugations. These give us two natural groups. Complex quaternions lead, by direct products of their closed subsystems, to the Dirac algebra with its many natural conjugations. This gives us many larger groups. But the many conjugations here give the SAME group. Isn't this fantastic??! The world seems to be built on mathematical pillars in a limited range. The Dirac algebra relates to electron clouds in atoms, as we shall discuss in a later chapter. Therefore, this master group in the Dirac algebra should be important in the design. It's subgroup, SL(2,C), should then also be important, as well as the other subgroups.

SL(2,C), which comes from complex quaternions, turns out to have formulations in it that fit with the observed effects of the vacuum on clocks and rods and light beams! The details are called Lorentz covariance and are not easy to understand. They are straightforward, however, even if tedious. You may wish to skim the rest of this chapter if not really interested in the mathematical nitty gritty. Do not completely skip over this, however. It is the poetry of reality!

MATHEMATICAL RELATIVITY

The Lorentz group is $AA^{\neq} \equiv 1\sigma_0$, where $A = a^{\mu}\sigma_{\mu} + b^{\mu}i\sigma_{\mu}$ and $A_{\epsilon\delta} = 1\sigma_0 + \epsilon^k i\sigma_k + \delta^k\sigma_k$ is a group member 'close' to the identity element, $1\sigma_0$. The numbers ϵ^k, k = 1,2,3, and δ^k, k = 1,2,3 are very small (like 10^{-20}) so that $\epsilon^1\epsilon^1 \approx 0$ ($10^{-20} \times 10^{-20} = 10^{-40} << 10^{-20}$), etc. The group has an infinity of members but is characterized by having 6 free parameters to specify these members. These 6 parameters turn out to be: (1) relatable to the three angles required to specify a rotation of the coordinates in three dimensional space (two angles to give the axis of rotation direction, and one angle to give the amount of rotation around that axis); and (2) related to the three speed components along the x,y,z axes of an object moving through space in a straight line. Physicists call these last 3 'boost' components and the first 3 'rotation' components. The pure rotations form a three parameter subgroup and are generally described by $AA^{\neq} = 1\sigma_0$ and also $A^{\neq} = A^*$. The rotation group's infinitesimal members are of the form $A_\epsilon = 1\sigma_o + \epsilon^k i\sigma_k$. They are 'pure' quaternions. Products of rotations, $A_\epsilon B_\epsilon = C_\epsilon$ give another rotation about a third axis through some angle. This one rotation produces the same final setup as the two, separate rotations do. Notice $B_\epsilon A_\epsilon = D_\epsilon$ gives a different third rotation in general from $A_\epsilon B_\epsilon$. (Check this by rotating a book 90° around its perpendicular axes with different orders of rotation.)

The speed boosts have infinitesimal members like $A_\delta = 1\sigma_0 + \delta^k\sigma_k$. They don't form a closed subgroup because $\sigma_1\sigma_2 = i\sigma_3$, whereas $i\sigma_1 i\sigma_2 = -i\sigma_3$. The $i\sigma_k$'s are 'closed' under multiplication. A product $A_\epsilon B_\delta = C_{\epsilon\delta}$ generally gives a mixture of boost and rotation instead of another pure boost, unless both boosts are along the same direction in space (say, the x axis, for example). This is all just mathematical fun and games! It need have no physics in it, but it does, somehow! The boosts, in particular, are related to length contractions, time dilations, and even the color dependence on beam direction for a moving source of light. We shall save the color dependence for the next chapter, since it really is a little blob phenomenon.

We must next bridge our way from the pure math above to predictions about distance and time for moving platforms of humanoid size. This won't be easy. We will need to display a 'representation' of the Lorentz group. By the way, Lorentz came upon this group in 1900. He discovered it to be a part of Maxwell's equations, which describe how moving charges make electric and magnetic effects. We shall see these equations in Chapter 6. They are not needed, however, to come up with the Lorentz group. We just came in through that side door in our history on our planet, as our physics knowledge evolved, because physicists didn't take Hamilton's discovery of quaternions seriously enough after 1843. Perhaps they did better on other planets out there in the cosmos.

Here we go then: An event is a 'flash' localized to a small volume of space and lasting for a short time. We describe an event in flat and static (hypothetical) spacetime by

$$x = x^{\mu}\sigma_{\mu} = x^{*} = ct\sigma_{0} + x^{k}\sigma_{k}$$

and call it a 4-vector of simple spacetime. Here c is the speed of light (thrown in here for dimensional uniformity), t is the time of the event, and $\{x^1, x^2, x^3\}$ is the cartesian location of the event {x, y, z}. The x^k notation saves space and that is its only purpose. Here $\mu = 0,1,2,3$ and we sum on repeated indices as we did earlier. (This is only a convenient short hand.) Next, we imagine 'inventing' a new description of the same event through the strange process of defining

$$x' = x^{\mu'}\sigma_{\mu} = A^{*}xA$$

where A is any Lorentz group member ($AA^{\neq} = 1\sigma_0$). First notice that $(x')^{*} = (A^{*}xA)^{*} = A^{*}x^{*}A^{**} = A^{*}xA = x'$, so x' does not have any $i\sigma_{\mu}$ parts. It is also a 'space' 4-vector. We start with $\{x^{\mu}\}$ and choose A. This produces $\{x^{\mu'}\}$. We can prove, by only mathematics, that (theorem) $S \equiv x'^{\neq}x' = x^{\neq}x = [(x^0)^2 - (x^kx^k)]\sigma_0$. We say S is invariant under the group of Lorentz transformations of events. This works if $AA^{\neq} = 1\sigma_0$. So what if S is an invariant of the Lorentz group? Well, ain't it pretty at least?

Notice that S is positive, zero, or negative, depending on the size of $x^0 = ct$ compared to $x^kx^k = x^2 + y^2 + z^2 \equiv r^2$. A flash at (0, 0, 0) at t = 0, and a flash at (x_1, y_1, z_1) at $t = t_1 > 0$ can be thought of as representing two events separated in space and time. Any projectile fired from the location (0, 0, 0) at the time t = 0 will reach (x_1, y_1, z_1) at some time t_1. If we choose t_2 to be the time interval that a light pulse would need to reach (x, y, z) then $t_1 < t_2$ means that the pulse from the first flash was still in transit when the second flash already occurred. Then $S < 0$. If the second flash occurs at $t_1 > t_2$, then the photon light pulse had already passed (x_1, y_1, z_1) when the second flash occurred there and $S > 0$. Since we now think that no physical pulses under human control can travel through space faster than light, whether $S > 0$ or $S < 0$ is significant. (That is the only reason.) $S < 0$, we say, is a 'space-like' distance and $S > 0$ is a 'time-like' distance, in spacetime. (Distance squared, actually, of course.) People have theorized that Tachyons exist and that they travel at $c < v < \infty$ only. Big blobs like us, if we could emit such pulses and could turn them on and off at our will, as in morse code, could produce some insane results, in principle (like receiving an answer before you ask the question). Therefore, most physicists believe, as a matter of faith, that such control is impossible. Little blobs, however, have a nasty way of sampling events far from where they are supposed to be confined in space and time. They sneak out far and fast in space and even go backward in time, but we cannot. There is a great distinction in nature between the worlds of big and small blobs. It is almost like two different worlds in one.

When t is very small and {x, y, z} very small, we still can have $S > 0$, $S < 0$, or $S = 0$. But down there, the rules seem to be completely different from the ones we big blobs have to go by. We cannot go faster than light or backward in time or 'know' the future, but they can!

Now let us get back to the x^μ representation. Choose $A = cos(\theta/2)\sigma_0 + sin(\theta/2)i\sigma_3$. Choose $x = 1\sigma_0 + 1\sigma_1 + 0\sigma_2 + 0\sigma_3$. Calculate $\{x^{\mu'}\}$ and S. Find that $x^{0'} = x^0$, and $S' = S$. Also, $\{x^{k'}\}$ is a point, $\{x^k\}$, rotated around the z axis through an angle θ. It is a tedious calculation, but it is also good practice for those so inclined to such challenges. Next, look up sinh and cosh and choose $x = \{x^\mu\}$ and $B = cosh(\theta/2)\sigma_0 + sinh(\theta/2)\sigma_1$ and calculate $\{x^{\mu'}\}$. Let $tanh\phi \equiv V/c$, so V determines ϕ, and rewrite $\{x^{\mu'}\}$. Now V's show up in the equation, indirectly. This 'is' a boost along the x^1 axis. Here, the new coordinates are related to the old ones by the result (prove this for yourself as outlined above):

$$t' = (t - \frac{V}{c^2}x)\gamma$$

$$x' = (x - Vt)\gamma, \quad \gamma \equiv (1 - V^2/c^2)^{-1/2}$$

$$y' = y, \quad z' = z$$

Please do try to work out the details, if you like, and see that it is really true. Again, so what? Well, consider the special case $x' = y' = z' = 0$. The event is at the prime frame origin. Calculate t'. Find $t' = t(1 - V^2/c^2)^{1/2} < t$. Look familiar? We guess that this somehow **physically** means that a clock, moving absolutely at speed V through the vacuum, will run slow by the amount shown above. This is actually confirmed by experiment! Lorentz has TRUTH for the real world and Lorentz comes from complex quaternions.

Next, consider the special case $t = 0$, $x' = L$, $y' = 0$, $z' = 0$. Calculate t' and x. Find $t' = (-V/c^2)x\gamma$ and, eventually, $x = L(1 - V^2/c^2)^{1/2}$ after some algebra. We guess that this **physically** means that a moving rod, which had a length L before being set into motion through the vacuum, acquires a shortened length x as it moves. Its length goes toward zero as its speed approaches that of light! (Note also how the previous case shows that time stops, as the clock approaches the speed of light.) Notice also that $t' \neq 0$ for this case of the moving rod, even though $t = 0$ here. What could this mean? We think it means that a 'well calibrated' coordinate system of clocks, riding on the rod, would find two explosions which really occur simultaneously at the ends of the passing stick ($t = 0$), with $x = L(1 - V^2/c^2)^{1/2}$ between them, would seem to have occurred with a time lapse t' for the clocks 'properly' calibrated and riding on the rod. For example, the two explosions could stop the clocks at the ends of the rod, near the explosions, for later inspection. Of course, this only tells us how to instruct the riders on the rod to calibrate their clocks 'correctly'. Since they are screwed up observers, their calibration is not important anyway, so $t' \neq 0$ is not as important as $x < L$. The loss of simultaneity, $t' \neq 0$, $t = 0$, is curious but I know of no profound implications.

We have seen here that the Lorentz group is so powerful that it **SUGGESTS** the world should have these peculiar manifestations for moving, big blob clocks and rods! The Lorentz group also shapes the laws of little blob physics. Somehow these laws for the subatomic world produce the results in the big blob world above. These big blob results seem to come directly from the Lorentz group structure as a purely mathematical 'machine'. It must remain 'magical' until we have successful little blob level explanations.

A second boost can be performed along the same x' axis, with a new speed parameter Q, relative to the boosted observers riding on the moving rod. The Lorentz transformations can be compounded to produce a curious and interesting result of 'relative' velocities. We in the Newton (absolute rest) frame see the aircraft carrier moving with absolute speed V. The people on the ship are screwed up observers, remember, but they still can measure a car, moving along the deck in the direction of the ship, to have some speed Q, using their lousy clocks and lousy measuring tapes. The real speed of the car, S, over the bottom of the ocean, is then found to be

$$S = \frac{V + Q}{1 + \frac{V}{c}\frac{Q}{c}} < V + Q$$

by using the Lorentz transformations twice. If $V = (4/5)c$ and $Q = (4/5)c$ then $S = (32/25)/(25/25 + 16/25) = (32/25)/(41/25) < c$. I throw this in to show that you cannot beat the speed of light in any way. Notice that Q = c gives $S = (V + c)/(1 + V/c) = c$. This again shows that the aircraft carrier frame has calibrated 'correctly' if they get the same numerical speed value for light that we know it has, with our well calibrated instruments in the Newton frame. Of course, the ocean floor is not really the Newton frame that is needed here, due to the earth's rotation and mainly due to its absolute translation through space, coming from the motion of the whole galaxy and the sun's motion around the galaxy.

We have ground through a lot of tedious algebra and trig, here, but we came up with some startling results. The simple idea $AA^{\neq} = 1\sigma_0$, as a general cornerstone of reality, leads to time dilation and length contraction as reasonable guesses for the real world. Sure enough, we find that they seem to be real. We still have no mechanism as to how the vacuum really does this. That solution is for you young scholars and your offspring to explore, in coming decades.

A big blob clock, in circular orbit around a huge, spherical rock, out in space, will run slow due to that motion, and it will run slow due to the curved space (gravity) produced by the rock. The motion effect we have already 'dealt' with above. The curved space effect also applies to a stationary clock, violently retrofiring its rocket to hold position at fixed r from the 'center' of the rock. Einstein's guessed-at-equations for how mass curves space were so complicated that he could not solve them for such a rock. Schwarzschild was already an expert in curved space, pure math, and found the solution soon after Einstein published his theory. The result is surprisingly simple to write down, but (as always) we have to guess like crazy to give it interpretation. We must guess what it says about real, eyeball observations that we big blobs can actually make.

The curved space around a stationary, spherical rock of mass M is described by the 'metric' formula (Schwarzschild's solution of Einstein's theory):

$$(ds)^2 = U^2(cdt)^2 - U^{-2}(dr)^2 - r^2(d\theta)^2 - r^2\sin^2\theta(d\phi)^2$$

where $U^2 \equiv (1 - R_s/r)$ and $R_s \equiv 2GM/c^2$ is the theoretical Schwarzschild radius for the rock. Notice first that $r \to \infty$ means $U \to 1 - 0$. Thus, empty space, far from the rock, is supposed to have

$$(ds)^2 = (cdt)^2 - [(dr^2) + (rd\theta)^2 + (r\sin\theta d\phi)^2]$$

$$= (cdt)^2 - [(dx)^2 + (dy)^2 + (dz)^2]$$

You recognize this as very similar to the S discussed above for Lorentz. It is essentially the same. The Lorentz group led us to S in a totally abstract way. We see it is the spacetime distance for flat space. The Schwarzschild solution is not really correct because there is NO flat spacetime in the universe! The space around the isolated rock should be thought of as full of thin sawdust, extending far out or even to infinity. The total mass of all the sawdust should greatly exceed the rock's mass and should be falling inward radially, near the rock, yet expanding outward at large distances, very far from the rock. The $(ds)^2$ would then never have the simple, empty space form above, as $r \to \infty$. In fact, this whole, idealized sawdust universe could be closed by the sawdust density, if high enough. See Chapter 9. You don't have to be a mathematician to see that this more realistic model is horrendous to solve. So the static, empty form, for $r \to \infty$, was chosen instead. Not to be realistic, but to be solvable. That was already hard enough. If the rock were also spinning, the curved space around it would be much more complicated, having axial symmetry about the spin axis. That mathematical problem wasn't solved until the 60's. It is called the Kerr solution. Next, we could put an electrical charge, Q, on the spinning rock and we would have an even harder problem to solve, and we would obtain a new space curvature, ds^2, solution. As $Q \to 0$, the last solution reduces to the Kerr solution. As the spin speed $\to 0$, the Kerr solution reduces to the Schwarzschild. As $r \to \infty$ the Schwarzschild solution behaves very similarly to the Newton gravity solution, where we imagine the force **F** of attraction to be acting in flat space. See Chapter 4. Very far away from M, the curved space becomes flat (and this matches the $\mathbf{F} \to 0$ as $r \to \infty$ result, in the old, Newton gravity view).

A clock, holding position at $r = r$, $\theta = 0$, $\phi = 0$ will be ticking with tick interval $d\tau$. The same clock, when slowly and gently moved to $r \approx \infty$, would be found to be ticking with tick interval $d\tau_\infty$. The next question is, of course, how do these compare? Suppose two identical clocks float in space at $r \approx \infty$, keeping time together. Then, one clock is moved slowly to $r << \infty$ and it stays there a long time. Finally, it is slowly and gently returned to position beside the other clock at $r \approx \infty$. How much time will show up as lost on the clock that has moved in and out? A simple question. The moving in and out will itself delay it some. We can neglect that if it stayed a very long time at $r << \infty$, by comparison. So we now need the curvature effect on time at r. We have to guess! All we have is the Schwarzschild solution to build on. It says in this case that

$$(ds)^2 = U^2(cdt)^2 - U^{-2}[0] - [0]$$

for a clock ticking at rest (due to rockets firing) at r. A similar clock at rest at $r \approx \infty$ would have $(ds)^2 \cong 1(cdt)^2 - [0]$ so that $ds = cdt$ for such a clock at ∞. We next make the big guess that (ds) in the Schwarzschild solution means $cd\tau$, the tick rate, for a clock at rest at any r. (This guess is later extended to also include even accelerating clocks, but we don't need to get into that here.) We then will predict that $(cd\tau_r)^2 = U^2(cd\tau_\infty)^2 - U^{-2}(dr)^2 - (rd\theta)^2 - (r\sin\theta d\phi)^2$ is part of the physical meaning of the Schwarzschild solution. We sill don't know what r really means and that is a can-of-worms also. It is roughly the distance to the center of the rock. Therefore, we finally get $d\tau(at\ r) = Ud\tau(at\ \infty) = d\tau_\infty(1 - R_s/r)^{1/2}$. If the rock were very dense, so that $r \approx R_s$ were possible before hitting it (as in a neutron star), then we could see dramatic effects on time at $r \approx R_s$. Remember, however, that there are also tidal forces that tend to pull the bottom of the retrofiring ship off of the top of the ship. In the old flat space view (Newton's original theory), these stretching forces are roughly

$$F_{stretch} \approx \frac{GMmL}{r^3}$$

where L is the radial direction length of the ship. ($L << r$ is needed for this formula to be very useful.) Here r is roughly from the center of the rock to the center of the ship. The r^3 in the denominator means that these forces grow rapidly as r shrinks. The stretching may rip the clocks (and observers) apart, if the weight force, pushing them into the floor (GMm/r^2), has not already squashed them, at 'small' r. The force pressing any mass m into the stationary ship's floor is roughly

$$F_{squash} = \frac{GMm}{r^2} = \frac{GMmL}{r^3}\left(\frac{r}{L}\right)$$

$$= F_{stretch}\left(\frac{r}{L}\right) >> F_{stretch}$$

Having r > > L is reasonable. We have for the general free fall situation, either (1) motion radially inward, or (2) sweeping 'sideways' while coming in close and out again, (in a close encounter, 'hyperbolic' arc—thinking in terms of the flat space approximation). Here the $F_{squash} = 0$, for free fall, but the $F_{stretch}$ is still there and can kill you before you hit the rock, if falling radially in. In a close encounter fly by, it can also kill you, though you still move in and back out to far away once more. (Your atoms and the ship's atoms may be a jumbled streak in the end when you get back to far away.) A neutron star has typically the mass of the sun and its surface is at about $2R_s$. Throw in the numbers here and calculate roughly some time slowing factors and stretching forces at $r = 2R_s$ or $10R_s$. Beyond about $10R_s$, the old flat space, Newton theory gives surprisingly good results, compared to the more difficult curved space theory. Of course, only the curved space theory makes any sense near R_s. (Quantum gravity may take over even before R_s is approached.)

A truly gigantic mass, like a million suns squeezed down to less than $2R_s$ in size, turns out to have a huge R_s, a very low density (something like water), and very weak tidal forces, even right at R_s! You would stay alive as you approach $2R_s$ until you still splatter onto the surface, if falling inward. However, the surface at $r \approx 2R_s$ would not be stable. The gravity would be so strong that the huge 'rock' would crush itself inside of R_s, perhaps very quickly, even if very cold. You then could apparently fall all the way to R_s without harm. What happens? The time dilation equation says that time stops at $r = R_s$ for clocks at rest. It is guessed that this also means that you will fall slower and slower and move in closer to R_s with ever slower motion as you near R_s. We, in the preferred frame, outside and far away, will 'see' you 'frozen' at $r > R_s$ indefinitely into the future. Some have guessed that you fall to $r < R_s$ while still young and alive but that is basically religious speculation—untestable! We get bored with watching you be 'frozen' at $r > R_s$ and move on to other things. If you should turn on your hypothetically VERY POWERFUL super-rocket, before we lose interest, we would see you slowly talking and slowly creeping outward at first and then talking ever faster as you creep farther away. When you again reach $r \approx \infty$, where we are, you will be much much younger than we are, and may have to be talking to our great great grandchildren, depending on how close you got to $r = R_s$. Such super-rockets don't exist, of course! You would have to be content with going in to only $r >> R_s$ and then powering back out. Your atomic clocks on board would show that you did age a little bit slower over all, due to the round trip in and out. What a fantastic world we have!

Remember however, none of this wonderful curved space stuff is compatible with the quantum results we have already tested to 10 digit accuracy. It may fail badly in close. The R_s 'location' may be a quantum world of little blobs. We need a quantum gravity theory that works, in order to find out the truth at R_s. That is for future generations to discover.

Recently, gravity theorists have begun to talk about the temperature inside a black hole or its entropy. Can we talk about the temperature inside an atom? NO. Can we talk about the entropy increase of an atom when it absorbs a photon and jumps to an excited state? No. If a black hole is a quantum blob, then it is more like the insides of an atom than like the insides of a star. I doubt very much that classical thinking can be used productively to explore or explain the behavior or properties of a black hole or even of a neutron star. These are nuclear worlds with the rules of nuclei working there, compounded by highly curved space. Such highly curved space must also exist inside nuclei and we have no idea of how to deal with that at present.

CHAPTER 6

Quantum Interactions

We have seen that big blobs move in the curved space produced by other big blobs. We call this gravity. Charged blobs seemingly push and pull at a distance and create photons (little blobs) when accelerated. Big blobs get shrunk by their motion through the vacuum, somehow, and their internal 'experience' of time is slowed. There is a special, comoving (Newton) reference frame for any locality in the expanding and open or closed universe. We are big blobs, so we only can have a description of nature that is big blob testable. Anything else not testable in the lab or cosmos is theology or fantasy. We must approach little blobs, not on their own terms, but on our big blob terms. This is extremely important and has not sunk in sufficiently on most physicists. They still think they are studying the mind of God. But really they are only following the trail of his/her/its droppings. We are very frustrated by this sense limitation, but we have absolutely no choice. No alternative. We are here, the world is there (and also inside us) and we can only preform tests we actually see to tell us if we are at all on even roughly the right track. That is all we can hope to do—get roughly on the right track, but a track highly biased by our big blob limitations. This puts constraints on what we can see, and puts limitations on our way of thinking; it even shapes how we start out in designing our theories.

We naturally tend to start out with locality and path and forward only time flow. All of this is garbage, but what else can we do? We must test only at the big blob level where nature will manifest itself to us in exactly these terms. We experience only events at specific localities and at specific times, running in succession and going only forward in time; classical moving stuff, in forward time evolution. Look at cosmology, or star formation, or planet geology, or even biological evolution or economics. All have this classical under pinning in common. Yet, DOWN THERE wild and wonderful things are going on that defy our imaginings. We only scrape at the surface, as best we can. It is that or just give up, 'turn on', and 'drop out'. Why should we give a damn what is inside electrons?

So, we guess and we guess. Then we calculate and calculate. Then we design and redesign. Then we measure and confirm. Out of all this sweat and tears comes a nucleus of ideas that seems to be on the right track. We now can use the past 100 years of such suffering and hindsight to bring you to the current frontiers of our quantum ideas! Again, with as simple math as is adequate. Quantum is 90% math, so you cannot go very deeply into it without math. The world is math brought to fruition—whatever that means. (Sounds good anyway.)

How do we start? Where else, with QUATERNIONS! We earlier saw groups and representations of groups in them. The Lorentz group, $LL^{\neq} = 1\sigma_o$ has an important 4-vector representation $P = P^{\mu}\sigma_{\mu} = P^{*}$. The position 4-vector $x = x^{\mu}\sigma_{\mu}$ is not really useful in curved space. Only $dx = dx^{\mu}\sigma_{\mu}$ is useful there and σ_{μ} is even replaced by $b_{\mu}(x) = b_{\mu}{}^{\nu}(x)\sigma_{\nu}$. The 16 numbers $b_{\mu}{}^{\nu}$ ($\mu = 0,1,2,3$; $\nu = 0,1,2,3$) describe the coordinate system and the spacetime curvature. We mention this only for completeness, but we shall overlook it in what follows, since it is very complicated and not fully developed yet. It is probably central to any quantum gravity theory of the future. (However, there may be more dimensions than 4.)

The 4-vector $P = P^{\mu}\sigma_{\mu}$ plays a central role in the quantum foundations. It also is useful in explaining the dependence of color on the direction of a flashlight beam, relative to the direction of the flashlight's motion through space. That is called the relativistic Doppler effect on waves. We have the Lorentz transformation

$$L^{*}PL \equiv P' = P'^{*} \text{ if } P = P^{*} \text{ and } LL^{\neq} = 1\sigma_o$$

Here $P = P^{\mu}\sigma_{\mu}$, and we have $P^{\neq}P = P'^{\neq}P'$ an invariant of the group. Photons of light have $P = (E/c)\sigma_o + P_x\sigma_1 + P_y\sigma_2 + P_z\sigma_3 = (E/c)\sigma_o + \boldsymbol{P}$ and $P^{\neq}P = \cdots = (E/c)^2 - |\boldsymbol{P}|^2 \equiv 0$. That this invariant is zero, for photons, is an empirical 'fact', of unknown origin but highly accurate. We say photons are massless and only travel long distances at the speed c.

The above view of photons is in terms of their particle nature. But they are 'wavicles'—both particles and waves. Their wave nature connects to the above through the facts that $E = h\nu$ and $|\boldsymbol{P}| = h/\lambda$, so that

$$(E/c)^2 - |\boldsymbol{P}|^2 = \frac{h^2\nu^2}{c^2} - \frac{h^2}{\lambda^2} = 0 \Rightarrow \left(\frac{\lambda\nu}{c}\right)^2 = 1 \Rightarrow \lambda\nu = c.$$

This last equation is like $d = vt = v\,1/f \Rightarrow fd = v$ which even works for boxcars of length d going by at frequency f on a train moving at speed v. Train cars have fixed length, d, and variable v; but photons all have the same v (= c) and variable length, λ. That is an empirical fact too that is not understood. The photons all travel through the vacuum at this universal 'speed limit'.

A general Lorentz transformation has the internal form $L = L^{\mu}\sigma_{\mu} + \mathscr{L}^{\mu}i\sigma_{\mu}$, $LL^{\neq} = 1\sigma_o$. Under it, $P \rightarrow P' = L^{*}PL = P^{\mu\prime}\sigma_{\mu}$. This is tedious to grind out, but possible, given L in detail. This gets complicated in a hurry, for the general case. Let us instead do the simplest case: a flashlight of color λ_o is accelerated to speed v along the +x axis and it is shining light in the +x direction. What color do we see for this light moving towards us? We already discussed the fact that its speed is still c, and independent of the v for its source. But the color will be 'blue' shifted. How much? We need a 'boost' Lorentz transformation along x of speed v for a photon P. There are complicated details like which $\pm x$, $\pm v$, $\pm P_x$ to choose in deciding the approximate L to match this physical situation. Roughly speaking we want (remember $E/c = |\mathbf{P}|$): $P' = L^{*}PL$ with $P = (E_o/c)\sigma_o + (E_o/c)\sigma_1$, $L = cosh(\phi/2)\sigma_o + sinh(\phi/2)\sigma_1$, where $tanh\phi = v/c$. Grinding this all out gives (as you can check—on the next page):

$$P' = \left[\cosh^2(\phi/2) + \sinh^2(\phi/2) + 2\sinh(\phi/2)\cosh(\phi/2)\right]\frac{E_0}{c}(\sigma_0) + \cdots$$

$$= \left[\cosh\,\phi + \sinh\,\phi\right]\frac{E_0}{c}(\sigma_0) + \cdots$$

$$= \left[\frac{c}{c\sqrt{1 - \dfrac{v^2}{c^2}}} + \frac{v}{c\sqrt{1 - \dfrac{v^2}{c^2}}}\right]\frac{E_0}{c}(\sigma_0) + \cdots$$

$$= \frac{1 + \dfrac{v}{c}}{\sqrt{1 - \dfrac{v^2}{c^2}}}\frac{E_0}{c}(\sigma_0) + \cdots = \frac{\sqrt{1 + \dfrac{v}{c}}\sqrt{1 + \dfrac{v}{c}}}{\sqrt{1 - \dfrac{v}{c}}\sqrt{1 + \dfrac{v}{c}}}\frac{E_0}{c}(\sigma_0) + \cdots$$

$$= \frac{E_0}{c}\sqrt{\frac{1 + \dfrac{v}{c}}{1 - \dfrac{v}{c}}}(\sigma_0) + \cdots = \frac{E'}{c}(\sigma_0) + \frac{E'}{c}(\sigma_1)$$

$$\cosh^2\phi - \sin^2\phi = \cdots = \frac{1 - \dfrac{v^2}{c^2}}{\left(\sqrt{1 - \dfrac{v^2}{c^2}}\right)^2} = 1, \quad \textit{as expected}$$

We readily see that

$$E = E_0\sqrt{\frac{1 + v/c}{1 - v/c}}.$$

For a receding flashlight, shinning backward, we would get ($v \rightarrow -v$)

$$E = E_0\sqrt{\frac{1 - v/c}{1 + v/c}}.$$

For slow speeds (v/c) $<<$ 1, we can use the marvelous theorem $(1 + x)^n \approx 1 + nx$ for any x $<<$ 1 to get the approximate result

$$E \approx E_0((1 + \frac{1}{2}v/c)(1 + \frac{1}{2}\frac{v}{c}))$$

$$\approx E_0(1 + \frac{v}{c})\ for\ v << c,$$

when the source is coming at us. Notice that the accurate formula gives E $\rightarrow \infty$ as v $\rightarrow$ c for a source coming at us. We would be fried by the x-rays and γ-rays coming from an ordinary flashlight if it were moving fast enough toward us! We make these predictions from only Lorentz math and they seem to fit the experiments.

The angle that a photon path has, compared to the x-axis, is not the same angle for people moving through space along the x-axis with the flashlight. That isn't surprising, but what they see isn't important. The Lorentz transformation machinery can convert from one angle to the other through the P^μ transformation details in three dimensions.

We saw in the curved space, Schwarzschild solution that time dilates for a clock that is moved inward to r $<< \infty$, 'near' a mass M. A stationary flashlight, holding at r, shining light outward toward r $\approx \infty$ has its light also change color as it pulls away. The effect is very small. How small? Talking about quanta in gravity is really a no-no, at present! We have no theory, but we can guess the right answer anyway. The trick is to use $T = 1/f$ for photons and GUESS that T behaves like dt for clocks, in the Schwarzschild case. We have already seen that

$$dt_{at\ r} = dt_{at\ \infty}\sqrt{1 - R_s/r}$$

Therefore, maybe

$$T_{at\ r} = T_{at\ \infty}\sqrt{1 - R_s/r} \qquad T = 1/f$$

Therefore, we surmise that maybe

$$f_{at\ r} = f_{at\ \infty}(1/\sqrt{1 - R_s/r})$$

Or, at ∞, we would see the numerical value $f_{at\ \infty}$ from an actual source with frequency f_o at r. Therefore, we see a red shifted color at infinity:

$$f_\infty = f_o\sqrt{1 - R_s/r} < f_o$$

By 'reciprocity', we surmise that f_o emitted at near ∞ will reach us holding position at a closer r, looking like

$$f_r = f_o/\sqrt{1 - R_s/r} \rightarrow \infty \ as\ r \rightarrow R_s$$

if we are really stationary at $r > R_s$. We would be fried by the ordinary incoming star light, if hanging out too near R_s, *i.e.*, on a cold, dead neutron star. Of course you have other problems there, like being stretched and crushed onto the surface. This frequency shift has been verified for star light reaching us from the sun, and even for light beams made here and traveling vertically a few stories in the atmosphere of the earth. This about covers what we know about photons in outer space. We also know a lot about how they move through big blob substances, such as lenses, but that phenomenon is very complicated at the microscopic level. Photons have an internal structure called polarization. There are four internal parts, A^μ, $\mu = 0,1,2,3$. These manifest themselves in bizarre behavior in polarized sheet transmissions and moving through the crystal calcite, and others. The speed of quanta of light is not strictly c either, in empty space at the micro-level. This is part of the general non-locality problem of all quanta there. What we need is an equation that at least crudely describes how moving charges can create light and electromagnetic fields. We must guess. Before we do that, we first turn to electrons. They too need an equation to crudely describe how photons can make electrons and anti-electrons. It works both ways! Each is a source for 'quanta' of the other. Photons are unique, for they equal their antiparticles, $\gamma = \overline{\gamma}$, whereas $e^- \neq e^+$, $p^+ \neq p^-$, and $n^o \neq \overline{n}^o$. Light is its own antiparticle, whereas anti-electrons, anti-protons, and anti-neutrons are distinct. These other particles also differ in having some mass: $(E/c)^2 - |\boldsymbol{P}|^2 = (E_o/c)^2 > 0$ and $E_o \equiv m_oc^2$ is a Lorentz invariant. So, antimatter stars would give off anti-photons if they existed, but we cannot distinguish them. (Theory suggests there are none out there.) Cosmic rays are mostly p^+, protons, and not p^-. Therefore, our galaxy is very probably all matter. The highest energy cosmic rays are p^+ and believed to come from beyond our galaxy. We must really guess though, about other galaxies. We guess conservatively, naturally. Antimatter probably got killed off in the Big Bang's early, high density days, we guess, therefore all galaxies are probably matter.

Anti-people, made of anti-matter, would be possible on anti-planets. The chemistry is the same as that for matter, as far as we know. As long as they stay away from our atmosphere, they are safe. We can talk to them easily using photons, but we cannot communicate safely with morse code beams of protons. Such protons would cause havoc when they hit their ship, unless we keep the intensity very low. Antimatter is a true wonder. Why is it there at all? Because the math needs it? We may never know.

All three particles, e^-, p^+, and n^o, have a similar internal structure as to angular momentum. This, so called spin 1/2 nature results in at least four separate, complex functions ψ^μ to describe 'one' of them or even a cloud of electrons and antielectrons. Because photons are their own antiparticles, all of the four A^μ parts are not independent, in contrast to the four ψ^μ parts for particles like e^-, p^+, and n^o. But our current thinking here is probably very naive. There is much more to be found concerning these fundamentals.

The p^+ and n^o particles have other internal parts, quarks, so we shall give up on them here. The curled up, tiny extra dimensions of space may be critical to understanding their internal nature. It seems the e^- field/particle cloud is the simplest, next to the photon field/particle. However, there are also neutrinos, ν^o, that may be even simpler than electrons. They have no charge, yet $\nu^o \neq \overline{\nu}^o$. They may have a very small rest mass, as discussed earlier. They are spin 1/2 particles like electrons. The photons are spin 1 particles and, because they have no mass, they travel through space with 'spin axis' only forward or backward (helicity $\pm$).

All this preliminary information is the result of the past few decades of stringent and heroic effort. The only physics that we presently understand fairly well is QED (quantum electrodynamics). This is a fancy name for photon clouds interacting with electron/antielectron clouds. That mathematical formulation was completed in the late 1940's. We have been stuck since then. Probably this is due to the limited scope of our vision. The next advance requires something even more radical than the quantum advancement required. I share my speculations on this future in the next chapter.

Before we tackle the job of finding the field equations, notice that

$(E_o/c)^2 \equiv (mc)^2$, therefore, $E_o = \pm mc^2$. We don't know why E_0 should be different from zero in the basic laws of our universe. Perhaps it is an arbitrary rule, but because it is not zero in our universe, we are here to be conscious and ask the question in the second place. The $+mc^2$ goes with e^- (matter) and the $-mc^2$ goes with e^+ (antimatter), in one way of looking at things. In another view, $E_o = +mc^2$ only but Et = ± and t = − (going backward in time for e^-) is the 'way' of having e^+ seem to exist at all! In other words, e^- can go both forward and backward in time but with only $+E_0$ and $+mc^2$. This somehow describes a bunch of electron/antielectron creations and annihilations. Feynman said that Wheeler once theorized that there is only one e^- in the whole universe—it just zig zags in time a whole bunch, to fill the universe! We don't know much here either. We just guess and do the best we can to get something that can make numerical predictions, for comparison with experiments involving big blob detectors, that we can see directly.

WAVE EQUATIONS

We next focus on the foundations of QED. These were historically found in bits and pieces, and without appreciation of Hamilton's 1843 quaternions. We shall here do it the way that they should have done it, perhaps, and actually did it on many other planets, once quaternions were invented.

The historical narrative of physics on this planet is well captured in the two volume set by Emilio Segrè. Very worth reading. The depressing thing that comes from that reading is how insignificant we all are in physics. A couple of very promising physicists died young. Their great discoveries were just made by someone else instead of them. Everything would go pretty much as it has, regardless of who lived to contribute and who died early. We are cogs in an inevitable machine cranking out scientific and technological progress (followed by eventual decay of our civilization). The only real question on a planet is one of timing! Certain people or civilizations could speed up the steps; maybe as much as 50 years, for individuals, and perhaps centuries for civilizations. Consider the Greeks and what they could have done if they had remained prosperous longer. We might have been on the moon 2,000 years ago!

Of course, classical ideas are discovered first. These begin to fail to explain things found empirically as technology advances, and a quantum theory then supplements them. (It does not really replace them because we can still only see big blobs to test the new theories). The classical particles had ordinary nonlinear, differential equations and the classical waves had partial differential equations, to naturally describe them. The quaternions are the foundation for both sets of classical equations and, on our planet, quaternions were discovered before vectors were. It would be so very interesting to see how that struggle occurred on other planets, which were going along the general path we have followed. The quaternions are naturally 'relativistic' in structure—suggesting space and time are parts of one thing, mathematically. They also naturally suggest that Lorentz symmetry fits the world. Combining these with the wonderful realizations that particles act like waves, we are led to explore partial differential equations involving quaternions and derivatives with respect to space and time in 3+1 dimensions of space and time.

The key step in the transition to quantum machinery is the replacement of 4-momentum in classical thinking with a 4-operator of spacetime:

$$P = P^{\mu}\sigma_{\mu} \rightarrow (i\hbar\partial^{\mu})\sigma_{\mu}$$

Here, $i = \sqrt{-1}$, and it is there as part of the wave nature of particles. (More about that later.) The ∂^{μ} is a shorthand:

$$\partial^0 \Rightarrow \frac{\partial}{\partial ct}, \partial^1 \Rightarrow -\frac{\partial}{\partial x}, \partial^2 \Rightarrow -\frac{\partial}{\partial y}, \textit{etc.}$$

These minus signs are just traditional. We sum on $\mu = 0,1,2,3$ as usual. This P will make mathematical sense when it operates on an unknown function, and the resulting equations are connected to a 'source' of that field function, made up of other fields. The quanta create each other, in a bazarre dance of comings and goings at the little blob level. The general pattern of little blob nature is

$$P\psi = f(\phi)$$

where ψ is the particle-field-cloud of interest and ϕ is a field source for ψ. The f symbolizes some 'function' of ϕ, structured such as to make the left and right sides 'compatible'.

Remember that this really should all take place in curved and stretching space. The basic guesses should be general enough to encompass this possibility, even if we don't yet know how to do the curved space quantum details. If you do not know the $\partial/\partial x$ derivative notation, that is not a real problem here. We are not going to solve any equations. Just understand that $P\psi$ means that if $f(\phi)$ is known, then ψ is determined by $f(\phi)$. If $f(\phi) = 0$, then there is still a solution, ψ. The particle nature of the quanta requires much modification of these simple equations anyway, to do real calculations that can be compared with experiment. We are concerned with the basic, skeletal structure.

Once that is grasped, the rest can eventually be worked out in detail. If we have the wrong skeletal ideas, then no details will come out right, no matter how clever we are or how hard we try.

Appreciating the significance of P, as the fundamental operator, is the first step. The second step is finding some constraints on the basic wave equation structure, to limit the guesses we must make for the basic wave equation ($P\psi = ?$).

We have empirically found that there are spin 1/2 and spin 1 particle/fields and these are radically different. The differences seem to be somewhat related to first order (one P) and second order (two P's) differential operators: $P\psi =$ and $PP\phi =$. Of course, we don't presently know why these distinctions and correlations exist. But they do. This complexity is very interesting, for it further expands the possibilities for wave equations. These limitations on equation structure, that seem to work, involve the Lorentz group once more. This is not surprising once we have discovered how the Lorentz group successfully suggests space and time peculiarities for big blobs of the kind discussed in the last chapter. However, do not get overly excited by this classical success. Big blob physics is very 'averaged out' and crude. What we actually see up here may be very over simplified compared to the subtle things going on at the little blob level.

So, our next step is to again look at the Lorentz group and its representations (manifestations). We have already seen that

$$P \rightarrow P' \equiv L^{*}PL,\ LL^{\neq} \equiv 1\sigma_0$$

is a useful idea for photon color dependence on light beam direction and source speed. If it is important to photons, then it could easily be important to other quanta. When P becomes a partial differential operator, we have things like $P\psi = 0$. Then $P \rightarrow P'$ is combined with $P\psi$, to give us $P'\psi' \rightarrow L^{*}PL\psi' = ?$, and $LL^{\neq} \equiv 1\sigma_o$. Clearly, $\psi' \equiv L^{\neq}\psi$ would be interesting. Then $P'\psi' = 0$ reduces back to $P\psi = 0$ as follows:

$$P'\psi' = (L^{*}PL)(L^{\neq}\psi) = L^{*}P(LL^{\neq})\psi$$

$$= L^{*}P\psi \equiv 0 \Rightarrow L^{\neq *}(L^{*}P\psi)$$

$$= L^{\neq *}(0) = 0 = (LL^{\neq})^{*}P\psi$$

$$= P\psi = 0$$

This is sort of a game we can play and it seems very useful in guiding our guesses at the fundamental equations!

Maybe God played the game also, in selecting the rules for the particular universe that we evolved in. That is not a physics issue. But for some reason, our universe seems to be limited in its rules by just such games. The basic equations, that we know so far to be very significant, all have this pretty pattern. The $P \rightarrow P'$ idea can be used to find $\psi \rightarrow \psi'$ and $\phi \rightarrow \phi'$ structures, such that a primed equation can be 'stripped' of all its L's. This is not really a transformation of any physical kind. It still applies to the equations that actually work in curved and unstable (expanding) space. It is just wonderful magic! It helps us zero in on the most likely equations that will actually work for our universe, to numerically meet our humanoid needs for modeling. Whether it is 'deep' in the real rules that are down there, we can never know. Probably it is not. It may be a crude approximation itself, just as everything else is, that we know.

But now we are ready to start guessing seriously and productively about the 3+1 spacetime equations for our little blob world of atoms. Naturally, the first guess is $P\psi = \phi \Rightarrow P'\psi' = \phi' = L^*PLL^{\neq}\psi = L^*P\psi = \phi'$. Therefore, we need $\phi' = L^*\phi$. It is, therefore, fundamentally different from ψ, with respect to L 'behavior'. We say ϕ is a source for ψ quanta. If $\phi = 0$, then we have 'free' ψ quanta. We say ψ has spin 1/2 because $\psi \rightarrow \psi' = L^{\neq}\psi$. The name choice is not obviously useful, but historically motivated. The $L^{\neq}$ and L^* transformations are associated by mathematicians with co- and contra- spinors, in case you have heard those words elsewhere. The coupling term, ϕ, can be more sophisticated. For example, consider

$$\phi = A\psi, \quad \phi \rightarrow \phi' = L^*\phi$$

Here A is the new source for ψ and it is directly 'tangled' with ψ in this source term. We clearly need $A \rightarrow A' = L^*AL$ here, to reproduce the ϕ' nature that was needed because of $P' = L^*PL$. We see that such an A field is 'like' P in its L structure. There should also be $P\phi =$ and $PA =$ equations as well, or perhaps $PP\phi$ or PPA equations. We readily see that $P^{\neq}P$ is nice because $P'^{\neq}P' = (L^*PL)^{\neq}(L^*PL) = L^{\neq}P^{\neq}L^{*\neq}L^*PL = L^{\neq}P^{\neq}(LL^{\neq})^*PL = L^{\neq}(P^{\neq}P)L$. Notice that $(P^{\neq}P)^{\neq} = P^{\neq}P^{\neq\neq} = P^{\neq}P$. Therefore, $P^{\neq}P$ has no σ_k parts. It can, therefore, commute through other quaternion pieces. Therefore, $P^{\neq}P \rightarrow P'^{\neq}P' = \cdots = (P^{\neq}P)L^{\neq}L = L^{\neq}L(P^{\neq}P)$. We have only assumed $LL^{\neq} \equiv 1\sigma_o$. What about $L^{\neq}L$? We would have to grind out all the components, and we would find $L^{\neq}L = 1\sigma_o$ also. (Theorem.) There is another useful field with the peculiar L structure $F \to F' = L^{\neq} F L$. Notice that $F \equiv \pm F^{\neq} \Rightarrow F' = \pm F'^{\neq}$ here. If $F = -F^{\neq}$ then it has only $\{\sigma_k, i\sigma_k\}$ parts in it. If it has $F = +F^{\neq}$, then it has only $\{\sigma_o, i\sigma_o\}$ parts and it commutes with other quaternions. In this latter case, we call it ϕ: $\phi \to \phi' \equiv L^{\neq}\phi L = \phi L^{\neq} L = \phi$. We say such a ϕ is a scalar, for historical reasons. A field $F \to F' = L^{\neq} F L$, $F = -F^{\neq}$, we call an antisymmetric tensor. (It needs a shorter name, obviously; how about 'antor'?) It is associated with the **E** and **B**, electromagnetic fields, in its parts. So let us call it a '6-antor' field.

We have, so far, found five types of fields that may be useful for modeling the real world:

$$\psi' = L^{*}\psi, \; \lambda' = L^{*}\lambda, \; A' = L^{*}AL,$$

$$F' = L^{*}FL, \; \phi' = \phi L^{*}L = L^{*}L\phi = \phi$$

Are there others? I don't know. These are all simple and natural, but they may not be enough. Nature is very complicated. It has, after all, evolved humanoids, using no further magic after the Big Bang, and evolution allows humanoids to be very intelligent and self-aware. We cannot expect our models to remain simple, as we dig deeper and deeper into the structure that is really there. It is incredible that we are here and that babies can grow from one cell! The basic 'free' quanta wave equations are, of course, $P\psi = 0$ or $P^{\neq}P\psi = P^{\neq}(0) = 0$, which follows directly from it. Replacing the zero on the right-hand side leads to great complexity! The guessing is not so clearly guided any longer. We seem to have the prime directive in nature that all the L's must be canceled out and $LL^{\neq} = 1\sigma_o$. There are also basic coupling strengths, ϵ, δ, $\cdots$, on the right hand side, when we put fields there to act as sources. These couplings are not known in size and they are adjusted so as to fit with some experiments, the best we can. We will probably never know why they have the values that we find in the lab. Are they arbitrary or unique? Let us now couple ψ and F and A together in a set of 'covariant' equations. The choices we make are not unique here. We are, however, able to guess that the following might be useful:

$$P\psi = \epsilon_1 A\psi \quad PF = \epsilon_2 J$$

$$P^{\neq}A - (P^{\neq}A)^{\neq} = \epsilon_3 F \Rightarrow F^{\neq} = -F$$

$$P \rightarrow P' = L^{*}PL \quad A \rightarrow A' = L^{*}AL$$

$$\psi \rightarrow \psi' = L^{\neq}\psi \quad J \rightarrow J' = L^{*}JL$$

To complete the set, we need the source J, or we need J expressed in terms of the other fields. *This is a very important problem*! Clearly, $J \rightarrow J' = L^{*}JL$ is required here. Its details still are not very clear, however. We see A is a source for ψ and F is a source for A, but J is a source for F. To close the circle, we would guess that ψ is a source for F, or maybe a combination of A and ψ is the source. To meet the L behavior needed, we can easily take: $(\psi^{\neq *}\psi^{\neq})^{*} = \psi^{\neq *}\psi^{\neq **} = \psi^{\neq *}\psi^{\neq} = (\psi\psi^{*})^{\neq}$. We check the L behavior:

$$(\psi'\psi'^{*})^{\neq} = [(L^{\neq}\psi)(L^{\neq}\psi)^{*}]^{\neq} = [L^{\neq}\psi\psi^{*}L^{\neq *}]^{\neq}$$

$$= L^{\neq *\neq}\psi^{*\neq}\psi^{\neq}L^{\neq\neq} = L^{\neq\neq *}(\psi\psi^{*})^{\neq}L = L^{*}[(\psi\psi^{*})^{\neq}]L$$

Just like J needs to be! This is not a unique choice, however. The right choice can only be found by guessing and then comparing with experiment. Notice that this source is quadratic in ψ, contrary to the A and ψ sources, which are linear in the external field coupling. This suggests that ψ is unique among the three fields here. We say that it is a fermion field, whereas A and F are boson fields. This means that ψ has more peculiarities, compared to 'classical' wave equations above. The parts in ψ, $\psi = \psi^{\mu}\sigma_{\mu}$, have totally weird functions, ψ^{μ}, compared to the field parts in F and A. These ψ^{μ} parts, we would easily guess, are ordinary complex number fields (not all independent, perhaps); but that is not the case in our universe! The ψ^{μ} parts come from a whole new mathematical thing that only shows up in the little blob world. They are so called Grassmann numbers, not ordinary numbers. Grassmann invented and played with weird numbers, a and b, such that $ab = -ba$ and $aa = -aa \Rightarrow aa + aa = 0 \Rightarrow 2aa = 0 \Rightarrow aa = 0/2 = 0$. Therefore, $a^2 = 0$ and $b^2 = 0$. It turns out that nature uses such numbers in the field equations in a very important way! They nicely prevent ψ quanta from ganging up together in the same 'state'. A consequence of this is called the Pauli Exclusion Principle. An electron cloud around a large nucleus has the electrons 'excluding' each other. The atom is then much larger and more complex, internally. It also results in white dwarf stars being unable to squeeze together beyond a certain, very high density and therefore they stop contracting. The more massive they are, the smaller they are, up to the Chandrasekhar mass limit, $1.4M_{sun}$.

It is not easy to deal with solutions of these coupled partial differential equations. The Grassmann number complication is not the only problem. These smooth looking, causal equations only crudely describe actual quanta and the quanta only behave in a probabilistic way. In fact, they behave in every possible way they can. The more bizarre the behavior, the less likely it is, but nothing is left out that is possible at all and that fits within the basic rules. "Anything that can happen, will happen, if you wait long enough."

We have only just begun, when we have guessed the above, basic equations of interaction for quanta, even if correct. The bosons naturally are much easier to deal with. The A and F quanta can bunch up and can, therefore, more easily display their wave-like characteristics to big blob observers like us!

Their wave nature was discovered, not surprisingly, before the fermion wave-like behavior of electrons, protons, and neutrons. The photon equations are called Maxwell's equations, and were discovered in the 1800's. The J source is electrons on the move. Depending on how they move, they make different patterns of F field which in turn make related patterns of A field. The A field acts back on the electrons to modify their motions. This is, more or less, quantum electrodynamics. It is easy to write down but very difficult to solve and interpret.

If there are no electrons in the vicinity, we can still have free A and F fields. We call this situation a beam of 'light'. It travels through stretching space independently, spreading out as time goes forward. The F field, with $F^{\neq} = -F$, is described by $F^k\sigma_k + \mathscr{F}^k i\sigma_k$ and these are, for historical reasons, written as $-E^k\sigma_k + cB^k i\sigma_k$. The equations for F, with J = 0, are said to have **E** generating **B**, through its changes, and **B** also generating **E**, through its changes, as dictated by the equation for F.

At this level of approximating nature, the A field seems redundant, but it is not. It turns out that $A = A^*$, which is compatible with $A \rightarrow A' = L^*AL$ and $A'^* = A'$. It, therefore, has only four functions, A^μ, in it. The six field parts in F are sources for these 4 field parts in A. These 4 parts in turn interact with the electrons as a source in their equation. The zero mass value for photons is believed to result in even these four A^μ's not being independent. There can be so-called Gauge transformations where $A^\mu \rightarrow A^{\mu'}$ but F is unchanged. This freedom of choice in A may not still exist at a deeper level of approximation! It is not natural at all. There must be a reason for A and F being both needed in the complete set of equations. See the next chapter for speculations on this.

We could take the source J to represent one proton, sitting at rest. We could find its F solution in the space around it. Next we could find the A solution generated by this F. Then we could put this into the ψ equation and try to solve for ψ, with the assumption that $\psi^{\mu} \rightarrow 0$ as $r \rightarrow \infty$. This 'bound state' boundary condition will result in the large ψ_{n} family of ψ solutions being so called eigenstates of the ∂^{o} operator, with discrete eigenvalues given by $i\hbar\partial^{o}\psi_{n} = (E_{n}/c)\psi_{n}$. The set $\{E_{n}\}$ is an infinite series of real numbers, each with a corresponding ψ_{n} 'state' to go with it. Transitions from ψ_{n} to $\psi_{n'}$ we guess are associated with 'slight' changes in the overall F field, which the ψ field and the central proton make, such that quanta of F are somehow created and ejected at 'speed' c from the vicinity. This is a difficult calculation and still it is only a crude approximation to all that is going on here.

After all this calculational work, we find only crude agreement with the actual photons that are detected in this situation. Electrons have another quantum coupling coefficient besides ϵ_{1}. We call this 'self' coupling, m_{e}. There is no place in $P\psi = \epsilon_{1}A\psi$ for this additional constant, yet it must get in there if we have the right equation. (We know this from classical physics!) We are seeing a breakdown here of our starting assumptions. The complex quaternions are not rich enough in structure to handle the real world. We need a more complicated $P^{\mu}\sigma_{\mu}$ structure. Just letting P^{μ} become complex: $P^{\mu}\sigma_{\mu} + Q^{\mu}i\sigma_{\mu}$ is not enough. There has to be real generalization of a radical kind to accommodate mass. We need a bigger number system. How do we find it? Not easily by any means. We have $\{\sigma_{\mu}, i\sigma_{\mu}\}$ so far. This system has $\{\sigma_{0}, -i\sigma_{1}, -i\sigma_{2}, -i\sigma_{3}\}$ and $\{\sigma_{0}, i\sigma_{1}, \sigma_{2}, \sigma_{3}\}$ as closed subsystems, among others like $\{\sigma_{0}, i\sigma_{0}\}$ or $\{\sigma_{0}, i\sigma_{1}\}$, etc. (The $i\sigma_{1}$ choice in the second subsystem above is only one of three 'equivalent' forms, since the i could instead be with σ_{2} or σ_{3}.) What if we generate a bigger number system from what mathematicians call a direct product $\otimes$ of these two subsystems? New things appear, like $ai\sigma_{2}\otimes b\sigma_{2}$, *etc.*, where a and b are real numbers. The new basis obviously then consists of $\{\sigma_{0}\otimes\sigma_{0}, \sigma_{0}\otimes i\sigma_{1}, \sigma_{0}\otimes\sigma_{2}, \sigma_{0}\otimes\sigma_{3}, i\sigma_{k}\otimes\sigma_{0}, i\sigma_{k}\otimes i\sigma_{1}, i\sigma_{k}\otimes\sigma_{2}, i\sigma_{k}\otimes\sigma_{3}, k = 1,2,3\}$. We have a new 16 part number system here, with the old 8 part system as a subsystem, as expected. Addition and multiplication are as expected: We combine left members to get a new left member, and do the same for the right members. Remember, we are doing this because objects in our universe can slow down and even be stopped completely—they have rest mass, $m \neq 0$. We obviously need to get m into our field equations. Once we act on this motivation, we must again remind ourselves that mass is really a big blob experience. At the little blob level, there may be wild and new things going on that only average to a single, lumped mass parameter that we see at the big blob level. So we proceed cautiously with an open mind here. The bigger number system that we have gone to may contain much more than the one mass parameter that motivated it in the first place!

The direct product notation is awkward but adequate. We can instead rename 16 basis elements $\{e_\mu, ie_\mu, f_\mu, if_\mu\}$. This set will have $\{e_\mu, ie_\mu\}$ isomorphic to $\{\sigma_\mu, i\sigma_\mu\}$ in its multiplication table, by definition. We need to choose the f multiplication rules so the full set behaves like the members of the direct product system. The rules that work, I found to be the following:

$$f_\mu e_\nu \rightarrow \sigma_\mu \sigma_\nu \rightarrow \sigma_\lambda \rightarrow f_\lambda$$

In other words, $(if_1)e_2 \rightarrow i\sigma_1\sigma_2 = (ii\sigma_3) = -\sigma_3 \rightarrow -f_3$. However, if the right side element is an f, we have a new rule:

$$f_\mu f_\nu \rightarrow (\sigma_\mu)^{\neq *}(\sigma_\nu) \rightarrow \sigma_\lambda \rightarrow e_\lambda$$

$$e_\mu f_\nu \rightarrow (\sigma_\mu)^{\neq *}(\sigma_\nu) \rightarrow \sigma_\lambda \rightarrow f_\lambda$$

In other words, $(if_1)f_2 \rightarrow (i\sigma_1)^{\neq *}\sigma_2 = --i\sigma_1\sigma_2 = -\sigma_3 \rightarrow -e_3$ and $(ie_1)(f_2) \rightarrow (i\sigma_1)^{\neq *}(\sigma_2) = -\sigma_3 \rightarrow -f_3$. To remember the pattern, think of f as like $-$ and e as like $+$. Recall that $(+)(+) = +, (-)(-) = -, (+)(-) = (-)(+) = -$. There is no real need for you to compare the details here with the direct product multiplication table, to see that they really are the same. I have, and they are! Even if they weren't, the ef system described here does work for the real world! We now have a new number system to try out on nature. The proof is in the pudding, as they say.

Before we can turn to the wave equation generalizations, we need the conjugation generalizations. We seek so called antiautomorphic ones as before: $A^{conj}B^{conj} = (BA)^{conj}$. We find there are now many, instead of two! They fit a pretty and symmetrical pattern. I've called three of them $\uparrow$ $\wedge$ $\vee$ and $\downarrow \equiv \uparrow \wedge \vee$ applied together (in any order). These change the signs of $\{e_\mu, ie_\mu, f_\mu, if_\mu\}$, in a specific pattern, just as * and $\neq$ do for the set $\{\sigma_\mu, i\sigma_\mu\}$. Playing with the possibilities leads to the workable rules:

$$(e + f)^{\uparrow} \rightarrow e^{*} + f^{\neq}, \quad (e + f)^{\wedge} \rightarrow e^{\neq} + f^{*}$$

$$(e + f)^{\vee} \rightarrow e^{\neq} - f^{*}, \quad (e + f)^{\downarrow} \rightarrow e^{*} - f^{\neq}$$

This is easy to memorize. It means that, for example, $(ie_1)^{\uparrow} \rightarrow (i\sigma_1)^{*} = -i\sigma_1 \rightarrow -(ie_1)$ but $(if_1)^{\uparrow} \rightarrow (i\sigma_1)^{\neq} = -i\sigma_1 \rightarrow -(if_1)$. Also, $(ie_1)^{\vee} \rightarrow (i\sigma_1)^{\neq} = -i\sigma_1 \rightarrow -ie_1$ and $(ie_1)^{\uparrow\wedge\vee} = (ie_1)^{\downarrow} \rightarrow (i\sigma_1)^{*} = -(i\sigma_1) \rightarrow -(ie_1)$, but $(if_1)^{\downarrow} \rightarrow -(i\sigma_1)^{\neq} = +i\sigma_1 \rightarrow (if_1)$. The pattern is easy to apply, but does it actually work? We really need to check all 16×16 pairs for each of the three conjugations! They all do work! There is an elegance here that cannot be denied, but it is complicated. For example: $[(if_1)(f_2)]^{\uparrow} \rightarrow [-\sigma_3]^{\uparrow} \rightarrow -e_3$ and $(f_2)^{\uparrow}(if_1)^{\uparrow} = (-f_2)(-if_1) = f_2(if_1) \rightarrow (\sigma_2)^{\neq *}(i\sigma_1) = -\sigma_2 i\sigma_1 = -i\sigma_2\sigma_1 = -i(-i\sigma_3) = -\sigma_3 \rightarrow e_3$. This checks one of the 16×16 combinations for the $\uparrow$ conjugation.

A computer program could easily express these rules and grind out all $3 \times 16 \times 16$ cases, individually. This new number system is, it turns out, isomorphic to the Dirac algebra of 4×4 'complex' matrices, first shown to be important to electrons in the late 1920's. We need to put i's in front of our 16 basis elements later, to duplicate the full 32 element Dirac algebra.

DIRAC GROUPS

We had two basic groups in the complex quaternion system: $AA^{\neq} = 1\sigma_o$ and $BB^{*} = 1\sigma_o$. Now we have at least three natural groups: $AA^{\wedge} = 1e_o$, $BB^{\uparrow} = 1e_o$, and $CC^{\vee} = 1e_o$. As before, the infinitesimal members, 'close' to the identity, are such that the conjugations change the signs of the basis elements associated with the infinitesimal parameters:

$$A_{\epsilon} = 1e_0 + \epsilon^{k}(ie_k) + \delta^{k}(e_k) + \lambda^{\mu}(if_{\mu})$$

$$B_{\epsilon} = 1e_0 + \epsilon^{k}(ie_k) + \epsilon^{0}(ie_0) + \delta^{k}(f_k) + \lambda^{k}(if_k)$$

$$C_{\epsilon} + 1e_0 + \epsilon^{k}(ie_k) + \delta^{k}(e_k) + \lambda^{\mu}(f_{\mu})$$

Each group has 10 parameters. They might be identical groups, so called isomorphic by mathematicians. They are in fact, identical! The group is called Sp(4) and A_{ϵ} clearly has the Lorentz group as a familiar subgroup! Remember, the e's alone are like the σ's. This larger group replaces Lorentz symmetry in a natural way for the little blob world! We shall continue to use the L notation, but $LL^{\wedge} = 1e_o$ is now a 10-parameter group built on L_ϵ above. By the way, Lie showed that two groups like this are identical, if they have the same number of parameters and corresponding commutators: $[\alpha, \beta] \equiv \alpha\beta - \beta\alpha$, for the infinitesimal basis elements. For example, check $[ie_1, if_\mu]$, etc. This is tedious to show, but a one-to-one correspondence can be found for each of the 10 parameters in each of the three groups. (It is a long process.) Having found that these correspondences hold means the groups are just one group.

The other group, $DD^{\downarrow} = 1e_o$ has infinitesimal elements like

$$D_\epsilon = 1e_0 + \epsilon^k(ie_k) + \epsilon^0(ie_0) + \delta(f_0) + \lambda(if_0)$$

Clearly 6 parameters, and it can be shown to be isomorphic to SU(2)⊗SU(2), in matrices. Therefore, this group ought to also have some importance in the equations, somewhere, but not the importance of Sp(4), which limits the basic equation structures. We complete this generalization guess by the natural extension guess $P = P^{\dagger}$, since it contains $(P^\mu\sigma_\mu)^* = (P^\mu\sigma_\mu)$ as a special case and $\dagger$ is the Hermitian conjugate in matrix representation. This leads us quite naturally to

$$P = P^{\dagger} = P^\mu e_\mu + P^4(if_0) + P^5(f_0)$$

We now have two new parameters, P^4 and P^5 to 'play' with. (We only wanted one!) What could they mean? We needed the new parameter so that mc could be accommodated, but we instead see two new slots. They both must be important somehow? Maybe $P^4 = -mc$, $P^5 = 0$ for some particles, and $P^4 = 0$, $P^5 = -mc$ for others, and $P^4 = 0$, $P^5 = 0$ for still others, like photons. We find in the world, leptons (light weight particles) and hadrons (middle and heavy weight particles). This is unexplained at present, but the above slots might be related to this observed split. There are serious problems with $P^4 = -mc$ and $P^5 = -m'c$, as you will quickly find when you try to solve seemingly simple problems like the hydrogen atom energy levels. But we find $m = 0$ or $m' = 0$ will work well for the hydrogen atom energy levels. The solution there shows us no preference in the choice between P^4 and P^5, I suspect.

It is reasonable that a distinction between P^4 and P^5 should show up at some deep level, and this generalization is, so far, too naive to cover that deeper level. It does work well for QED, and that is all we really know at present. See the next chapter for further speculations.

We don't need to make much change in the set of equations that we guessed before for ψ, A, and F. We simply have to extend P and put in P^4 or P^5 (replaced by mc) for the appropriate quanta in $P\psi$. For QED, we need only $m_e c$ for either P^4 or P^5, in $P_\psi \psi$ and $m_\gamma c = 0$ for $P_\gamma F$ and $P_\gamma{}^{\neq} A$. (Recall that photons are presumably massless—for unknown reasons.) We of course replace $P^{\neq}$ by $P^{\wedge}$ or $P^{\vee}$ and P^* by $P^{\uparrow}$ or $P^{\downarrow}$. We next need to consider $J^{\uparrow} = J$ and whether J should now have more than four parts. The traditional QED guess here has been $J = \psi^{\uparrow} e_\mu \psi e^\mu$, but this is ugly, ugly! No other term brings in 'bare' basis elements like the e_μ here. (This creates problems in curved space also.) I follow the beauty instead and guess that $J \equiv (\psi\psi^{\uparrow})^{\wedge} = J^{\uparrow}$ is the way to go here. (Comparison to experiment is the only and final judge of what is right.) Those comparisons remain to be done.

Since the $\{e_\mu, ie_\mu, f_\mu, if_\mu\}$ algebra has a 4×4 matrix equivalent, we see that ψ has a 1×4 matrix equivalent in $P_\psi \psi = \epsilon A \psi$. But J above may involve a ψ that is more complicated than this. If so, then F is more complicated and A is also more complicated than usually assumed. QED may only scratch the surface of the structures that lurk here!

The choice $P^{\neq} \rightarrow P^{\wedge}$ or $P^{\vee}$ has not yet been settled in the generalized guess. I would be inclined toward $P^{\wedge}$ since + is 'superior' to −, but only experimental testing of the theory's details can really settle such speculations. Since the $P^{\uparrow}$ conjugation turns out to be the familiar hermitian conjugation in matrix language, clearly it is preferred to $P^{\downarrow}$. Also, $P^{\wedge}$ is equivalent to $f_o P^{\uparrow} f_o$ in matrix language. This cumbersome form is used effectively in the standard textbooks on QED, where f_μ is called γ_μ.

Notice that $\{\sigma_0, -i\sigma_k\} \otimes \{\sigma_0, i\sigma_1, \sigma_2, \sigma_3\}$ led to $\{e_\mu, (ie_\mu), f_\mu, (if_\mu)\}$ with real coefficients in both cases. That means 16 free parameters. But 4×4 complex matrices have 32 free parameters, when complex coefficients are allowed. That algebra is the Dirac algebra. We saw $P^\mu \sigma_\mu = (P^\mu \sigma_\mu)^*$ and $P^\mu \rightarrow \hbar\partial^\mu$ are basic to massless quantum. In $P = P^\mu(e_\mu) + P^4(if_o) + P^5(f_o) = P^{\uparrow}$ we really considered only real coefficients. Generalizing to complex coefficients seems trivial at first, but it opens important, new insights! The first problem is how to define things like $[i(ie_3)]^{\uparrow}$. This is not obvious and is important.

Playing around with the Dirac equation for years and years, finally led me to be convinced that the most 'physical' way to generalize is to have all three conjugations, ↑, ∧, and ∨ change the sign of any i coefficient that is outside of the parentheses. The two i types are very distinct. Thus $[i(ie_3)]^{\uparrow} = (-i)[ie_3]^{\uparrow}$, *etc*. The amazing result of this is that ↑, ∧, ∨, and ↓ all become even more alike than before! None seems special at all now. The complex Dirac algebra is obviously $\{(e_\mu), i(e_\mu), (ie_\mu), i(ie_\mu), (f_\mu), i(f_\mu), (if_\mu), i(if_\mu)\}$ with 32 elements and real coefficients. The i's out front commute through the hypercomplex basis elements and do not affect the old rules of multiplication: $[i(ie_1)][i(if_2)] = ii[(ie_1)(if_2)]$, *etc.*

We now find that $P = P^{\uparrow}$ can have up to 16 components! We find that $\uparrow$, $\wedge$, $\vee$, $\downarrow$ all change 16 signs and leave 16 alone. Any pair of these conjugations still changes 8 signs when applied together. This automatically generalizes the results we had before for 'Dirac'. The $LL^{\wedge} = 1$ group now has 16 parameters, rather the 10, and I call it the master group of Dirac. The subgroup $L^{\wedge} = L^{\vee}$ now has 8 parameters instead of 6. Study of these two extra parameters indicates they suggest that there is no natural length scale in nature and that ψ or $e^{i\theta}\psi$ both give the same physics results, for any θ value. It is nice to see these properties right along side SL(2,C) as natural parts of the Dirac algebra. In the general case, we have

$$P = P^{\mu}(e_{\mu}) + P^4(if_o) + P^5(f_o) + Q^{\mu}i(ie_{\mu}) + R^k i(f_k)$$

$$+ S^k i(if_k) = P^{\uparrow} = P_{16}$$

We can then analyze $P^{\wedge}P = (P^{\wedge}P)^{\wedge}$ which could have no more than 16 elements also. If we guess that nature still 'likes' $P^{\wedge}P \propto (e_o)$, so that $L^{\wedge}(P^{\wedge}P)L = P^{\wedge}PL^{\wedge}L = P^{\wedge}P = \textit{invariant}$, then we find that there are three distinct possibilities (at least)

$$P = P^{\mu}(e_{\mu}) + P^4(if_0) \quad L^{\wedge} = L^{\vee}$$

$$P = P^{\mu}i(ie_{\mu}) + P^4(if_0) \quad L^{\wedge} = L^{\vee}$$

$$P = P^0(e_0) + P^k i(f_k) + P^4 i(ie_0) \quad L^{\wedge} = L^{\downarrow}$$

There is a peculiar 3-part P, $P^k i(if_k) = P = P^{\uparrow} = P^{\wedge}$ with $L^{\wedge} = L^{\uparrow}$. There $P^{\wedge}P = P^k P^k(e_o)$. I don't know what this represents, if anything. There are two more 3-part P equations also, similar to this, but we shall not discuss them further here.

The $P^{\mu}i(ie_{\mu}) + P^4(-if_o)$ operator is closest to the original Dirac operator: $P_D = i\hbar\partial^{\mu}(f_{\mu}) - mc(e_o)$. It is 'identical' if $P^{\mu} \rightarrow \hbar\partial^{\mu}$ instead of $i\hbar\partial^{\mu}$, as we used for the $P^{\mu}\sigma_{\mu}$ physics before. This P^{μ} structure is a basic guess, at best, and the old Dirac equation works well in QED calculations. Therefore, for physics in complex Dirac algebra, we might guess $P^{\mu} \rightarrow \hbar\partial^{\mu}$ and P^{μ} is a real coefficient.

Multiplying the P_D operator above from the left by (if_o) produces $(if_o)P_D = i\hbar\partial^\mu(ie_\mu) - mc(if_o)$, so '$\gamma = -f$'.

This question of whether $P^\mu \rightarrow i\hbar\partial^\mu$ or $P^\mu \rightarrow \hbar\partial^\mu$ and $P^{\mu*} = \pm P^\mu$ is perplexing. The real Dirac and the complex Dirac systems seem to require different viewpoints. Also, the complex Dirac gives us three (actually four) Dirac equations, all with $P = P^\dagger$ and covariance under $LL^\wedge = 1$. This needs further investigation. The octonion extensions in the next chapter will be based on only real Dirac, but again that would all need to be generalized to complex Dirac as well. Nature just ain't simple.

COMPLEX PAULI ALGEBRA

The i in quantum physics is notorious. Some give it mystical significance in their speculations. It is very important in its implications for the physics. Let us look again at the Pauli algebra, $\{\sigma_\mu, i\sigma_\mu\}$, with 8 basis elements and two natural groups, $AA^{\neq} \equiv 1\sigma_o$ and $BB^* \equiv 1\sigma_o$. We first consider only real coefficient. We naturally have $P\psi = 0$, $P = P^\mu\sigma_\mu \equiv \hbar\partial^\mu\sigma_\mu$ and no room for a mass parameter in P. I discovered in the early 70's that $P\psi = \psi^{\neq *}Mc$ is a covariant generalization and $M = M^\mu\sigma_\mu = M^*$ is certainly possible. But it wasn't until the early 90's that I fully realized that this equation is incompatible with the Klein-Gordon equation: $P^\wedge P\psi = m^2c^2\psi = P^\mu P_\mu\psi = (i\hbar\partial^\mu)(i\hbar\partial_\mu)\psi$, which then leads successfully to the non-relativistic limit and Schrödinger's quantum ideas. Also in the early 90's, I discovered from Marx that van der Waerden had a similar equation to this in the late 20's, when Dirac found his version. Dirac is equivalent to $P\psi_a = mc\psi_v$ & $P^{\neq}\psi_v = mc\psi_a$. But Waerden had to include complex coefficients to get a normal looking mc parameter. The larger algebra has 16 basis elements really, $\{(\sigma_\mu), (i\sigma_\mu), i(\sigma_\mu), i(i\sigma_\mu)\}$, where again the i's outside are totally distinct from the i's inside. The importance of this distinction is easily overlooked. Remember then that we are really dealing here with the algebra $\{e_\mu, (ie_\mu), i(e_\mu), i(ie_\mu)\}$ and $[i(e_\mu)]^\wedge = i^\wedge(e_\mu)^\wedge$ is like $[i(\sigma_\mu)]^{\neq} = i^{\neq}(\sigma_\mu)^{\neq}$, so $i^{\neq} = -i$ here. Being 'unable' to clearly separate the i's here, held me back for two decades!

Now let us look at what really is in the Pauli algebra and the complex Pauli algebra. To start, let us prove that $P\psi = \psi^{\dagger\wedge}Mc$ is covariant in either the Pauli or Dirac algebras, for $P \rightarrow P' \equiv L^\dagger PL$, $LL^\wedge \equiv 1e_o = L^\wedge L$:

$$P'\psi' = \psi^{\dagger\wedge}M'c' = (L^{\dagger}PL)(L^{\wedge}\psi)$$

$$= (L^{\wedge}\psi)^{\dagger\wedge}Mc = L^{\dagger}\psi^{\dagger\wedge}Mc$$

$$= L^{\dagger}PLL^{\wedge}\psi = L^{\dagger}P\psi$$

The last L's are removed by $L^{\dagger\wedge}$ left multiplication on both sides. Notice that $M = M'$ (guess) is totally unspecified so far here. If we are in Dirac instead of Pauli, then P has 5 parts and there is a second kind of mass in the equation besides M. Next we look at the K-G equation 'requirement' (guess) $P^{\wedge}P\psi = {m_{exp}}^2c^2\psi$. We get, by operating on both sides:

$$P^{\wedge}P\psi = P^{\wedge}\psi^{\dagger\wedge}Mc = (P^{\dagger})^{\wedge}\psi^{\dagger\wedge}Mc$$

$$= (P\psi)^{\dagger\wedge}Mc = (\psi^{\dagger\wedge}Mc)^{\dagger\wedge}Mc$$

$$= \psi(M^{\dagger\wedge}M)c^2 = ? = (M^{\dagger\wedge}M)c^2\psi$$

Notice that we used $P = P^{\dagger}$ here and we need

$$\pm M^{\dagger\wedge}M = m_{\mathbf{exp}}^{2}$$

to reach the known physics of non-relativistic quantum. Remember too that $P^{\wedge}P$ is a 5 part thing in Dirac, so there is another mass parameter there too. For Pauli, the P has no internal mass part. Further note that $P^{\wedge}P$ has $P^0P^0 - P^1P^1 - P^2P^2 - P^3P^3 \equiv P^{\mu}P_{\mu} = (\hbar\partial^{\mu})(\hbar\partial_{\mu}) = -(i\hbar\partial^{\mu})(i\hbar\partial_{\mu})$. This minus sign is very important!! It dictates which parts of ${m_{exp}}^2$ must be positive and larger than the negative parts. This is all new stuff in the mid-90's, but it should have been developed in the early 30's! It may still prove useless for real world physics, but it should be old stuff. (The final chapter summarizes my latest insights and has more on this sign problem.)

We should not go much further into this here. I wanted you to see the elegance in it though. We finish this discussion by examining the Pauli subalgebra:

$$P\psi = \psi^{\dagger\wedge}Mc, \; P^{\wedge}P\psi = \psi M^{\wedge\dagger}Mc^2$$

$$P^{\wedge}P = (P^{\mu}P_{\mu})e_0 = (\hbar\partial^{\mu})(\hbar\partial_{\mu})e_0$$

$$= -(i\hbar\partial^{\mu})(i\hbar\partial_{\mu})e_0$$

$$M_I \equiv M^{\mu}e_{\mu}, \; M_I^{\dagger\wedge} = M^{\mu}e_{\mu}^{\wedge}, \; M_I^{\dagger\wedge}M_I =$$

$$(M^0M^0 - M^1M^1 - M^2M^2 - M^3M^3)e_0$$

$$M_{II} \equiv (iM^{\mu})e_{\mu}, \; M_{II}^{\dagger\wedge} = i^{\dagger\wedge}M_I^{\dagger\wedge} = +iM_I^{\dagger\wedge},$$

$$M_{II}^{\dagger\wedge}M_{II} = iiM_I^{\dagger\wedge}M_I = -M_I^{\dagger\wedge}M_I$$

Therefore, we have finally for such an M_I

$$P^{\wedge}P\psi = M^{\wedge\dagger}Mc^2\psi$$

$$(i\hbar\partial^{\mu})(i\hbar\partial_{\mu})\psi = (-M^{\dagger\wedge}M)c^2\psi \equiv m_{exp}^2c^2\psi$$

You can now easily see why van der Waerden needed the 16 element complex Pauli algebra for old fashioned physics. He needed M_{II} in order for m_{exp} to be real, if $M^kM^k \equiv 0$. Thus he required complex coefficients. I can see, however, that we only need M^o small compared to at least one M^k element to get m_{exp} real. The exotic M parts, M^k, must dominate in the Pauli algebra $\{(e_\mu), (ie_\mu)\}$. This is new stuff, mathematically, for physics; but does nature use it? Have we missed a simple but exotic kind of spin 1/2 particle with quaternion mass? (Rest states require complex Pauli.) Your generation will find out, one way or the other.

There are interesting questions about how many parts or functions are in ψ when $\psi^{\dagger\wedge}$ is included. A lot more for sure than when M = 0. Again more internal structure with probably new interpretations and surprises. Isn't it amazing how the math itself guides and dictates the physics. The math here is not really as pretty to my eye with $\psi^{\dagger\wedge}$Mc instead of, or in addition to, P having internal mass. But it adds much richness of structure if it is in the real word design. It will look prettier to your generation if it proves to have lots of new insights about the parts inside protons and neutrons.

We have covered the 'waterfront' by now. There is little left but hard nosed, detailed calculations of special cases for the wave equations as they stand, and comparison with experiment. Of course, the lumpy quantum nature needs to be further guessed at in detail. These smooth equations that we have displayed need to somehow become statements about lumps, i.e., little blobs that register as clicks in our big blob equipment. That is what the propagator approach to QED does in detail. The calculations for 'click' prediction get very tedious and they rely completely on ϵ_2 and ϵ_3 being very small couplings! If they were not small, experimentally, we would have no QED theory at all! It is essentially a perturbation expansion, springing from the ideas 'contained' in the field equations above. The details were all worked out by 1950, and they are displayed in modern versions, like Chang's 1990 summary text. There is no need to duplicate all that here. Instead, we push on now beyond the basics of QED. We can push the $L^{\dagger}PL$ form invariance to new heights by looking for even more exotic expansions of the Dirac number system itself, which may be only an approximation to an even richer structure. For example, we have assumed that P^4 and P^5 would just be numbers, taken from experiment. Why these particular numerical values? Maybe there is a deeper theory that turns these into dynamic operators, like the other four parts, and perhaps there are even more variables, besides the 6 we have discovered so far in P. We have already come a long way, but we still know very little. We cannot predict mass values yet for muons, electrons, protons, etc. That is the ultimate GOAL! Until we can do that, we still have only a crude theory of nature.

CLASSICAL BLUES

We've got these blues because we are big blobs and can only see other big blobs. The beautiful equations and number systems, that we have been discussing here, apply only to the world of little blobs. Clearly, the details in any development are not unique. We always have to guess. Naturally, we want to know which guesses are the right ones. But we cannot directly look at little blobs. So, we enter the 'mine field' of quantum measurement. How to interpret the big 'clicks' we do see in terms of the little blobs that are in the system producing these clicks or flashes.

The key to resolving all the various conflicting views here is to face the fact that we must guess in the lab also. We certainly have guessed at the theory! We then calculated lots of detailed consequences. All this calculation was still at the little blob level. Then some of us lumber into the lab as huge hulks, made of some 10^{27} little blobs, or more, and try to study interactions by a few little blobs. What a hopeless joke. We might as well give it up and go fishing.

We'll never get very far.

Most humans do not even start such study and many of the rest quickly give up, but the diehards remain in the labs, soldering iron and lathe 'in hand'. The challenge is just too irresistible for the few. Clever and sophisticated things have been done. The most important are clearly the bubble or spark chambers and the accelerators that go with them. They allow us to see a series of big blobs, with about 10^{20} little blobs in each, as a series of 'points' in 3-space. We interpret these as 'tracks' of little blobs, interacting gently with billions of other little blobs along the way. This is our basic experimental guess about micro-reality.

A little blob runs into a big detector and sets billions of little blobs into new behaviors. We boil this down to thinking of it as 'one thing'. We 'caught a little blob', we would say. If nothing at all happens that we can see, then we say that the little blob went right through the equipment undetected, or maybe it never was there at all. We have to guess and guess, over and over. We think in terms of these big blobs and whether they get altered in a directly observable way, (even if no one actually does look) due to the little blobs within them.

The little blob wave equations, with 'lumpiness' added to them, must somehow be turned into calculations of what changes will instead occur in these big blobs of 10^{23} such little blobs. What a stupid thing to have to do. But we have no other choice. We cannot do better. Better means asking God what really happens down there. But we probably couldn't relate to the answer anyway.

So—here is how it goes: We set up big blobs in a special arrangement. Then we later see these big blobs in a new arrangement. We have to GUESS at which arrangement of little blobs we first 'produced' in the initial set up of the big blobs. This is the 'big jump'. We have to also GUESS, from the final setup, what set or sets of little blobs we could have 'produced' in the final setup—another big jump, similar to the first. We repeat the setup conditions many times, because the outcome is only probabilistic. One 'run' is not enough usually. We use the theory to calculate between the initial and final theorized (guessed) states of little blobs, that we first made and later found, at the little blob level (going both ways in time in the calculations). We then have to hope that the little blob details we guessed at really did fit with the big blob states we actually observed at the beginning and at the end. We cannot initially guess downward to the little blob level, calculate, then finally guess upward to yield specific, testable, big blob outcomes that humanoids can see with their (10^{23} little blob) eyeballs. We must guess downward from BOTH ends, then calculate the odds in the middle, using both time directions in the process.

Failure to realize these limitations and the unavoidable 'disaster' of our measurements because of our limitations, are what cause all the interpretation problems for quantum physics. We want to think that things go on at the little blob level like they do on our big blob level. We want to say that little things move along in the equipment. We say there are sources, then collisions, then detections. We cannot say this because it is all bullshit. There is equipment at t_1 and there is equipment at $t_2 > t_1$. In between, the little blobs are all over the equipment at once and running around both directions in time from sources to sinks (the sinks are themselves sources for backward time propagation).

We must break out of our old thinking and accept all this strangeness as OK for the real world; a world that we can never experience directly. Well, your generation must break out. Mine must just die off and get out of your way. Just as our ancestors did for us. Time flows forward only on the average over long intervals. There was no overall time evolution for the universe's Big Bang at times less than some t_o, where t_o is the first classical time that physically makes any sense after $t = 0$, when the bang of creation occurred, whatever that means—if anything. We essentially are classical, we think classical, we see classical. The real world just ain't classical. Take it up with God if you object.

Classical thinking theorists have tried to give things like $t = 10^{-40}$ seconds physical meaning for time evolution. They need a 10^{80} swift expansion of the universe at such times to 'explain' how the universe is so flat, homogeneous, and simple. (Maybe God just did it that way with the magic wand of creation.) This small time evolution thinking really is garbage for the universe, even long after 'reaching' nuclear density, as the universe expanded. It was still in a totally quantum gravity and otherwise state there, where time flow forward was barely meaningful as some kind of drift forward. When we reach the helium synthesis stage, supposedly in the first few minutes of the universe's life, we finally might be seeing a universe where classical time flow of the overall structure just begins to make sense. We cannot know much more until we have a quantum theory of gravity.

CHAPTER 7

Very Little Blobs: Extra Dimensions – Non-associativity?

I have made the continuing point ad nauseam that we humanoids experience a highly averaged version of the reality that makes us up. We scientists do (we hope) only physics, not theology. Therefore, we rely only on experimentally testable hypotheses and guesses. All testable experiments involve two time 'slice' experiences for us humans: (1) preparation of some big blobs at time t_1, and (2) detection of some other big blobs at $t_2 > t_1$. Usually, we have $t_2 >> t_1$, compared to sub-atomic time frames. In this time interval, all kinds of strange things 'go on' at the little blob level. These conspire in complicated ways to produce the big blob situation actually seen by us at t_2. We can never know what really took place between t_1 and t_2. That is an important realization! Saves us a lot of philosophizing about fantasies! Don't even try to 'know' the truth of what went on. Just guess and calculate, then compare with actual observations at t_2. The most difficult thing to accept is that time does not necessarily flow only one way between t_1 and t_2. It does flow one way for **our** conscious experience, but we self-aware beings are only big blobs. We cannot directly sample 'below' that averaged experience. We can still calculate both ways in time, however, in theorizing about what happened between t_1 and t_2, which led to the humanly observed outcome at t_2. Whatever works is valued by us, no matter how startling it is, or improbable, compared to our averaged experience. It appears that we need to specify, in advance, both the t_1 and the t_2 conditions, actually seen by us, to then calculate the odds of this outcome being observed. It also appears that observed outcomes are not definite. The same t_1 conditions, duplicated again at $t_3 < t_4$, will produce different results at t_4, even though $t_4 - t_3$ is the 'same' as $t_2 - t_1$. This has nothing to do with $t_3 \neq t_1$. We could instead build many duplicate set ups, and activate all of them at the same t_1, then observe all set ups at t_2. They would be different. A GOOD theory is one which gives the right odds for these various outcomes. It appears that the little blob world has 'roulette wheels' in it, so that outcomes are only probabilistic. This is beautiful since it means that the universe is not a movie with a pre-written script. (Even God must wait to see the details on each planet as they evolve.) Only the average, over all behaviors are predictable from the Big Bang; *e.g.*, what percentage of planets will blow themselves up when they hit their atomic bomb stage of evolution, *etc.*, *etc.*

While we obtain from this the advantage of not feeling like robots, we have much more difficulty in putting together a successful, relativistic quantum theory. We must learn to calculate probabilities. Yet, how can this be? We had causal, partial differential equations at the base of our theory in the previous chapter! This conflict has driven us mad for years. It is usually called the quantum measurement problem—the 'collapse' of the wave function. It has not been resolved and cannot be. We must face our limitations and give up the idea of time evolving at the subatomic level. Therefore, collapse is a meaningless idea. We have not been willing up until now to give up time flow, hence the confusion still reigns. We must learn to work both directions in time and think only in terms of macro-jumps in flowing time; from humanly experienced set up to humanly experienced observation. In the dark, meantime, there are ghosts and goblins, if you like, doing all sorts of wild and wonderful things.

For example, little bits can tunnel through walls, or even move around inside the walls where they never could be found if they were big blobs. Whole atoms diffract from very thin crystals in a way that requires that the beam atoms 'know' the positions of all the atoms in the crystal; yet they don't rip up the crystal in going through. A more dramatic example of strange non-locality for little bits is neutron diffraction in two strips of crystal, very close together. By having spin flip detectors mounted behind each of these strips, a few centimeters apart, we can determine which crystal strip a neutron 'went through', as it went by the region and got diffracted. Notice the error I have made here already. I assumed it 'goes by'. We cannot assume it moves in time like we do!

With the spin flip detectors off, we cannot tell which strip the neutron 'experiences' and so we get a complicated pattern of detection arrival, for the many neutrons detected one at a time. With both spin flipper on, we again cannot determine which crystalline strip was 'used' by the neutrons to reach the final detectors. The pattern of arrivals is, therefore, unchanged. However, with only one flipper on, we can 'know' that a neutron that hits the final detectors with its spin direction flipped 180°, went through the crystal side which had the flipper turned on behind it. The flipper does not do anything radical here, as evidenced by both flippers being on. (That on condition does not disturb the final arrival pattern.) Nature behaves very differently if each neutron can be at both crystals at 'once', compared to circumstances where it must be only at one or the other. This marvelous experiment has actually been done and we have to get used to this strange result.

In tunneling, we find the wall thickness alters the odds of the tunneling taking place. Again, the particles on one side of the wall 'experience' the total wall, in the dark. With the lights on, the particles get measured by scattering the light (just like the spin-flippers being turned on above measured the particle's position). If the wavelength of the light is much longer than the wall thickness, then the 'light' scattering gives only a diffuse localization for the particles. 'We' cannot tell which side of the wall they are on, when near the wall, even with these lights on brightly. They, therefore, will then tunnel through the wall with greater than zero odds. This is just the way it is, as experienced by macro-beings like us. The bigger the blobs and the thicker the walls, the lower the odds of any tunneling at all, even in the dark. So we fortunately don't fall through the floor from the second story bedroom when we turn out the lights at night; we can successfully walk across to the bed. But it really could happen, once in a great while, if you sleep on the second or higher floor or have a basement. The odds are just incredibly small. The faster an object moves, the more it also seems to be able to do the unexpected, such as tunneling. Whether this has anything to do with length contraction, i.e., the big blob shrinking in the direction of its motion (due to that motion), is not presently known. I suspect they are all connected at the very smallest scale of reality.

The many quantum experiments, like tunneling and diffraction by two crystal strips, sample what to us seems like the very small scale of nature. It is, however, only the intermediate scale size of nature. We see that little blobs behave very strangely. However, this is not the end. There is a much smaller scale, such as that inside protons. There, we may well find new and radically different things, that make tunneling seem like child's play. In the rest of this chapter, I shall share some personal speculations about how the Dirac and Maxwell equations, that are the corner stones of QED, might be further generalized using 'quaternion' thinking. This is untested stuff, so the reader may wish to skip it (except those bright, young ones who aspire to join the hunt for the break through needed to go beyond QED and to someday explain why $m_p = 1800m_e$).

We certainly expect radical things to show up. Going to quantum mechanics replaced classical path motion by ψ, a wave that is all over the place at once and involving both directions in time. The next level below that could easily require higher dimensions and new kinds of mathematics. We shall introduce both below. This is only one man's attempt, and it may be totally useless. There is only one right answer for nature, and we cannot expect to find it easily. Most searchers will search their entire lives in vain. It cannot be otherwise. Perhaps whole generations will search in vain as well. The answer is already there, but it may be beyond our imagination.

So relax and enjoy my ravings below. They are beautiful in a sense, at least, even if possibly not right for our universe. (Some other universe, perhaps!)

OCTONION PHYSICS

Shortly after Hamilton opened the door with quaternions, mathematicians started inventing lots of generalizations and variations, just for the fun of it. Systematic study eventually revealed that $\{\sigma_0\}$, $\{\sigma_0, i\sigma_0\}$, $\{\sigma_0, -i\sigma_k, k = 1,2,3\}$, called R, C, Q, respectively, are very special compared to say $\{\sigma_0, -i\sigma_1, \sigma_2, \sigma_3\}$ or $\{\sigma_\mu, i\sigma_\mu\}$, or even $\{e_\mu, ie_\mu, f_\mu, if_\mu\}$. The 'norm' for R, C, and Q is positive definite and the 'norm' of a product of two such hypercomplex numbers is the product of the individual norms. We don't need to go into all that here. Take their word for it that this is an elegant and rare property. They eventually proved that there was only one other number system, octonions, that shared this property. Octonions were peculiar, however, because they failed to follow the usual associative law: $(AB)C = A(BC)$. Such strange mathematics is not useful in QED at all, so most physicists have ignored octonions. The late professor Gürsey jointly led the charge of a small group that searched for practical octonion applications in particle physics. I don't know of any real successes, however. They do not seem to have found any important ones, but I may be wrong because things have a way of coming back later to be important. Again, as always, most fundamental ventures will be unsuccessful. I played with octonions too in the 1970's but gave up on them. The non-associativity was just to radical and frustrating for me. But, having found that my other efforts also led to dead ends, I gave them a second shot. After all, they are the closest thing to quaternions and quaternions give us QED in a natural way, as we have outlined in the previous chapter.

The octonions have 8 parts: $O_0, O_1, \cdots O_7$. Their multiplication rules are complicated and usually displayed in a table (one which has to be constantly referred to). In my first efforts with them, I put together a simple rule for multiplying them, just as I had done for the Dirac algebra. It is not as simple a rule, since the math is non-associative. But it can be applied with a little care, even for high school students, as we now show: We start with $\{\sigma_0, i\sigma_k, \sigma_k j, i\sigma_0 j\}$, $k = 1,2,3$. We obviously have 8 basis elements here. The subset $\{\sigma_0, -i\sigma_k\}$ is the 'pure' quaternion subsystem, and we know all about it already. The j is the new thing here. We define (and memorize):

$$\sigma_k j = -j\sigma_k,\ j^* = -j,\ j^{\neq} = -j,\ (i\sigma_\mu j)^* =$$

$$j^*\sigma_\mu^* i^*,\ (i\sigma_\mu j)^{\neq} = j^{\neq}\sigma_\mu^{\neq} i^{\neq},\ ij = ji,\ jj = -1$$

The multiplication rules are as before except for $(\sigma_1 j)(\sigma_2)$, $(\sigma_1 j)(\sigma_3)$, $(\sigma_2 j)(\sigma_1)$, $(\sigma_2 j)(\sigma_3)$, $(\sigma_3 j)(\sigma_1)$ and $(\sigma_3 j)(\sigma_2)$. In other words, there is trouble only when a j is on the left and no j is on the right, and the σ_μ have only $\mu = 1,2,3$, but the μ values are NOT alike. In these cases and only in these cases, we put in a minus sign when removing the parentheses. This destroys the associativity of the algebra! For example: $(\sigma_1 j)(\sigma_2) = -(\sigma_1 j\sigma_2) = -(-i\sigma_3 j)$ and $(\sigma_1 j)(\sigma_3) = -(\sigma_1 j\sigma_3) = -(+i\sigma_2 j)$. More examples follow:

$$(\sigma_1 j)(\sigma_2 j) = (\sigma_1 j\sigma_2 j) = (-i\sigma_3 jj) = (i\sigma_3)$$

$$(\sigma_1)(\sigma_2 j) = (\sigma_1\sigma_2 j) = (i\sigma_3 j)$$

$$(\sigma_1 j)(\sigma_0 j) = (\sigma_1 j\sigma_0 j) = (\sigma_1 jj) = -(\sigma_1)$$

$$(\sigma_0 j)(\sigma_1) = (\sigma_0 j\sigma_1) = -(\sigma_1 j)$$

Of course, we can expand this easily to the basis $\{e_\mu, (ie_\mu), (e_\mu j), (ie_\mu j)\}$ which is equivalent to the complex octonion algebra. We also need the Dirac algebra, $\{e_\mu, (ie_\mu), f_\mu, (if_\mu)\}$ as a subsystem of any expanded number system, since that system is sufficient for the very successful QED theory of electrons and photons. The obvious generalization (guess) is $\{e_\mu, ie_\mu, f_\mu, if_\mu, e_\mu j, ie_\mu j, f_\mu j, if_\mu j\}$, but we have to define the multiplication rules before this 32 element set is well defined. The following guess is reasonable. Combine elements by first using the Dirac system rules, then using the octonion rules. An f on the right thus causes a $(\cdots)^{\neq *}$ to be taken on the left. The final form will eventually be an e or f, depending on the ee, ef, *etc.*, rules, as before. Meantime, we use the octonion rules and quaternion rules to combine and simplify in the middle. Examples will make this clear:

$$(e_1 j)(e_2) \rightarrow (\sigma_1 j)(\sigma_2) = -(\sigma_1 j \sigma_2) = (i\sigma_3 j) \rightarrow (ie_3 j)$$

$$(e_1 j)(f_2) \rightarrow (\sigma_1 j)^{**}(\sigma_2) = (\sigma_1{}^{**} j^{**})(\sigma_2) = (-\sigma_1(+j))(\sigma_2)$$

$$= -(\sigma_1 j)(\sigma_2) = --(\sigma_1 j \sigma_2) = -(i\sigma_3 j) \rightarrow -(if_3 j)$$

$$(e_1 j)(f_2 j) \rightarrow \cdots = (\sigma_1 j \sigma_2 j) = -(\sigma_1 \sigma_2 jj) = +(\sigma_1 \sigma_2)$$

$$= (i\sigma_3) \rightarrow (if_3)$$

$$(f_1 j)(f_2) \rightarrow (\sigma_1 j)^{**}(\sigma_2) = \cdots = -(i\sigma_3 j) \rightarrow -(ie_3 j)$$

$$(e_1)(f_2 j) \rightarrow (\sigma_1)^{**}(\sigma_2 j) = -(\sigma_1 \sigma_2 j) = -(i\sigma_3 j) \rightarrow -(if_3 j)$$

This is a pretty pattern. It clearly contains all the old QED stuff as a sub-system. But we are in new and uncharted waters here. Naturally we turn to the conjugations for further insights. We had three, $\dagger$, $\wedge$, $\vee$, for the Dirac algebra and one basic group $LL^{\wedge} = 1e_0$, called Sp(4), with 10-parameters, or a 16 parameter group for complex coefficients. We naturally guessed that $P \rightarrow P' = L^{\dagger}PL$ is basic to wave equations in QED. We now search for generalization of these conjugations, and then we can see what $P = P^{\dagger}$ looks like for this larger, non-associative algebra! We also will generalize $LL^{\wedge} = 1e_0$ as well, and see how many parameters it has, by investigating elements near the identity, L_{ϵ}. The 'group' is non-associative now, so we have a whole new ball game here as well. (I call non-associative groups, *toups*.) Non-associativity ruins all kinds of neat results, so nature better show us that octonions are used, or most of us will lose interest in a hurry! Even if we do lose interest, there will still be mathematicians who are looking for challenges, and who will push ahead and eventually explore all their abstract ramifications.

We must guess at the generalized conjugations. A natural thing to try is what we had before:

$$(\ )^{\dagger} = e^* + f^* \qquad (\ )^{\wedge} = e^* + f^* \qquad (\ )^{\vee} = e^* - f^*$$

These will be well defined for $i\sigma_{\mu}j$, defined above. It turns out that this REALLY works to give new and generalized, antiautomorphic conjugations! We now have $3\times32\times32$ combinations to check out!! A computer would come in very handy here. Let us try some examples:

$$(e_1 j)^\dagger \rightarrow (\sigma_1 j)^* = j^*\sigma_1^* = -j\sigma_1 = +\sigma_1 j \rightarrow +e_1 j$$

$$(f_2)^\dagger \rightarrow (\sigma_2)^* = -\sigma_2 \rightarrow -f_2$$

$$[(e_1 j)(f_2)]^\dagger = \cdots = [-(if_3 j)]^\dagger \rightarrow -(j^*\sigma_3^* i^*)$$

$$(f_2)^\dagger (e_1 j)^\dagger = (-f_2)(e_1 j) = \cdots = --(if_3 j)$$

I am pretty sure that the whole set will check-out OK. So, what then does $P^\dagger = P$ imply? We readily see that such a P has the structure

$$P = P^\mu e_\mu + P^4(if_0) + P^5(f_0) + P^{3+k}(e_k j)$$

$$+ P^9(ie_0 j) = P^\dagger$$

We have a 10-vector, with the old 6 parts of Dirac and four new, non-associative parts. Of these new parts, 3 are similar and the fourth is different. Are these four new parts space and time 'shadows'? Too early to tell. People are currently exploring this exciting matter. There is little guidance, however, for guessing here.

We do have one guide post here. That is the second order differential operator that derives from the basic operator P. We have seen that this P NATURALLY seems to be a 10-vector, from $P = P^\dagger$. The so called Klein-Gordon equation traditionally had an operator like $[(i\hbar\partial^\mu)(i\hbar\partial_\mu) - (mc)^2]\psi = 0$. Of course, we would not accept this old approach any more. But we need to start with P and generate a suitable 'PP' operator of some kind that is more or less like the K-G operator of old.

Recall from Chapter 6 that in 4-space $P^{\neq}P$ is a Lorentz invariant and $P^{\neq}P = \cdots = (P^\mu P_\mu)\sigma_0$. So the natural guess here is $(P^\wedge P)\psi = 0$. This almost comes from $P\psi = 0$, but P now is a 10-vector and non-associative. Can we still get all the cross terms to cancel out now?

The new basis $\{e_\mu, ie_\mu, f_\mu, if_\mu, e_\mu j, ie_\mu j, f_\mu j, if_\mu j\}$ is actually insufficient for physics because the successful Dirac description requires $P^\mu \rightarrow i\hbar\partial^\mu$. The i

$= \sqrt{-1}$ means that we will need complex coefficients in general. That requires that we further generalize the three conjugations to act on these complex coefficients. I have explored this in some depth and it appears that the best guess is that ALL THREE conjugations act the same on the complex coefficients. This common action is the usual complex conjugation. This is very important for examining ($P^{\wedge}P$). Table 2 shows the multiplication table for the Dirac basis elements of the 10-vector $P \equiv P^{\dagger}$. Notice that (f_0) is quite peculiar. It is the only term that will not cancel out in the cross terms! We can get rid of the cross terms only if the (f_0) coefficient is pure imaginary (or zero) and the other 9 coefficients are real, as expected in $P = P^{\mu}e_{\mu} + P^4(if_0) + \cdots = P^{\dagger} = P^{\mu *}(e_{\mu}) + P^{4*}(if_0) + P^{5*}(f_0) + \cdots$. In other words, P^5 must be zero, if $P = P^{\dagger}$, so we have only 9-vectors; or P^5 is imaginary and $P \neq P^{\dagger}$. If P^5 is somehow always very small, then $P \approx P^{\dagger}$. This is probably a very important matter and relates deeply to the very concept of mass. See the next chapter. With $P^{4*} = P^4$ and $P^{5*} = -P^5$ we wind up with $P^{\wedge}P$ in the form of $P^{\wedge}P = P^{\mu}P_{\mu} - [P^4P^4 - P^5P^5 + P^6P^6 + P^7P^7 + P^8P^8 + P^9P^9]$. In Dirac approximation, the bracket term, $[\cdots]$, above becomes $\pm(mc)^2$ with m taken from classical experiments involving the bending of proton trajectories in a magnetic field. The $-P^5P^5$ form here, where $P^{5*} = -P^5$ reminds us of something like $P^5 = im'c$. The concept of tachyons existing was guessed long ago due to $E = mc^2/(1 - v^2/c^2)^{1/2}$. If $v > c$ then the replacement $m \rightarrow im$ will compensate and E can still be real; it can participate in energy conservation. Tachyons were guessed to have $c < v < \infty$, whereas ordinary mass has $0 < v < c$. I think it likely that big blob tachyons are not possible but little blob tachyons might still exist as part of the sub-proton world. The loss of $P = P^{\dagger}$ as an exact idea is very radical, however, so maybe $P^5 = 0$ all of the time in all cases, for some reason. Much remains to be explored here. Puzzles like this usually lead to great advances when resolved.

The infinitesimal 'group' element has $L_{\epsilon}{}^{\wedge}$ change the sign of all basis elements besides e_0. We see that

$$L_{\epsilon} = \epsilon^k(ie_k) + \delta^k(e_k) + \lambda^{\mu}(if_{\mu})$$

$$+ \theta^{\mu}(e_{\mu}j) + \phi^{\mu}(ie_{\mu}j) + \alpha^k(if_kj) + \beta^0(f_0j)$$

with 10+12 parameters now. Notice how 'tight' all of this is, and notice we have assumed only real coefficients in L. There are few places where we can make arbitrary choices. We can also check out $A^{\vee}$ and $B^{\dagger}$ and we find again that they also have 22 parameters for the infinitesimal group members. Probably they are still isomorphic to L^ above, as 'groups'. The toup $CC^{\dagger} = 1e_0$ generalizes SU(2)⊗SU(2) and we find that (with only real coefficients):

$$C_{\epsilon} = \epsilon^{\mu}(ie_{\mu}) + \delta^1(f_0) + \delta^2(if_0)$$

$$+ \lambda^0(e_0j) + \lambda^k(ie_kj)$$

which now has 6+4 parameters.

These toups and P should be the foundation for expanded wave equations and their quantization. Future calculations can eventually compare with experiment and see if nature makes use of any of this pretty mathematics. If not, it is just pretty poetry.

Looking back at the coupled ψ, F, and A equations of QED, notice that only pairs of fields show up multiplied together. This nicely eliminates any ambiguity when the individual elements are non-associative! Therefore, these three old equations are still well defined in this generalization of the underlying number system. In the usual, so called Lagrangian formulation of QED, one guesses at simplified 'Lorentz' invariants, then uses a mathematical 'crank' to crank out the field equations. We have instead started with the equations by direct guessing. This Lagrangian idea won't be so easy now, since Lagrangians usually have three elements in the scalars, such as $\psi^{\dagger}P_{\psi}\psi$. Our ψ, F, A couplings can be simplified, to eliminate F, unless there is another experimentally determined coupling constant between F and A. In any case, our direct equation-forms eliminate any ambiguity as to where the parentheses should go, once we start combining the three separate equations! Lagrangians are in for a lot of trouble, I think, at this deeper level. Time will tell.

You now can readily see all sorts of complications and subtleties, that non-associativity brings into physics, **IF** it is really there. Even $P\psi = 0$ does not lead to $(P^{\wedge}P\psi) = 0$, but only $P^{\wedge}(P\psi) = 0$.

We humanoids do exist, and we would not do so if the rules were too simple. But are they this complicated? Could be. I wouldn't be at all surprised. They will need to be checked out by the next generations, unless someone soon finds an alternative approach that is obviously better. (There is none better up to now, 1996.)

Notice that the old Lorentz symmetry, used in predicting how moving objects shrink in length and slow in internal time, has been greatly expanded from such crude ideas for large, three-dimensional bodies. We are now into 10-

spacetime or higher here, with $LL^{\wedge} = 1e_0$ and $P' = L^{\dagger}PL$, 'form co-variance'.

We have not yet brought in the curved space that is certainly there at the large scale average. It must be there in the micro-world as well. Certainly all the extra dimensions in P, besides the usual 4 large ones of space and time, must roll up on themselves since we don't see them. They may even expand and contract over time, but probably remain subatomic. We don't really know that they will stay subatomic, or were always subatomic in the distant past. We only see the world as it is now and used to be, over the past billions of years, using telescopes. At very early or very late times, these other dimensions may become large. If so, then they may eventually affect humanoid consciousness, if it exists well into the distant future. (There was no intelligence alive in the beginning, at the hot soup, Big Bang, but there could be some left alive near the final, Big Crunch.) As the 'large' dimensions of space contract, near the end of time, the extra ones may grow. We just don't know! Could such conscious beings then 'think' in terms of more than 3 space dimensions if these dimensions become large enough? Could 'anyone' even stay alive with such a radical change in the structure of space and its subsequent effects on chemical bonding and neurons? I doubt it. Such changes in atomic electron cloud properties would likely lead to a breakdown of organic chemistry and the death of any remaining humanoids. The laws of increase in entropy will likely have eliminated life in the universe long before the Big Crunch becomes noticeable. (The crunch comes at about 400 billion years in the future, minimum). Stars like the sun live about 10 billion years. All will probably have come and gone in the next 100 billion years or so. Humanoid life would be hard pressed to stay alive after that, hovering around small stars which can live 100's of billions of years.

We still have not explained how moving in space causes large objects to shrink in the direction of motion. But we have seen that there may be 10-space instead of 4-space, even 16-space perhaps, so we should not expect to be able to explain this strange behavior with our limited, current knowledge. It remains one of the wonders of the world, like tunneling and non-locality for small blobs.

Big blobs, such as us, experience time flowing forward. Our brain flashes take about a millisecond, so we sample time on about that time frame of successive, forward events. If we big blobs turn off the lights and push off from the interior walls of our space capsules, while wearing our space suits and with no air in the capsules, we cannot know really where we are. I truly believe that quantum non-locality is telling us that there is a finite, though minuscule, probability that even we huge creatures could tunnel through the ship walls and appear, later, outside when the lights are again turned on. This may be silly in practice, but not in principle! Unfortunately, if only one neuron is thrown by itself at the ship's wall in the dark , it still has only a minuscule chance of tunneling through (unless thrown at near the speed of light). So, no conscious being could tunnel, even if zillions of near duplicate set ups were run at the same time, out in space. Any volunteers? What if we are thrown at the wall at near the speed of light? How does that change the odds for big blobs to tunnel?

One neuron still has some 10,000 atoms in it. One atom alone has a pretty good chance of tunneling through a thin aluminum foil, if going very fast. One electron or one neutron has even a better chance. The thinner the foil is rolled out, the greater the odds. It is interesting to ask if the relativistic squashing, observed for large objects, also applies to small objects. Can they tunnel easier, if squashed, due to high speed? Of course, small objects are not really squashed because they have no edges in the first place! They are 'all over the place' at once in some sense. But the penetration power generally goes up with increased speed. This is easy to see in an electron microscope, (my old stomping grounds). You can see through thicker foils at higher accelerating voltages for the electrons. The potential energy from the high voltage field, eV, goes into the kinetic energy of the accelerated electrons as follows:

$$eV = \frac{mc^2}{\sqrt{1 - v^2/c^2}} - mc^2 \rightarrow \frac{1}{2}mv^2 \ (slow)$$

This gives the classical speed, v, of such particles, considered as particles. Considered as waves, they have a wavelength

$$\lambda = \frac{h}{P} = \frac{h}{\left(\dfrac{mv}{\sqrt{1 - v^2/c^2}}\right)} \rightarrow \frac{h}{mv} \ (slow)$$

No one can see objects smaller than the wavelength, λ, of the 'radiation' used to try and see them. For the electron microscope, typically operating at V = 100,000 volts, we could even 'see' inside atoms, in principle. But the magnetic, focusing lenses are not that good. They cause enough distortion that we can only see whole atoms as fuzzy blobs on the final viewing screen. Inside of the atoms, the 'parts' are non-local anyway, and our usual viewing concepts probably fail. Could we really ever image a nucleus as a well defined, fuzzy ball in only three dimensions of space? Even in principle? Good question. Inside protons, we may have more dimensions, so surely we lose the classical, big blob concept of 'imaging' at a small enough scale of nature.

The electron 'is' a spinning wad of charge that is spread out as ψ. It shows us no internal parts, yet it has an angular momentum. There could also be charged, rotating 'stars' possible, in principle, and these big blobs give a known gravity curvature solution, in Einstein's gravity theory, for the curved space around them. The charge, Q, and spin, S, can't be too high for these stars or the solution breaks down. Perhaps this is where the whole star would become a quantum blob. How much Q or S is really possible, is not known for a spinning black hole. It may well be that electrons, if not protons, are just such so-called naked singularities, i.e., black holes with excessive Q and S compared to mass. An interesting idea. We are at the current frontier here. Gravity must next be brought into $P = P^{\uparrow}$ in 10-space or higher and Einstein's 4-space gravity equation generalized and added to the QED generalized equations. After all, a universe could in principle consist of only light, negative electrons, and gravity. There would have to be rules for such a universe, with huge net charge. It would likely expand like crazy and the gravity would be negligible, unless the gravity constant, G, were very large for some reason there. We don't know at all why G is so small for our universe. We may not theoretically 'penetrate' inside the proton until we first have gravity merged with QED. This may even require 10-space, as above, or even more dimensions. Time will tell.

In 1993, while 'completing' this chapter, I found some new things in the Dirac algebra, as discussed earlier. The algebra really needs to be generalized to complex coefficients, as stated earlier. The conjugations take complex conjugation on all coefficients in front, so $[i(\cdots)]^{\uparrow} = -i(\cdots)^{\uparrow}$, *etc.* This seems trivial at first, but it has wide implications for the basic physics! Let us review those results:

The basic, complex coefficient algebra, before octonions are brought in, is $\{(e_\mu), i(e_\mu), (ie_\mu), i(ie_\mu); (f_\mu), i(f_\mu), (if_\mu), i(if_\mu)\}$ with 32 basis elements. The conjugations $P = P^{\uparrow}$, $P = P^{\wedge}$, and $P = P^{\vee}$ now all give 16-vectors, as does $P = P^{\uparrow\wedge\vee} \equiv P^{\downarrow}$. The group $LL^{\wedge} \equiv 1(e_0)$ has $(\ )^{\wedge}$ change the sign now of 16 basis elements, so this is a 16 parameter Lie group. Restrictions such as $L^{\wedge} = L^{\vee}$ then reduce L to 8 parameters, not 6 parameters as before. The two extra parameters seem to be associated with phase symmetry ($\psi \to e^{i\theta}\psi$, causing no change of any importance) and with scale invariance ($x^\mu \to \alpha x^\mu$, causing no change of any importance). This may mean that the so-called Planck length, 10^{-33} cm, is not a physically significant distance for the real world.

The $P^{\wedge}P$ picture now, for P possibly being a 16-vector, leads to P having only 5 components, IF no cross terms are to survive. Adding in octonions would expand this even further, of course. But without octonions, the 5th dimension $P^4(if_0)$ or instead $P^4 i(f_0)$, is naturally associated with the usual kind of mass. The P^P table then eventually shows that there are two basic mass equations:

$$[P^{\mu}(e_{\mu}) + P^{4}(if_{0})]\psi = 0 \equiv P_{5A}\psi$$

$$[P^{\mu}(e_{\mu}) + P^{4}i(f_{0})]\psi = 0 \equiv P_{5B}\psi$$

These differ in P^P leading to (+ − − − −) for the first, and leading to (+ − − − +) for the second. One of these cases has no Schrödinger, non-relativistic limit. It appears to be a really new equation. There are also two different versions of each equation, with different basis elements for the 5-vector P. There are also, it seems now, 3-part P's with P^P having no cross terms. These cannot mix with the other 5-vectors, or cross terms survive. Their meaning is currently unknown.

As mentioned earlier, I generalized an old idea I published in 1974 in the Am. J. Phys. that suggests a second form of mass! We guess two P operators give

$$P\psi = \psi^{\dagger\wedge}Mc \qquad P\psi = \psi Mc$$

(At the end of 1993, I learned from Egon Marx that van der Waerden had a similar equation in the 1920's.) The first form above then uses P^P to get a K-G equation. The second uses instead PP to get the K-G equation. Here P is the 5-vector which already has mass as the 5th dimension! The Mc is then a new kind of mass. Both it and the 5th component mass mix together in the K-G equation and thus in the experimental non-relativistic mass number we have always thought was the only mass number! New possibilities; these are discussed further in the next chapter.

The Mc can have up to 32 basis elements and some reason must be found for cutting this down in size. For example, maybe $(Mc)^{\dagger\wedge} = (Mc)$. The form covariance $P \to P' = L^{\dagger}PL$ and $\psi \to \psi' = L^{\wedge}\psi$, with $LL^{\wedge} \equiv 1e_0$, leaves Mc unspecified, to a large extent. It may have small extra parts that are neglected in the K-G equation and subsequent non-relativistic limit. When we bring in octonions, M can have even more components! There is much to be explored here. Nature may be much more complicated than we have imagined up until now.

The P^P analysis for cross terms in possible 16-vectors shows that (f_3) could also be a mass term and, therefore, we might need more conjugations. I then found that every basis element can generate a 'new' conjugation, as shown earlier. We already have $(f_0)A = A^{\uparrow\wedge}(f_0)$, $(if_0)A = A^{\uparrow\vee}(if_0)$, and $(ie_0)A = A^{\uparrow\downarrow}(ie_0)$. These provide no problem for the usual, associative Dirac algebra. We can do the following:

$$[(f_0)A] = [A^{\dagger\wedge}(f_0)] \Rightarrow [(f_0)A](f_0) = [A^{\dagger\wedge}(f_0)](f_0)$$

$$= A^{\dagger\wedge}[(f_0)(f_0)] = A^{\dagger\wedge} \Rightarrow (A^{\dagger\wedge})^{\dagger} = A^{\wedge} =$$

$$(f_0)^{\dagger}[A^{\dagger\wedge}(f_0)]^{\dagger} = (f_0)[(f_0)^{\dagger}A^{\dagger\wedge\dagger}] = (f_0)[(f_0)A^{\wedge}] = A^{\wedge}$$

and

$$(f_0)[(f_0)A] = (f_0)[A^{\wedge\dagger}(f_0)] = [(f_0)(f_0)]A = A$$

$$\textit{Let } B \equiv A^{\wedge} \textit{ then } B^{\wedge} = (f_0)[B^{\dagger}(f_0)]. \textit{ Similarly,}$$

$$B^{\vee} = (if_0)[B^{\dagger}(if_0)],\ B^{\downarrow} = -(ie_0)[B^{\dagger}(ie_0)]$$

We, therefore, define a new conjugation by $B^{\vartriangle} \equiv -(f_3)[B^{\uparrow}(f_3)]$: Now let B be (e_0), $i(e_0)$, (e_1), $i(e_1)$, $\cdots$, and see which signs change. We find a pattern as follows: Let $(\ )^{\perp}$ change (e_3), (ie_3), (f_3), and (if_3) by definition. $(\ )^{⧗}$ changes (e_0), (e_3), (ie_0), (ie_3), (f_0), (f_3), (if_0), and (if_3). As before $(\ )^{*}$ changes $i \to -i$. Then $(\ )^{⧗*}$ changes (f_0), (f_3), and (if_1), (if_2). We then find that (theorem):

$$(e + f)^{\vartriangle} = e^{\perp} + f^{⧗*}$$

$$(e + f)^{\triangledown} = e^{\perp} - f^{⧗*}$$

where $B^{\triangledown} \equiv -(if_3)[B^{\uparrow}(if_3)]$. The conjugations come in pairs and maybe single out $(\ )^{\uparrow}$ as a 'special' conjugation (hermitian conjugation for matrices).

We can prove in general that $(AB)^c = B^c A^c$, for an associative algebra, where $A^c \equiv b^{-1}[A^{\uparrow}b]$. We must show that $b^{-1}[(AB)^{\uparrow}b] = [b^{-1}[B^{\uparrow}b][b^{-1}A^{\uparrow}b]$. This is obvious for any associative algebra, since $(AB)^{\uparrow} = B^{\uparrow}A^{\uparrow}$.

If we bring in octonions, we have to check that

$$(e + f)^{\triangle} \equiv e^{\perp} + f^{⧗*}$$

still gives $(AB)^{\triangle} = B^{\triangle}A^{\triangle}$, just as we did before for $(e + f)^{\uparrow}$, $(e + f)^{\wedge}$, and $(e + f)^{\vee}$. This can be done with a computer that can check every pair individually. Similarly for $(e + f)^{\triangledown} \equiv e^{\perp} - f^{⧗*}$.

How many such conjugations are really useful remains to be seen. Whether $(\cdots)^{\uparrow}$ should be singled out as it was here, also needs to be further investigated. I suspect that, at the $(e + f)^{\text{conj.}}$ level, we find it is not special. No particular conjugation is special, what-so-ever, I suspect. Any $P = P^{\text{conj.}}$ changes 16 of 32 Dirac basis elements. The combination $LL^{\text{conj.}} = 1$, $L^{\text{conj.}} = L^{\text{conj.}'}$ still changes 8 signs. This is the pattern and its universality shows some deep restrictions on the laws of nature, I suspect.

This chapter now ending may seem to be beyond the scope of Younger students for whom this book was written but that need not be true. We still have no matrices. The rotation group in 3-space can be understood. It is simple algebra. The idea of Lie groups with free parameters comes directly from this example. The magic of $L^{\wedge}$ changing the sign of all terms that go with free parameters can be checked for the special case of rotations in 3-space, but this is admittedly abstract. We have made no use of the number of group parameters anyway here. The idea of form covariance is essential, and the group $LL^{\wedge} = 1(e_0)$, was used for selecting such $P\psi$ forms.

There is much more to be done here and you students will need more math to follow these developments of the future. I hope you will be more motivated to learn the math and thus be able to share in the joy of the discoveries ahead!

CHAPTER 8

The Three Kinds of Mass

Let us first summarize the concepts explored so far. We have a simple algebra called Pauli's algebra (complex quaternions) with basis elements $\{e_\mu, (ie_\mu)\}$. It has three useful conjugations at least, $\dagger$, $\wedge$, $\vartriangle$, where $e_0{}^\wedge \equiv e_0$, $e_k{}^\wedge \equiv -e_k$, $(ie_0)^\wedge \equiv (ie_0)$, $(ie_k)^\wedge \equiv -(ie_k)$, $(e_\mu)^\dagger \equiv (e_\mu)$, $(ie_\mu)^\dagger \equiv -(ie_\mu)$, and $\vartriangle$ changes the signs of only e_3 and (ie_3). If $P \equiv P^\dagger = P^\mu(e_\mu)$ or $P \equiv -P^\dagger = P^\mu(ie_\mu)$, then $P\psi = 0$ yields a basic 'wave' equation, called the Weyl equation, where $P^\mu = \hbar\partial^\mu$. Notice there is no mass term, mc, here at all yet. Notice also that there are no complex coefficients, whereas Weyl's equation really has $i\hbar\partial^\mu$.

The natural groups so far are:

$$AA^\dagger = 1(e_0)\text{: } A \rightarrow 1e_0 + \epsilon^k(ie_k) + \delta(ie_0)$$

$$BB^\wedge = 1(e_0)\text{: } B \rightarrow 1e_0 + \epsilon^k(ie_k) + \delta^k(e_k)$$

$$CC^\vartriangle = 1(e_0)\text{: } C \rightarrow 1e_0 + \eta^1(e_3) + \eta^2(ie_3)$$

The group B is called SL(2,C). It is the Lorentz group so we focus on it. Let $P \equiv P^\dagger \rightarrow P' = L^\dagger PL$ with $LL^\wedge \equiv 1$. Then $P' = P'^\dagger$ also. Notice that $P\psi \rightarrow P'\psi' = L^\dagger PL\psi'$, therefore, $\psi' \equiv L^\wedge\psi$ is natural if $LL^\wedge \equiv 1$. Now reconsider $P\psi = \psi^? Mc$. Assume $Mc = M'c'$, but what is $\psi^?$? We must choose $\psi^?$ such that all the L's can be removed from the $P'\psi' = \psi'^? M'c'$ equation. Look at $(L^\wedge\psi)^{\dagger\wedge} = L^{\wedge\dagger\wedge}\psi^{\dagger\wedge} = L^{\dagger\wedge\wedge}\psi^{\dagger\wedge} = L^\dagger\psi^{\dagger\wedge}$. (The antiautomorphic property makes this work beautifully.) The equation $P\psi = \psi^{\dagger\wedge}Mc$ is obviously 'form covariant' for the group $LL^\wedge = 1(e_0)$, which here is SL(2,C). This is where Dirac should have started in the late 20's. I discovered this equation in the early 70's (Am. J. Phys.) and no one did anything with it, then or since. To be self contained here, we repeat some earlier material.

I learned from Egon Marx, as I said, that van der Waerden discovered a similar equation in the late 20's, when Dirac was discovering his alternative. However, Waerden had to go to the complex coefficient extension of Pauli so that he could get M to be ordinary mass, as we discussed before and will more below. Notice that Mc is put in arbitrarily here and its being on the right of ψ means that $M = M^\mu(e_\mu) + m^\mu(ie_\mu)$ is possible here. The covariance is unaffected!

Next consider operation by P^ from the left. We get:

$$P^{\wedge}(P\psi) = (P^{\wedge}P)\psi = P^{\wedge}(\psi^{\dagger\wedge}Mc) = P^{\wedge\dagger\dagger}(\psi^{\dagger\wedge}Mc)$$

$$= (P^{\dagger}\psi)^{\dagger\wedge}Mc = (+P\psi)^{\dagger\wedge}Mc = (\psi^{\dagger\wedge}Mc)^{\dagger\wedge}Mc$$

$$= \psi^{\dagger\wedge\dagger\wedge}Mc^{\dagger\wedge}Mc = \psi(M^{\dagger\wedge}M)c^2$$

So $(P^{\wedge}P)\psi = \psi(M^{\dagger}{}^{\wedge}M)c^2$ and this is the needed Klein-Gordon equation for the Schrödinger, non-relativistic limit. Clearly $M^{\dagger}{}^{\wedge}M$ must reduce approximately to one, real mass parameter squared in the non-relativistic limit. This shows how to correctly introduce mass into the simplest, relativistic equations, ad hoc. We do this simply because we see mass for huge objects around us in our universe!

Before we move on to more complicated algebras, we need to carefully examine the Klein-Gordon equation and its non-relativistic limit, for it reveals startling things about mass in relativistic quantum. The conventional Klein-Gordon equation is

$$(i\hbar\partial^{\mu})(i\hbar\partial_{\mu})\psi = m^2c^2\psi$$

where ∂^{μ} means partial derivative with respect to t, x, y, or z, holding the other variables constant (for the moment). We had to use $P = P^{\mu}(e_{\mu}) = \hbar\partial^{\mu}(e_{\mu})$ and the i is missing here. This means one of two things: (1) nature demands the i, so the real coefficient Pauli algebra is NOT physical at all! It is then a purely mathematical fantasy, just like the quaternions above are; or (2), nature uses this simple algebra and, therefore, the M appearing here corresponds to physical things in the real world of little blob quanta. With this background perspective then, trust me that the usual physics of Schrödinger, for slow moving quanta, requires

$$\hbar\partial^{\mu}\hbar\partial^{\mu} = \hbar^2[\partial^0\partial^0 - \partial^1\partial^1 - \partial^2\partial^2 - \partial^3\partial^3] = -(m_{\exp}c)^2$$

If we do not reach this form, then our new equation is not one that yields reasonable predictions for slow motion quantum of the usual kind. Such particles, if they could exist, would be like nothing we have experienced before. There is a cosmological problem of missing dark mass, in galaxies, which seems to exert 10 times the pull of the known matter in the galaxies, on their neighbors. Some exotic mass things like these could possibly be involved there.

Let us look again at the Klein-Gordon structure from $\hbar\partial^{\mu}(e_{\mu})\psi = \psi^{\dagger\wedge}Mc$, allowing only real coefficients;

$$P^{\wedge}P\psi = \cdots = \hbar\partial^{\mu}\hbar\partial_{\mu}\psi = \psi M^{\dagger\wedge}Mc^2$$

For a non-relativistic limit to exist, $M^{\dagger\wedge}M$ must commute with ψ, and $M^{\dagger\wedge}M$ must be negative. This suggests $M = M_R{}^{\mu}(e_{\mu})$ or $M = M_I{}^{\mu}(ie_{\mu})$ and, in either case, we get $M^{\dagger\wedge}M = (M^0M^0 - M^kM^k)(e_0)$. Thus, we cannot avoid $(M^1M^1 + M^2M^2 + M^3M^3) > M^0M^0$ if we are to have a non-relativistic limit. Therefore,

$$(m_{\exp})^2 = (M^kM^k - M^0M^0)$$

We don't know why m's exist at all in nature, so $M^k \neq 0$ is possible.

Next consider the intermediate ('twilight zone') algebra, between Pauli and Dirac: $\{(e_{\mu}), (ie_{\mu}), i(e_{\mu}), i(ie_{\mu})\}$, with 16 basis elements. The same equation $P\psi = \psi^{\dagger\wedge}Mc$ with $P = +P^{\dagger}$ now can have, in general,

$$P = P^{\mu}(e_{\mu}) + Q^{\mu}i(ie_{\mu})$$

with real P^{μ} and real Q^{μ}. The group $LL^{\wedge} \equiv 1(e_0) \rightarrow L_{\epsilon} = 1(e_0) + \epsilon^k(e_k) + \delta^k(e_k) + \phi i(e_0) + \ell i(ie_0)$ shows us 8 parameters now! The new ones, ϕ and ℓ, relate to phase invariance and spacetime scale invariance. This new L is the basic group of nature, as we have already seen in the Dirac algebra. But do we really need the Dirac algebra at all? We might have $P\psi = \psi^{\dagger\wedge}Mc$ for free electrons/positrons, and $PF = 0$, $F = -F^{\wedge}$, for free photons. These equations can then be easily coupled within this twilight algebra. We might have all there is in the QED physics right here!? We could also extend this algebra by adding in the octonions, with only 'e's', as well, if need be. Why then are there any 'f' basis elements needed for the real world?

Two possibilities: (1) they are not needed—Dirac went overboard in his generalization from Pauli, or (2), they are needed because of other quark-like considerations or because there is another kind of mass. (The possible quark tie-in will have to be developed later by others.)

The second kind of mass possibility comes from guessing that $P = P^{\dagger}$ but also that $P^{\wedge}P$ is proportional to (e_0), so it commutes. These basic guesses leave us with 5 parts in P, as Table 1 shows, for the full Dirac algebra. There are several ways of writing these P's with 5 parts. These several ways seem to all come down to basically just two independent forms, with metrics $(+ - - -)$ and $(- + + + -)$. If another mass is somehow associated with the 5th component, as a constant, instead of thinking of this dimension as an extra, rolled up, tiny dimension inside protons and electrons, then the final sign, ±, is very important in the metric. The (+ − − − −) case precludes having a non-relativistic limit for this new mass part. However, if this is only a minor part of overall mass and Mc dominates, in $P^\wedge P\psi = \psi M^\uparrow{}^\wedge Mc^2$, then both 5-space metrics are allowed and we have discovered here two types of Dirac-like, spin 1/2 wave equation! If both mass parts, m and M, are essential to nature, then we see why the full Dirac algebra is needed.

Table 1 shows something else that is seemingly new as well. (All of this was, by the way, recently published in Spec. in Sci. & Tech. and Nuovo Cim.—as well as rejected by Phys. Rev. and lost or thrown away by Rev. Mod. Phys.). There seem to be three-piece objects, P, for which $P^\wedge P$ is proportional to (e_0). These may have important physical applications also. If so, then this is another justification for the full Dirac algebra being used in nature for the fundamental things there. However, these may have 2 additional pieces that I missed seeing.

In the Dirac algebra, $LL^\wedge \equiv 1(e_0)$ gives a 16 parameter group, but there is, perhaps, a second requirement, such as $L^\wedge = L^\vee$ or $L^\wedge = L^\uparrow$, which again reduces us to this same, fundamental, 8 parameter group that comes so naturally from the twilight algebra. The Dirac algebra also has new conjugations, $\downarrow$, $\vee$ and $\triangledown$, so possibly other new structures, still unknown to us. This powerful notation will open many doors quickly and I am afraid my generation won't leave you youngsters much to discover here, once they jump on this material in force.

There is still the Mc question. Does it exist? How many terms in M? How big are they? Why do they have these particular values that seem to work? Why three mass types? There is much here to puzzle your generation until you die off as well.

Nature is so complicated! But again, evolution could probably not have spawned us humanoids if it were any simpler! What other wonders lie ahead, DOWN THERE?

MORE DETAILS ON THE THREE MASS TYPES

You may want to skip over this material unless you just really love physics and math. This detail is for the really dedicated among you who want to see all that I know.

The Twilight Algebra $\{(e_\mu), (ie_\mu), i(e_\mu), i(ie_\mu)\}$ solves the mass problem we ran into for the Pauli algebra $\{(e_\mu), (ie_\mu)\}$. We have the new equation possibility now, which van der Waerden almost discovered long ago,

$$P\psi = \hbar\partial^{\mu}(e_{\mu})\psi = \psi^{\dagger\wedge}iMc$$

Since, $(i)^{\dagger\wedge} = --i = i$ when outside the (···) of a basis element, this leads to $iM^{\dagger\wedge}iM = -(m_{exp})^2$. Thus $M = M^{\mu}(e_{\mu}) = M^0(e_0) + 0$ is NOW POSSIBLE. This is a spin 1/2 wave equation with ordinary mass. Egon Marx says this is not Dirac's equation and van der Waerden's similar equation has gauge invariance problems that Dirac's equation does not. Don't worry about what that means if you never heard of gauge invariance. The important point is that this ordinary mass, M^0, is not Dirac's ordinary mass m_D which appears as the 5th dimension in his spin 1/2 equation. Both of these equations lead to a satisfactory Klein-Gordon equation, so why should nature use one and ignore the other completely? Marx used the Waerden equation as a basis for a quark theory, in the early 1990's (see Journal of Math. Phys. and Spec. in Sci. & Tech.).

Of course, we can combine the Waerden equation and Dirac equation into one new equation with both kinds of mass appearing for the same quanta, once we recognize m_D is a 5th 'dimension'. This is exotic and very exciting. Maybe we missed this in the 20's and needed it to go beyond QED in the 50's and 60's, ····. Or, it may again just be mathematical fantasy.

We guess that $P_{5I} = P^{\mu}(e_{\mu}) + P^4(if_0)$ with metric $(+ - - - -)$ from $P^{\wedge}P$ and Table 1. Then we have

$$P_{5I}\psi = \psi^{\dagger\wedge}Mc, \quad P^{\wedge}P = (|P^0||P_0| - |P^k||P_k| - |P^4||P^4|)e_0$$

as a new basic equation possibility, along with

$$P_{5I}\psi = \psi^{\dagger\wedge}iMc$$

Notice that the usual approach, $(i\hbar\partial^{\mu})$, will not work here, since $|P^0||P^0| = |i\hbar\partial^0||i\hbar\partial^0| = +\hbar^2\partial^0\partial^0$ and the i here cannot have any effect. But iM gets $(iM)^{\dagger\wedge} = +i(M^{\dagger\wedge})$ and eventually this i changes the sign for $(m_{exp})^2$.

We have already seen that $M^{\dagger\wedge}M$ leads to $M^0M^0 - M^kM^k$ without the i beside M and $(m_{exp})^2 = -M^{\dagger\wedge}M$. Notice that $-|P^4||P^4|$ will go across to the other side to combine with $M^{\dagger\wedge}M$ or $iiM^{\dagger\wedge}M$ and it becomes $+|P^4||P^4|$, regardless of the signs in P^4 itself as a mass parameter. Thus $(m_{exp}c)^2 = -[M^{\dagger\wedge}Mc^2 + |P^4|^2]$ or $(m_{exp}c)^2 = -[iM^{\dagger\wedge}iM + |P^4|^2]$. The $|P^4| \equiv mc$ *or imc*, as a second mass, definitely has the wrong sign here and we can get a non-relativistic limit only if iM is non-zero and dominates, or if M is non-zero and its M^k parts dominate. In no case can M = 0, if we are to get a non-relativistic limit here of the usual kind. Thus $P_{5I}\psi = 0$ is only an exotic kind of quantum in the Dirac algebra. A new thing to explore. Then what does Dirac's original equation look like? Table 1 has the answer. Consider the other simple and somewhat less natural 5-vector possibility there:

$$P_{5II} = P^{\mu}i(ie_{\mu}) + P^4(if_0) = P^{\dagger}_{5II}, \quad P^{\wedge}P = (-|P^0||P_0| + |P^k||P_k| - |P^4|^2)(e_0)$$

$$P^{\wedge}P\psi = P^{\wedge}(\psi^{\dagger\wedge}iMc) = \cdots = \psi(-M^{\dagger\wedge}Mc^2)$$

$$= [-(\hbar\partial^{\mu}\hbar\partial_{\mu}) - |P^4|^2]\psi = [-(-(m_{exp}c)^2) - |P^4|^2]\psi$$

$$\rightarrow (m_{exp}c)^2 = +|P^4|^2 - M^{\dagger\wedge}Mc^2$$

Thus, $|P^4|^2$ now has the right sign and can be the whole mass, and M = 0 is finally possible. This is where QED starts and M is never considered. We chose iM in the above analysis, but could have chosen M instead, which here would give for $P^4 \equiv 0$,

$$(m_{exp}c)^2 = +(M^0M^0 - M^kM^k)(e_0)$$

so M^0 could also be the mass, instead of P^4. In this case, we don't need any f's at all, once again, since (if_0) then disappears and it was the only f, unless M has some f's in it. We know NOTHING about M, remember. We are then back to the Waerden-like equation in the twilight algebra, with no i in M, but the i appears now in P, rather than with M. In case you missed it, in all this, Dirac's original equation is, essentially,

$$P_{5II}\psi = 0, \quad P_{5II} = (\hbar\partial^{\mu})i(ie_{\mu}) - mc(if_0)$$

$$P^{\dagger} = P \rightarrow L^{\dagger}PL, \quad P^{\wedge}P = [-\hbar^2(\partial^0\partial^0 - \partial^1\partial^1 - \partial^2\partial^2 - \partial^3\partial^3) - (mc)^2](e_0)$$

My apparently new variation is

$$P_{5I}\psi = 0, \quad P_{5I} = \hbar\partial^{\mu}(e_{\mu}) - mc(if_0)$$

$$P^{\dagger} = P \rightarrow L^{\dagger}PL, \quad P^{\wedge}P = [+\hbar^2(\partial^0\partial^0 - \partial^k\partial^k) - (mc)^2](e_0)$$

which does not have a classical limit, without the $\psi^{\dagger\,\wedge}$iMc term existing on the right, and it allows both mass types in the one equation. The pure Pauli algebra, with only real coefficients, has only exotic mass M^k dominating. Very strange indeed.

In all these cases, the natural guess for coupling to the electromagnetic field quanta is $P \rightarrow P - $ 'e'A, where 'e' is some real or imaginary coupling constant to be fitted by experimental findings later.

Before we end this development, you younger readers might be wondering what Dirac and Waerden did originally, since they did not have this elegant mathematical machinery before them at the start, like we did here. I have not seen van der Waerden's work. So cannot comment further. It seems left out of modern text books! Dirac suspected, wrongly, that the $\partial^0\partial^0$ second time derivative in $(-\hbar^2\partial^{\mu}\partial_{\mu})$, for the relativistic Schrödinger (later called Klein-Gordon) equation, was not physically reasonable. So he wanted ∂^0 only and relativity showed that space and time need to be treated equally. He guessed, therefore, that

$$P\psi = mc\psi$$

essentially, and realized P must be a complicated operator. Next, he essentially tried to get to the Klein-Gordon equation by

$$PP\psi = Pmc\psi = mcP\psi = mcmc\psi$$

If we choose $P \equiv P^{\mu}\gamma_{\mu} = (i\hbar\partial^{\mu})\gamma_{\mu}$, then

$$PP = (P^{\mu}\gamma_{\mu})(P^{\nu}\gamma_{\nu}) = P^{\mu}P^{\nu}\gamma_{\mu}\gamma_{\nu} = ?\ldots? = (mc)^2$$

$$= (-\hbar^2)(\partial^0\partial^0 - \partial^k\partial^k)$$

He thus 'needed' $\gamma_{\mu}\gamma_{\nu} = 0$ for $\mu \neq \nu$, $\gamma_0\gamma_0 = 1$, $\gamma_1\gamma_1 = -1$, $\gamma_2\gamma_2 = -1$, $\gamma_3\gamma_3 = -1$; but that is too radical: it means $\gamma_1\gamma_2 = 0$, *etc.* There was a less radical alternative, because $P^{\mu}P^{\nu} = +P^{\nu}P^{\mu}$:

$$PP = P^{\mu}P^{\nu}\gamma_{\mu}\gamma_{\nu} = \frac{1}{2}(P^{\mu}P^{\nu}\gamma_{\mu}\gamma_{\nu} + P^{\mu}P^{\nu}\gamma_{\mu}\gamma_{\nu})$$

$$= \frac{1}{2}(P^{\mu}P^{\nu}\gamma_{\mu}\gamma_{\nu} + P^{\nu}P^{\mu}\gamma_{\nu}\gamma_{\mu})$$

where we used the common trick of renaming $\mu \rightarrow \nu$, $\nu \rightarrow \mu$ in only the second term. This does not change anything because both indices are summed over. But now $P^{\nu}P^{\mu} = P^{\mu}P^{\nu}$ is alright in the second term, because $\partial^0\partial^k = \partial^k\partial^0$ for well behaved wave functions ψ, when taking derivatives. In other words, it doesn't matter which order you do the derivatives in, the answer is the same. This leads to

$$PP = \frac{1}{2}(P^{\mu}P^{\nu}\gamma_{\mu}\gamma_{\nu} + P^{\mu}P^{\nu}\gamma_{\nu}\gamma_{\mu})$$

$$= P^{\mu}P^{\nu}[\frac{1}{2}(\gamma_{\mu}\gamma_{\nu} + \gamma_{\nu}\gamma_{\mu})]$$

Therefore, the new 'operators' γ_{μ} needed the following pattern

$$\begin{aligned}
\gamma_0\gamma_0 + \gamma_0\gamma_0 &= 2 \\
\gamma_1\gamma_1 + \gamma_1\gamma_1 &= -2 \\
\gamma_2\gamma_2 + \gamma_2\gamma_2 &= -2 \\
\gamma_3\gamma_3 + \gamma_3\gamma_3 &= -2 \\
\gamma_0\gamma_k + \gamma_k\gamma_0 &= 0 \qquad \text{`}\gamma_{\mu}\gamma_{\nu} + \gamma_{\nu}\gamma_{\mu} \equiv 2g_{\mu\nu}\text{'} \\
\gamma_1\gamma_2 + \gamma_2\gamma_1 &= 0 \\
\gamma_1\gamma_3 + \gamma_3\gamma_1 &= 0 \\
\gamma_2\gamma_3 + \gamma_3\gamma_2 &= 0
\end{aligned}$$

Notice that he chose PP instead of $P^{\wedge}P$ or $P^{\dagger}P$, ··· If he had been properly steeped in quaternion math for ordinary relativity, he would have known that Lorentz has the form

$$P = P^{\mu}(e_{\mu}) \rightarrow P' = P^{\mu'}(e_{\mu}) = L^{\dagger}PL = P'^{\dagger}$$

$$LL^{\wedge} = 1, \;\; P^{\wedge}P = (P^{\mu}P_{\mu})e_0 = P'^{\wedge}P'$$

and he would probably have tried P^P instead of PP. In fact, if all the physicists had been taught this elegant math for relativity, I bet that Schrödinger would have even written down

$$P\psi_a = mc\psi_v \quad P^\wedge\psi_v = mc\psi_a$$

along side of $P^\wedge P\psi_a = (mc)^2\psi_a$ and $PP^\wedge\psi_v = (mc)^2\psi_v$ as well as his non-relativistic equation

$$P^0\psi = \frac{P{\cdot}P}{2m}\psi$$

where $P^\mu = i\hbar\partial^\mu$ and $P = P^\mu(e_\mu)$. (The right notation shows the right physics to be discovered.) The set $\{\psi_a, \psi_v\}$ has nothing multiplying it from the right and no $\psi^{\dagger\wedge}$ conjugations, so they are special objects (called spinors). The ψ we have used, with $\psi^{\dagger\wedge}$ in its equation, has more internal pieces. In fact, Dirac's equation $P_{5II}\psi_D = 0$ has a ψ_D which is equivalent to the $\{\psi_a, \psi_v\}$ set. For covariance, $\psi_D \rightarrow L^\wedge\psi_D$ and $\psi_a \rightarrow L^\wedge\psi_a$ but $\psi_v \rightarrow L^\dagger\psi_v$, as you can easily see for yourself. Thus, Dirac's equation was contained in the twilight algebra alright ($P^\mu = i\hbar\partial^\mu$ here), but it required two coupled wave function pieces there. Notice again that the i could be placed with the mc terms here, and $mc\psi_v \rightarrow \psi_v\, mc$ allows mc to be quaternion mass also!

Since I claim $P_{5II} = P^\mu i(ie_\mu) + P^4(if_0)$ is Dirac's equation with $P^4 = \pm mc$, we need to work backward toward where Dirac started with $mc\psi$, which is $\pm mc(e_0)\psi$. We multiply the above P_{5II} through by (if_0) from the left, and get

$$(if_0)P_5 = P^\mu i(if_0)(ie_\mu) + P^4(if_0)(if_0)$$

$$= (\hbar\partial^\mu)i(-f_\mu) + (\pm mc)(e_0)$$

Thus

$$(if_0)P_5\psi = 0 \Rightarrow (\hbar\partial^\mu)[i(-f_\mu)] = \mp mc\psi \equiv P_{DIRAC}\psi$$

$$\Rightarrow P_{DIRAC} \equiv (i\hbar\partial^\mu)\gamma_\mu = (i\hbar\partial^\mu)(-f_\mu) \Rightarrow \gamma_\mu = -f_\mu$$

You can verify that $\{-f_\mu\}$ has the required multiplication properties of $\{\gamma_\mu\}$ and that completes the bridge. If $P^4 = +mc$, then $\gamma_\mu = +f_\mu$. The $\pm$mc sign is not fixed yet here. Actually, $P^4\psi = \pm mc\psi$ can be viewed as an operator for matter and antimatter ψ states! We were earlier concerned with the $+(mc)^2$ or $-(mc)^2$, which is independent of the choice ($\pm$mc), in our non-relativistic approximation efforts. That non-relativistic approximation will eventually determine the sign needed in $P^4 = \pm mc$ for matter, and for antimatter (which cannot appear in a non-relativistic setting since it annihilates with matter).

We can see here that Dirac sort of backed into the world of hypercomplex numbers. He knew about {(e_μ)}, (ie_μ)} and was 'forced' to invent essentially {(e_μ), (ie_μ), i(e_μ), i(ie_μ), (f_μ), (if_μ), i(f_μ), i(if_μ)} to accommodate his mass type, mc, which goes naturally with i(ie_μ) and (if_0) for 5-vectors.

We have seen there are several 5-vector forms and they come in two types, $(+ - - - -)$ and $(- + + + -)$, essentially. The former is a new 'physics' and the later is Dirac's physics of QED. Both of these can be extended with $\psi^{\dagger\wedge}$Mc, covariantly, and this may be more new physics.

We have many conjugations in the full, 32 element Dirac algebra. We chose $P = P^\dagger$ for historical reasons. I am almost certain that it does not matter which we start with. We then chose $P^\wedge P \propto (e_0)$. Again arbitrary. We then need $P' = L^\dagger PL$ so that $P' = P'^\dagger$ and we need $LL^\wedge \equiv 1(e_0)$ so that $P'^\wedge P' = P^\wedge P$. But this L is a 16 parameter group! Huge! (Maybe SU(3)$\otimes$SU(3), and SU(3) was, after all, the beginning basis of the quark theory.) Our P then is $P_{5I} = P^\mu(e_\mu) + P^4(if_0)$ or $P_{5II} = P^\mu i(ie_\mu) + P^4(if_0)$. Now notice that $L^\dagger(if_0)L = (if_0)(L^\dagger)^{\dagger\vee}L = (if_0)L^\vee L$, but $LL^\wedge \equiv 1$. This common mass term, $P^4(if_0)$, is invariant ONLY if $L^\vee = L^\wedge$ and then L is only an 8 parameter group. I am virtually certain now that this group must replace Lorentz, its 6 parameter subgroup for two reasons: (1) P^4 should possibly be invariant from big blob, relativistic, moving clock experience, and (2) single mass parameter physics 'required God' to go to the twilight algebra, and there the 8 parameter group $LL^\wedge = 1$, naturally replaces the $LL^\wedge = 1$ group, SL(2,C), that was natural in the real coefficient Pauli algebra. Nature seems to require a certain complexity of algebra to achieve a simplicity of mass, and thus have huge creatures like us exist, to study our 'parts'. We are here and we have a simple, average mass concept for big blobs.

I leave you with these mass possibilities to ponder and explore. This is only the tip of a giant iceberg. We have only begun the adventure. There must be a reason for the mass values that we find in the lab for e^-, p^+, μ^-, n^0, π, *etc.* You youngsters must find it, as well as quantum gravity which may be an essential part of the mass answer.

Notice that the 5th component mass goes with a 5th dimension basis vector element (if_o). In curved spacetime, we may still have mc as a meaningful idea but the basis element may become location dependent just like $i(ie_\mu)$ or (e_μ) do, in the gravity theory. Thus gravity is essentially 5 dimensional, with an unusual 5th part. I recently published in Spec. in Sci. & Tech., 1994, the way to view gravity theory in this hypercomplex number perspective. That needs to be extended to 5 dimensions next, perhaps. I have included it later in this book although it is really complicated. You will have to get very comfortable with calculus before you tackle the problem of curved space quaternions and Einstein's gravity equation in terms of them. That chapter awaits you when you are ready for it.

As we shall see later, it appears that classical physics is best done with the (e_μ) basis but quantum, equivalently, needs the $i(ie_\mu)$ basis. Thus classical and quantum gravity will be done similarly, I suppose, using the one basis for classical and the other basis for quantum. Both of these go with the 5th dimension (f_0) or instead (if_0), and this component will be invariant in classical gravity but may become position dependent in quantum gravity. The 5th dimension may even be an operator element and, it is wild to say but possible that somehow 6-vectors may exist in quantum gravity with both of these f_0 elements present in very asymmetrical ways. We just don't know much at all yet about this world inside protons and its gravity details. Quantum gravity may be the last frontier eventually, or it may not be done independent of the strong interaction and both are related deeply somehow. Time will tell, but very slowly I suspect.

CHAPTER 9

Expanding but Flat Space

The actual general relativity equations are rather complicated, even for simple situations. We have qualitatively discussed two of these situations: The Schwarzschild solution is for the empty space around a spherical mass. The curvature there drops off to zero as $r \rightarrow \infty$. The other case is that of a 'saw-dust' universe, starting at $t = 0$ with infinite density of packing for the dust everywhere, but with tremendous speed of 'ejection' by means of the space itself violently stretching. This is like a trampoline with dust sprinkled on it everywhere. Have the trampoline already greatly stretched and held by people around the edge. Move the edge outward and film the scene from above. The saw-dust flakes move apart gradually. Run the film backward and you see how the universe becomes dense-packed as $t \rightarrow 0$. No flakes are actually moving across the trampoline here. They just move with it. Next, glue the dust specks on a balloon and see an approximation of what happens in a closed universe. We shall look into several flat space models in some detail. I will try to be clear and simple here, but that is difficult for such complicated stuff. I'll give it my best shot.

Let us start with an almost infinite, flat trampoline. As time goes forward the edge moves outward but slower and slower, just as if gravity were slowing it. The stretching rate starts fast and dies away gradually, as $t \rightarrow \infty$, but never quite stops existing. Keep the camera above, and centered over one particular flake of saw-dust, at a random location far from the edge—not at a special place. This is like our galaxy cluster and the other flakes are like other galaxy clusters moving out from us. Relative to that flake, the others move away with their speed of relative recession proportional to their distance at that time. Later on, any one flake is farther away and it is moving out slower, on and on forever.

Notice that if we let the trampoline instead shrink, as we 'walk inward' while pulling outward, then we see how the early universe looked. At some time the dust flakes will begin to touch and slide over each other, piling up. We call this the coupling time, or decoupling time if it happens as we are letting the trampoline stretch.

Now clear off the dust on a section of the trampoline and stop the motion. Paint parallel lines and a triangle on it. Start the motion again and see that the parallel lines stretch in width but remain parallel as they get farther apart also. The triangle similarly grows in area and line width, but the sum of the angles remains equal to 180°. This is a flat but stretching space!

If we imagine going back in time, by letting the trampoline shrink, we see that it remains nearly infinite in size, if large enough to begin with, but eventually the edges do come in and the dust is piled high on what is left of the surface.

Our universe may be just like this: flat and stretching and huge but finite. Or, even if our trampoline itself is really infinite, the dust sprinkled on it may still only cover a huge but finite section of it at the beginning. We can never know for sure unless we are close enough to the edge of the dust area to see some evidence of the 'void' in one direction. This infinity possibility is then a religious question, not a physics question, so long as no evidence of a special direction for the edge shows up in our telescopes.

Now imagine a three dimensional distribution of uniform dust that extends to infinity or nearly so in all directions around our favorite dust particle. Nothing exists but the space and the dust and Einstein's gravity law. The average density of the universe right now, at t_0, is about 10^{-30} grams/cm^3. That is, about 10^{-30} times the density of water; it equals about 10 hydrogen atoms per cubic meter. If we have saw-dust, instead of hydrogen, with 10^{19} atoms per small flake, then we have one flake per 10^{18} cubic meters $= (10^6 \text{ met})^3 =$ about one flake per cube 1000 km (600 miles) on an edge. That is a thin universe!

These little imagined flakes do pull on each other, through mutual gravity, with their influence dropping off as the distance squared. Now imagine that the space itself is stretching and taking the dust flakes with it. Just like on the trampoline.

Pick a particular flake, any flake, in your mind, and 'look' at the flakes receding outward from it. Imagine a sphere of flakes of radius $\ell_0 \equiv 10^3$km. It would then contain only the one flake, on the average, or a few flakes at most. So consider a much bigger sphere, say, 10^{12}m. It will have a volume of $(4\pi/3)\ell_0^{\,3} = (4\pi/3)(10^{12}m)^3 = 4.19\times 10^{36}m^3$. We can use this flat-space volume formula since our space is flat, even if expanding—again like the trampoline. Notice that our meter standard is not stretching here. We imagine that a real ruler, floating among the dust particles, will keep its length as time goes on here, just like the flakes. The atomic forces overpower the stretching space effects inside, which must still be there. This assumption may run into trouble if space were contracting, rather than expanding, when the space shrinking becomes very violent!

We will have $10^{36}/10^{18} = 10^{18}$ saw-dust flakes in this hugh sphere. Their collective mass can be estimated, since 10^{19} atoms of wood will have typically 10 protons and neutrons per atom and each of these has a mass of about

10^{-27}kg. Thus the mass is about $10^{18}\times10^{19}\times10^{-26} = 10^{11}kg$. Their gravity at the sphere surface would be only

$$g = \frac{GM}{\ell^2} = \frac{(6x10^{-11})(10^{11})}{(10^{12})^2} = 6\times10^{-24}m/s^2$$

Not much of a gravity pull. What speed would a flake need at this 'surface' to be moving radially away from our favorite flake at the center, such that it never quite falls back? But first, what about the influence of the other flakes, outside the sphere, on that 'fleeing flake'? So long as the interaction drops off as $1/\ell^2$ exactly, their collective pull adds up to zero exactly! Amazing but true and it is because a sphere has a surface area of $4\pi\ell^2$. These ℓ^2 parts can cancel exactly, it turns out. So, the issue of whether this universe expands forever or not hinges on the speed of this little flake relative to the flake at the center of this arbitrary sphere. What then is the escape velocity for this flake? So long as all speeds are less than about (1/10)c, we can use f = ma and the conservation of energy that follows from it. Thus:

$$E_O \equiv K_O + U_O = K_f + U_f = \frac{1}{2}mv_O^2 + \frac{-GMm}{\ell_O}$$

To just escape, we need $v_f \approx 0$ and $\ell_f \approx \infty$. Therefore, $v_{Oesc} = \surd(2GM/\ell_O)$ $= \surd[(2(6\times10^{-11})(10^{11})/(10^{12})] = \surd(10^{-11}) \approx 10^{-5}m/s$. Not very fast at all. Define $v_O \equiv H_O\ell_O$, where H_O is called the Hubble constant (even though it decreases with time as the expansion slows down). We get $H_O = 10^{-5}/10^{12} = 10^{-17}sec^{-1}$ and $T_O \equiv 1/H_O = 10^{17}sec$. We then use $3.15\times10^7sec = 1$ *year* to find $T_O \approx 3.2\times10^9$ or 3 billion years, give or take a few billion. This is roughly the time for the dust universe to have reached our low density situation. This would be correct except that past expansion was faster than at present and we have neglected this so far. The actual age will be shorter in general.

But you can see that even Newton could have come to this general conclusion, had he thought that the universe might be infinite, thin, unstable and thus expanding rapidly enough to possibly never quite stop expanding forever. Just how thin, he probably could not have guessed. All this takes no fancy curved space mathematics. Our H_O value, $10^{-17}sec^{-1}$, corresponds to a current expansion at the rate of 10^{-5}m/s at a distance out of 10^9 meters. What is this speed for, say, a distance now of 10 billion light years (10 bℓy)? We get:

$$1\ \ell y = c(1\ year) = (3\times10^8 m/s)(3.15\times10^7 s) \approx 10^{16} m$$

$$1\ b\ell y = 10^9\times10^{16} = 10^{25} m$$

Therefore, the dust recession speed there could be, at this time:

$$v_1 = H_O \ell_1 = 10^{-17}(1/s)\times10^{25} m = 10^8 m/s$$

but $c = 3\times10^8$m/s so this is greater than (1/10)c which is a 'no no' for our non-relativistic energy calculation. We cannot let ℓ get too large here. We can still handle 100 mℓy or closer alright. That is 1000× our galaxy's diameter. But can we just blindly use $v_1 = H_O\ell_1$, since we only calculated $v_O = H_O\ell_O$? You have to start over with ℓ_1 and find v_1. Notice that

$$\frac{v_1}{v_O} = \frac{\sqrt{2GM_1/\ell_1}}{\sqrt{2GM_O/\ell_O}}$$

but $M_1 = \rho(4\pi/3)\ell_1^3$ and $M_O = \rho(4\pi/3)\ell_O^3$. Therefore, $v_1/v_O = \ell_1/\ell_O$ and $v_1 = H_O\ell_1$ is, in fact, true for a $(1/\ell^2)$ gravity law! Newton could have also found this result. He would have predicted Hubble's distance law: the farther the galaxy cluster the faster it is going away from us right now. But we cannot really see it right now. We see it as it used to look, since the light might take 100's of millions of years to get to us. This isn't such a problem for closer clusters, which are used to estimate H_O, experimentally. Remember, our H_O value came from measured ρ_O and assumed minimal escape. Our universe may not have v at ℓ corresponding to this special condition. Then it either escapes even faster or it stops and falls back in later. (I wonder what theological interpretation Newton would have put on all that!)

I think you will agree that we have kept it pretty simple so far. To take the next step into the details, we need some basic calculus. We have for the case of barely escaping, at time t_O,

$$\frac{1}{2}mv_O^2 - \frac{GMm}{\ell_O} = 0 - 0 = 0$$

and $v_O \equiv d\ell_O/dt$ is the meaning of speed here: a small step $d\ell_O$ happens in a small time dt, right now at t_O. Notice that the mass, m, of the one saw-dust flake cancels out. We must then somehow solve

$$v_{Oesc}^2 = \left(\frac{d\ell}{dt}\right)^2 = \frac{2GM}{\ell}$$

which is a non-linear, first order differential equation. We can easily isolate dℓ and ℓ on one side as

$$\ell^{1/2}d\ell = (2GM)^{1/2}dt$$

This says you go dℓ higher in dt if you start at ℓ and have speed enough to barely escape, eventually. As this flake moves out, the sphere below it grows, but it still contains the same M and that M acts the same, gravitationally, as if it were all at the center flake. Thus this same calculation also applies to escaping from any planet, if you start at ℓ_O from the center and ℓ_O is equal to or greater than the radius of the planet.

The $\Sigma x^n dx = (x^{n+1})/(n+1)$, as dx → 0, theorem can be used here to get:

$$\frac{2}{3}\ell^{3/2} = (2GM)^{1/2}t + k$$

$$\ell^{3/2} = \frac{3}{2}(2GM)^{1/2}t + k'$$

$$\ell = (\frac{3}{2})^{2/3}(2GM)^{1/3}t^{2/3}$$

if the constant term, k, is zero. It is zero for cosmology. It cannot be zero for escaping from a planet. There, we would have

$$\ell^{3/2} = \frac{3}{2}(2GM)^{1/2}t + k'$$

with t = 0 corresponding to $\ell = \ell_O$. Therefore, $\ell_O{}^{3/2} = 0 + k'$, so $k' = \ell_O{}^{3/2}$. Then, while escaping from above the planet, we have

$$\ell = [(\frac{3}{2})(2GM)^{1/2}t + \ell_O^{3/2}]^{2/3}$$

for the time to reach a distance ℓ from the center, starting at ℓ_O distance at t = 0. This very special escape case can be easily solved! Notice that $\ell \to \infty$ as $t \to \infty$. For cosmology, M is trapped in the sphere $(4\pi/3)\ell^3$, so as $t \to 0$, $\ell \to 0$, and the density goes through the roof! But we must stop calculating the past history when the saw-dust begins to touch a lot, whatever that means. Maybe we could go back in time to a time when there was one flake per cubic meter. We had $\ell_O = 10^{12}m$, $M = 10^{11}kg$, and one dust flake per $10^{18}m^3$, which requires that $\ell_O \to \ell_e = 10^{-6}\ell_O$. Notice that

$$\frac{\ell_e}{\ell_O} = \frac{t_e^{2/3}}{t_O^{2/3}} = 10^{-6} \Rightarrow t_e = 10^{-9}t_O$$

Since our constant of integration, k, is zero for cosmology, the current time, t_O, is roughly $1/H_O$ but not exactly. We can now actually calculate H as a function of time directly

$$H = \frac{v}{\ell} = \frac{(2GM)^{1/2}}{\ell(\ell^{1/2})} = \frac{(2GM)^{1/2}}{\ell^{3/2}} = \frac{(2GM)^{1/2}}{\frac{3}{2}(2GM)^{1/2}t}$$

$$\therefore H = \frac{1}{\frac{3}{2}t} \Rightarrow t_O = \frac{2}{3}\left(\frac{1}{H_O}\right) = \frac{2}{3}(10^{10}y) = 7\ by$$

Therefore, $t_e = 10^{-9}(7\times 10^9 y) = 7\ years$. We could calculate the past and future expansion behavior, starting a few years after the bang and extending on into the future forever. Isn't this wonderful? And Newton could have done it all, conceptually! We have rounded off our numbers very crudely and you might like to go back through all this with better accuracy. The essentials are clear, however.

THE FALL BACK UNIVERSE

We were very successful with the barely escaping universe. We got the same answers that Friedmann got long ago by solving Einstein's curved spacetime gravity equation (GR). However, $ds^2 \neq (cdt)^2 - (dx^2 + dy^2 + dz^2)$ for this case because the space is really stretching. That condition is qualitatively different from non-stretching. The spacetime is actually curved even though the (x,y,z) SUB-space is flat. Einstein does really give us more information.

What if we have a house sized rock in among the dust, and what if a very much smaller, marble sized rock is orbiting that big rock? The orbiting marble would pick up or knock aside some dust flakes occasionally. The local flakes would be attracted to the big rock and rain down onto it, from some distance. Beyond that, they would drift outward with the general expansion. Would the orbiting marble spiral outward?? What about the rock and marble at t = 0, the Big Bang? If we don't worry about that, then how do we place the dust flakes now, at t_O, in the general vicinity of the rock? Any way we want to? Very interesting questions but too hard for me to answer. I wonder if anyone knows yet. This ever expanding dust at infinity should be the boundary condition for the Schwarzschild problem, not a static space far away. But stretching space is really hard to deal with! Everything is gradually changing with time for this case.

Let us try the 'fall-back' case from our simple, flat space point of view. How far can we get this time? We have v_O and ℓ_O as before but $v_f = 0$ at $\ell_f < \infty$, because $v_O < v_{esc}$. The speed is otherwise arbitrary. Energy conservation looks like:

$$E_O \equiv \frac{1}{2}mv_O^2 + \frac{-GMm}{\ell_O} = E_f = \frac{1}{2}mv_t^2 + \frac{-GMm}{\ell_t} = \frac{-GMm}{\ell_f}$$

where ℓ_f is unknown and depends on M, v_O, and ℓ_O. The M here is still all the mass in the sphere $(4\pi/3)\ell_O^3$. Einstein's theory says, instead, that the (x,y,z) space is now curved and closed, so that the volume formula is altered. But we push on anyway.

Our energy equation can be rearranged to become ($v_f \equiv 0$):

$$\left(\frac{v_O}{c}\right)^2\left(\frac{\ell_f}{2GM/c^2}\right) + 1 = \frac{\ell_f}{\ell_O}$$

If we choose ℓ_O and ℓ_f, then we find the v_O needed to eventually reach ℓ_f before falling back. Let $2GM/c^2 \equiv \ell_M$. It is the 'black hole' radius associated with a mass M, here a sphere of dust. Our energy equation then can look like

$$\left[\frac{d\,\ell_o\left(\frac{\ell_f}{\ell_M}\right)^{1/2}}{d(ct)}\right]^2 + 1 = \frac{\ell_f\left(\frac{\ell_f}{\ell_M}\right)^{1/2}}{\ell_o\left(\frac{\ell_f}{\ell_M}\right)^{1/2}}$$

Now define new and temporary variables

$$y_o \equiv \ell_o\left(\frac{\ell_f}{\ell_M}\right)^{1/2}, \quad x_o \equiv ct, \quad 2\eta \equiv \ell_f\left(\frac{\ell_f}{\ell_M}\right)^{1/2}$$

and our equation becomes (with η a constant)

$$\left(\frac{dy}{dx}\right)^2 + 1 = \frac{2\eta}{y}$$

General Relativity says instead:

$$\left(\frac{dR}{d(ct)}\right)^2 + 1 = \frac{4G''M''/3\pi^2c^2}{R}$$

where R is a scale factor proportional to our ℓ distance. The 'M' is interpreted as the total mass of the closed space which is finite. The solution $\ell(t)$ or R(t) is easiest to find by knowing that this y(x) equation fits a rolling wheel, where y is the vertical height and x the horizontal distance of a point on the rim. The wheel has radius η and rolls through angle θ, starting with $x_i = 0$, $y_i = 0$, $\theta_i = 0$. The solution is:

$$y = \eta(1 - \cos\theta), \quad x = \eta(\theta - \sin\theta)$$

$$\frac{dy}{dt} = \eta\omega\sin\theta, \quad \frac{dx}{dt} = \eta\omega(1 - \cos\theta)$$

$$\left(\frac{(dy/dt)}{(dx/dt)}\right)^2 = \left(\frac{dy}{dx}\right)^2 = \dots = \frac{\sin^2\theta}{(1 - \cos\theta)^2}$$

$$1 = \frac{(1 - \cos\theta)^2}{(1 - \cos\theta)^2} = \frac{1 - 2\cos\theta + \cos^2\theta}{1 - \cos\theta)^2}\left(\frac{2\eta}{2\eta}\right)$$

You can now readily check that this is indeed the solution. We then have for our space-time 'cycloid':

$$\ell_t\left(\frac{\ell_f}{\ell_m}\right)^{1/2} = \eta(1 - \cos\theta), \quad ct = \eta(\theta - \sin\theta)$$

$$\ell = 0,\ t = 0,\ \theta = 0, \quad \eta = \frac{1}{2}\ell_f\left(\frac{\ell_f}{\ell_M}\right)^{1/2}$$

From the wheel, we see that $y_{max} = 2\eta$, therefore, $\ell_{max} = \ell_f$, which is consistent with the simple, non-relativistic flat picture. At $\theta = 2\pi$, we get $\ell_t = 0$ and this is the 'end of the world'. That occurs at $ct_{end} = \eta(2\pi - \sin 2\pi) = 2\pi\eta = \pi\ell_f^{3/2}/\ell_M^{1/2}$. We still have not decided upon ℓ_f and ℓ_M yet. We have two free parameters, whereas, General Relativity has only one, 'M'. Our Hubble law is:

$$d\ell_t\left(\frac{\ell_f}{\ell_M}\right)^{1/2} = \eta(--\sin\theta d\theta) = \eta\sin\theta d\theta$$

$$cdt = \eta(d\theta - \cos\theta d\theta) = \eta(1 - \cos\theta)d\theta$$

$$v \equiv \frac{d\ell_t}{dt} = c\left(\frac{\ell_M}{\ell_f}\right)^{1/2} \frac{\sin\theta}{1 - \cos\theta}$$

$$\frac{v}{\ell} = \frac{c\sin\theta}{\eta(1 - \cos\theta)^2} \equiv H_t$$

Notice that $H \approx (c/\eta)(\theta/\theta^2/2) \to \infty$ as $\theta \to 0$, i.e., $t \to 0$. Also, $H = 0$ at $\theta = \pi$, *i.e.*, $t = (1/2)t_{end}$. The expansion is stopped. The ℓ(t) expansion is very similar for both $\ell_f = \infty$ (flat, eventual escape) and $\ell_f \leq \infty$, during the initial rise from 'zero'. Only later does ℓ(t) fail to rise almost as fast as the escape case. It eventually flattens and then decreases.

Let us now link H and t to see how the current age relates to the current H value here. We find that

$$H = \frac{1}{t}\left(\frac{\sin\theta(\theta - \sin\theta)}{(1 - \cos\theta)^2}\right)$$

which tells us nothing about the age of this universe. We believe our universe has roughly

$$H = \frac{20km/\sec}{m\ell y} = \frac{1}{15by} = \frac{1}{4.725x10^{17}\sec} = 2.116x10^{-18}(1/\sec)$$

Suppose that M, contained in a sphere of radius ℓ_O, is 10^{11}kg, as in the escape model before. Then

$$\ell_M = \frac{2GM}{c^2} = 1.482x10^{-16}m$$

We know that ℓ_f is a fundamental free parameter. Let us just choose $\ell_f \equiv 4\ell_O$

so that this universe stops when four times as expanded as now. This means that

$$\ell_O = (4\ell_O/2)(1 - \cos\theta) \Rightarrow \theta = 1.047rad = 60^o$$

$$\eta = \frac{1}{2}\ell_f^{3/2}/\ell_M^{1/2} = 4\ell_O^{3/2}/\ell_M^{1/2} = ?$$

We need to choose the sphere that contains the mass M. Let us instead choose $H_O \equiv 2.116\times 10^{-18}sec^{-1}$. Then

$$\eta = \frac{c\sin\theta}{H(1 - \cos\theta)^2} = 4.911x10^{26}m \Rightarrow$$

$$\ell_O = \left(\frac{\ell_m \eta^2}{16}\right)^{1/3} = 1.307x10^{12}m \quad (for\ \ell_f \equiv 4\ell_O)$$

$$t_{end} = \frac{2\pi\eta}{c} = 1.029x10^{19}\text{sec} = 326by$$

$$v_O = H_O\ell_O = 2.766x10^{-6}m/s \quad (v_e = 10x10^{-6}m/s)$$

$$t_O = \frac{\eta}{c}(\theta - \sin\theta) = 2.966x10^{17}\text{sec} = 9.42by$$

$$\ell_{\max} = \ell_f = 4\ell_O = 5.229x10^{12}m \quad at\ \theta = \pi rad = 180^o$$

$$\rho = M/[(4\pi/3)\ell_O^3] = 1.06x10^{-26}kg/m^3 = 1.06x10^{-29}gm/cm^3$$

We can then double check all of this with the original conservation of energy equation:

$$\left(\frac{v_o}{c}\right)^2\left(\frac{\ell_f}{\ell_M}\right) + 1 = ? = \left(\frac{\ell_f}{\ell_o}\right) = 4$$

$$\left(\frac{2.766x10^{-6}}{3x10^8}\right)^2\left(\frac{5.229x10^{12}}{1.482x10^{-16}}\right) + 1 = 3.999$$

At present H, density, and t are all in the right range here. This could be our universe! So everything fits; energy is indeed conserved. We cannot deal with $(v_O/c) > (1/10)$ and this sets the smallest ℓ_O value we can deal with. Using the energy equation above for v_t and ℓ_t we can find ℓ such that v, for this same flake, later approaches c. We are considering here the case $\ell_t \rightarrow 0$ as $t \rightarrow t_{end}$. Setting $(v_t/c) = 1/10$ gives us

$$\ell_{tmin} = \frac{\ell_f}{\left(\frac{1}{10}\right)^2 \frac{\ell_f}{\ell_M} + 1} = 1.48x10^{-14}\ m = 15\ Fermi$$

One proton has a diameter of 1 Fermi and our dust flakes will be touching long before we reach this small size. So excessive speed is not a problem in this model. It is non-relativistic over the entire middle life of the universe, so long as we don't go too far out in space at any one time, to look at dust flakes, nor near the beginning or end, when H is very large. For $\ell_O = (1/4000)\ell_f$, we find $t_O = 270{,}000y$. This is the decoupling time when radiation 'stopped' hitting the atoms, and our universe was about 1000 times 'smaller' than now.

This is about as far as we can go in non-relativistic, flat space modeling. Let us finally try to relate our constants to the one in General Relativity. We stated that it gives

$$\left(\frac{dR}{dct}\right)^2 + 1 = \frac{4G''M''/3\pi^2c^2}{R}$$

But R can physically be replaced by αR where α is any constant. Only ratios of R are said to be really of interest: $R(t)/R_O(t_O)$. Our energy equation can be written as

$$\left(\frac{dR}{dct}\right)^2 + 1 = \frac{\ell_f\left(\frac{\ell_f}{\ell_M}\right)^{1/2}}{R}$$

if $R_t \equiv \ell_t(\ell_f/\ell_M)^{1/2}$. This then requires that $4G'M'/3\pi^2c^2 = \ell_f(\ell_f/\ell_M)^{1/2}$ where $\ell_M \equiv 2GM_{\ell O}/c^2$. Our ℓ_O at t_O was arbitrary and so $M_{\ell O}$ is not a special mass value. This is about as far as we can take this analogy. I see no way to motivate the 'M' value relating to (ℓ_f/ℓ_M).

I hope you have enjoyed this very down to earth survey of cosmology. It should make your reading of more advanced accounts somewhat more intuitive.

RELATIVISTIC EXTENSION

Before totally ending this flat space discussion, I would like to share an interesting observation with your 'teachers'. We know relativistic $F = ma$ is more accurate than non-relativistic $F = ma$. Therefore, we have been limited to $v/c < (1/10)c$ in our discussion so far for cosmology. Yet we still got the exact General Relativity result for the equation of expansion, $v^2/c^2 = R_M/\ell$. Therefore, if we replace $K = (1/2)mv^2$ with the more accurate $K = mc^2(1 - v^2/c^2)^{-1/2} - mc^2$, then we MUST replace the $U(\ell)$ potential energy also with some more accurate form. I worked backward from the $\ell(t) = ((3/2)\sqrt{(R_M)}ct)^{2/3}$ solution to get the needed $U(\ell)$ potential. It is

$$U(\ell) = \frac{-mc^2}{\sqrt{1 - R_M/\ell}} + mc^2$$

Then I looked again at the energy conservation equation and this became totally obvious, if $v^2/c^2 = R_M/\ell$ is still to hold. We have This quite obviously means that, if $E = 0$ then $v^2/c^2 = R_M/\ell$ will suffice. There may be another choice for $U(\ell)$ that also works, since our differential equation is non-linear, due to v^2 and $(1/\ell)$.

This potential energy is very interesting and is very complicated, though it looks innocent enough. The non-relativistic limit $v/c << 1$ means $R_M/\ell << 1$. Then $(1 + x)^n \approx 1 + nx$, for $x << 1$, leads us to

$$K_O + U_O = K_t + U_t = E$$

$$= \left(\frac{mc^2}{\sqrt{1 - \frac{v^2}{c^2}}} - mc^2 \right) + \left(\frac{-mc^2}{\sqrt{1 - R_M/\ell}} + mc^2 \right)$$

$$U(\ell) \approx -mc^2(1 - (-1/2)R_M/\ell) + mc^2$$

$$= -mc^2 R_m/2\ell = -(mc^2 2GM/c^2)/2\ell$$

We see the usual potential as a crude approximation. Does this mean that gravity is 'not' really $1/r^2$, as a force law, and the question of whether the pulls of the stars outside the radius ℓ sphere still cancel is now open? The force law comes from the derivative of the potential,

$$F(\ell) = -dU(\ell)/(d\ell)$$

$$= -mc^2(-1/2)(1 - R_M/\ell)^{-3/2}(-R_M(-1)\ell^{-2})$$

using the chain rule of derivatives. This simplifies considerably and also gives us the non-relativistic case, where $R_M << \ell$. We find that

$$F(\ell) = -\frac{GMm}{\ell^2}(1 - R_M/\ell)^{-3/2}$$

$$= -\frac{GMm}{\ell^2}\left(1 - \frac{2GM/c^2}{\ell}\right)^{-3/2}$$

$$F(\ell) \approx -\frac{GMm}{\ell^2}\left(1 + \frac{3GM/c^2}{\ell}\right)$$

There is an extra pull proportional to $(1/\ell^3)$. This will cause planets to move faster than expected if in a circular orbit and it is not hard to actually calculate that extra amount. It will also cause elliptical orbits to precess forward by about the same amount as for a circle if nearly circular. We don't need curved space to get the famous precession of the perihelion of Mercury effect! But do we get the right amount of precession? Compare this with the U(r) potential in an earlier chapter which does give the correct result. We cannot get the same $U(\ell)$ to meet these two needs it seems! But here we really have m just above the filled sphere M, not a point M!

The bad side of this neat potential energy is that we lose all the wonderful simplicity of Coulomb's law of electricity. There

$$F = \frac{kQ_a q}{\ell^2}, \quad U_a = \frac{kQ_a q}{\ell}$$

is exact for the force on q, due to a small Q_a when sitting ℓ away from q. Two Q's produce $U = U_a + U_b \Rightarrow F = F_a + F_b$ *if* Q_a and Q_b are on the same line as q. We call this superposition and it is a great simplifier. We are so used to it that it is hard to imagine the complications of losing it. For

$$-F_m \approx \frac{GMm}{\ell^2} + \frac{3G^2M^2m/c^2}{\ell^3}$$

we lose $F \propto M^1$ and $F_M \neq -F_m$ anymore. They are approximately equal and opposite (the first terms are) since R_M/ℓ is very small. At the surface of the earth, this is about $10^{-2}m/10^4m = 10^{-6}$ so things are as before to a part in a million. But not quite the same! When both M_a and M_b pull on m along a line, we can perhaps calculate the force due to each by itself easily, but when we put both M_a and M_b into the picture, the M values depend upon how close M_a is to M_b! They attract each other and the closer they are together the smaller is the effective mass of the combination! So if they are on the same side, along the line, and approach each other, we have problems as they approach. This shows clearly that gravity is a big mess! This mess exists even for the cruder $1/\ell^2$ force law approximation, for gravity. Even there, $M_a + M_b$ is not the right mass when a and b are close to each other. There is really no superposition, so we cannot even integrate the scalar potential. We naively might think we can jump from a large source to a small source and have

$$U \approx -\left(\frac{GMm}{\ell} + \frac{3G^2M^2m/c^2}{2\ell^2}\right) \Rightarrow$$

$$dU = -\left(\frac{Gm}{\ell}dM + \frac{3G^2m}{2c^2\ell^2}(dM)^2\right)$$

We then run up against $\int (dM)^2/\ell^2$ which ain't ordinary calculus! The relativistic situation is even worse looking:

$$dU = \frac{-mc^2}{\sqrt{1 - \frac{2G}{c^2\ell}dM}} + mc^2$$

But our U was only from a large sphere of uniformly spread out mass, pulling on m just at its surface. We cannot assume U is unchanged if the M is compressed! That requires new guesses!

I think that even undergraduate engineering students should see this. It breaks the mind set of electricity being about the same as gravity. It gives a hint that gravity might be very special, and thus require a whole new generalization of our approach to nature from electricity. It does not suggest curved space is the answer, of course.

Two black holes, in a line, 'pulling' on a third one on that line and 'pulling' on each other, is the simplest 'superposition' framework within which to really deal with the full non-linearities of curved space gravity. Chandrasekhar has gotten some amazing, exact solutions. The world just ain't simple. We only approximate it successfully in our theory. Your generation has to develop curved space quantum field theory, it seems.

SPEED DEPENDENT GRAVITY

Our speculations on flat space gravity are incomplete since we did not have any speed dependence for the force. A spinning planet has more internal energy and, since $E = mc^2$, it thus should have slightly more mass and more gravity pull. Similarly, a spinning space station in orbit should have more mass and thus be attracted more strongly toward the planet it circles, to compensate for its increased inertia. This speed dependence makes some calculations easier and others much harder. (Speed dependent forces are avoided in many advanced mechanics books.)

The first issue, before speed is considered, is the space dependent part of the gravity pull. We try to match, as much as possible, the results obtained by General Relativity theory. Therefore, for motion far from the surface of a massive sphere, we shall assume the gravity force is

$$F(r,v)_m = \frac{mc^2}{r\sqrt{1 - v^2/c^2}}\left[\frac{R_s}{2r}\left(1 - \frac{R_s}{r}\right)^{-3}\right]$$

This works out quite well as you will soon see. The exponent, -3, is chosen to match the precession of the perihelion of Mercury that is observed experimentally. The speed term is the usual energy associated with the moving mass, m, that is being pulled on. The source of that pull, M, is found inside the R_s term, which we shall guess to be

$$R_s = \frac{2GM}{c^2\sqrt{1 - V^2/c^2}}$$

where V is the speed of M, when it is moving. It won't be in the sample calculations to follow.

We shall first tackle the orbit problem for this force law. For a circle, $a = v^2/r$, toward the center of the circle, and v is constant in magnitude. The relativistic $F = dp/dt$ equation then gives us

$$\frac{-mc^2}{r\sqrt{1 - v^2/c^2}}\left[\frac{R_s}{2r}\left(1 - \frac{R_s}{r}\right)^{-3}\right] = \frac{-mc^2(v^2/rc^2)}{\sqrt{1 - v^2/c^2}}$$

Simplifying now gives

$$\beta^2 \equiv \frac{v^2}{c^2} = \frac{R_s}{2r}\left(1 - \frac{R_s}{r}\right)^{-3}$$

for the orbit speed, v, at a distance r from the center of M. The non-relativistic case corresponds to r > > R_s. We can easily determine the orbit speed needed at any r, and we find: $r = 2.4R_s \Rightarrow \beta = 1.024$ which is slightly greater than c and not allowed. Therefore, the closest allowed stable orbit is at slightly greater than $2.4R_s$. This compares with GR theory which predicts a similar orbit limit at $1.5R_s$ for the closest orbit.

Radial motion is more difficult. Earlier, we used energy conservation to get a relationship between v and r. However, our speed dependent term in the new force law makes us have to strike out anew, using the old results as a guide only. For radial motion the $\boldsymbol{F} = dp/dt$ equation gives us the basic guess:

$$\frac{-mc^2}{r\sqrt{1 - v^2/c^2}}\left[\frac{R_s}{2r}\left(1 - \frac{R_s}{r}\right)^{-3}\right] = \frac{d}{dt}\left(\frac{mv}{\sqrt{1 - v^2/c^2}}\right)$$

and this time v is a decreasing and unknown function of time. (Those without calculus background may prefer to browse the rest of this section, for it must get technical in order to get a specific answer, v(r), for the motion outward.)

We use the calculus 'trick' $df/dt = (df/dv)(dv/dt)$. We find in our case:

$$\frac{d}{dv}\left(v\left(1 - \frac{v^2}{c^2}\right)^{-1/2}\right) = \dots = \left(1 - \frac{v^2}{c^2}\right)^{-3/2}$$

which is surprisingly simple, when the dust settles. Next, we multiply both sides of the force equation by dr and use $(dv/dt)dr = dv(dr/dt) = dv(v)$. Then, to get all the v stuff on one side, we move the $\sqrt{(1 - v^2/c^2)}$ term across from the force term. This gives us

$$\frac{-c^2}{r}\left[\frac{R_s}{2r}\left(1 - \frac{R_s}{r}\right)^{-3}\right]dr = \left(1 - \frac{v^2}{c^2}\right)^{-3/2} v \left(1 - \frac{v^2}{c^2}\right)^{1/2} dv$$

$$= \frac{v\,dv}{1 - \frac{v^2}{c^2}}$$

We can then integrate both sides, in principle, and we find the general integral form. We want this to reduce to $(1/2)mv^2 - GMm/r$ in the low speed limit, and this helps us choose the arbitrary constant for physical interpretation of the equation:

$$mc^2 \ell n\left(\frac{1}{\sqrt{1 - v^2/c^2}}\right) - \frac{mc^2/4}{\left(1 - \dfrac{R_s}{r}\right)^2} + any\ const. = 0$$

$$mc^2 \ell n\left(\frac{1}{\sqrt{1 - v^2/c^2}}\right) + \left[\frac{mc^2/4}{\left(1 - \dfrac{R_s}{r}\right)^2} - \frac{mc^2}{4}\right] = same(v_O, r_O)$$

where $v \to v_O$, $r \to r_O$ on the right hand side in the same expression.

This equation replaces the usual energy conservation equation, for radial motion at least. The first term is a kinetic energy type term and, using $\ell n(1 + x) \approx x - x^2/2 + \cdots$ for small $x << 1$, we can verify that it becomes $(1/2)mv^2$ for low speeds. The other term is only position dependent and it is a kind of potential energy. However, the force follows from it in the form

$$F_r = \frac{1}{\sqrt{1 - \dfrac{v^2}{c^2}}}\left(\frac{-dU}{dr}\right)$$

as you can check by differentiating.

We now have the needed connection between speed and distance, r, on the way up and back down. This will work well for a hammer thrown upward on the moon, and maybe for the whole universe expanding and recontracting. The barely escape case corresponds to reaching $v \to 0$ as $r \to \infty$, so the constant is zero in that special case, just as before. However for cosmology, the M is a filled sphere right up to the mass m, rather than a 'point' mass, as for Mercury and the sun.

We can now examine how the escape speed increases as we approach R_s. We need to solve for v/c in the 'energy' equation. Obviously we are better off specializing to the barely escape case, to keep things simple and to compare with General Relativity. A lot of straight forward algebra leads to

$$\left(\frac{v}{c}\right)^2 \equiv \beta^2 = 1 - e^{-\frac{1}{2}\left[(1 - R_s/r)^{-2} - 1\right]} \approx \frac{R_s}{r}$$

where the large r simplification has also been shown, and follows from $exp(x) = 1 + x + x^2/2! + \cdots$ for $x << 1$. This speed formula is more complicated than Einstein's theory for the expansion and it should have come from an energy equation that takes into consideration the pull of all the other flakes both inside and outside our sphere. The mass of any flake depends on the 'pull' of all the other flakes around it and that greatly complicates things here.

We can directly examine the escape speed at any $r > R_s$, from a point mass M. We find $\beta = .9908$ at $r = 1.5R_s$, and $\beta = 1 - .625\times10^{-2215}$ at $r = 1.01R_s$. We can also verify analytically that $r \to R_s$ gives $v \to c$ for escape. This is the same as General Relativity. So far, we are doing very well compared to GR. So let us try something really radical—gravitational red shift for color changes, as light pulls away from a mass M.

We have to guess of course, since photons are quantum and don't have a variable speed. Let us guess that the kinetic energy term corresponds to the photon energy, hf. For the potential energy term, let us assume mc^2 becomes hf. This gives us an hf in each term and we get

$$hf_O - \frac{hf_O/4}{\left(1 - \frac{R_s}{r_O}\right)^2} + \frac{hf_O}{4} = hf_f - \frac{hf_f/4}{\left(1 - \frac{R_s}{r_f}\right)^2} + \frac{hf_f}{4}$$

If f_f corresponds to $r_f \equiv \infty$, then we get a formula for f at r for photons coming in from, or going out to, infinity:

$$f_\infty = f_O\left(\frac{5}{4} - \frac{1}{4}\frac{1}{\left(1 - \frac{R_s}{r_O}\right)^2}\right)$$

Notice that $r = 1.8091R_s$ gives $f_\infty = 0$ for any f_O frequency. Thus light cannot escape from this distance or closer. This is not consistent with the escape speed for a mass m being less than c at this distance. But using the $r >> R_s$ approximation and simplifying, we get

$$f_O \approx f_\infty + f_\infty \frac{R_s}{2r}$$

From this we can construct

$$\frac{f_O - f_\infty}{f_\infty} = \ldots = \frac{R_s}{2r} = \frac{\Delta f}{f_\infty} \approx \frac{\Delta f}{f_O}$$

This approximate form is the only one that has been tested experimentally, and it is the same as the approximation for General Relativity.

We cannot get time dilation naturally. We have seen that $\sqrt{(1 - R_s/r)}$ appears naturally in the special relativity approach to cosmology and somewhat in the above theory. Therefore, we could maybe guess that $\Delta t_O = \Delta t_\infty \sqrt{[(1 - R_s/r)(1 - v^2/c^2)]}$ for a clock at r moving with speed v. Speed through space slows clocks and the gravity curving of space also slows clocks, it seems for our universe.

I wonder what kind of deflection we would get for a golf ball moving at 0.999999c past the edge of the sun, using **F** = dp/dt and the F(r,v) law we have been using here. If it is at all close to the photon path bending of Einstein's theory, then we will have captured almost all of the gently-curved-space results. The **F** equation is not Lorentz covariant. It is like Coulomb's law compared to Maxwell's equations. It would be only part of a Lorentz covariant system of equations, with something analogous to magnetism getting added in. Since we still don't have a quantum gravity theory, we don't know if curved space will survive into the next level of refinement. It has an elegance that this formulation lacks and that prejudices me to expect that it will survive, even though that is very complicated stuff, compared to the theory here displayed for your amusement.

We have seen some interesting possibilities for the pull of gravity near the surface and far beyond the surface of a mass M. Enough to be convinced that gravity (and thus cosmology) is a very special case in nature. It is not then surprising if it requires a radical formulation compared to quantum electricity. There is much left for your children to discover here, I think.

The attraction of two bodies of comparable size, or even a third orbiting the two, is very interesting for curved space and for flat space gravity. We shall explore this briefly next.

BALANCING CHARGED MASSES AND GRAVITY

There is another consequence of General Relativity that we cannot duplicate with a complicated, flat space force law for gravity. This is static equilibrium for charged masses, where the electrical repulsion balances the gravitational attraction. This is a very interesting problem, for it nicely shows the complexity of an M value depending on the other m's around it.

In flat space Newtonian and Maxwellian physics, three masses can have zero force between any pair if $\sqrt{(k)}Q_i = \sqrt{(G)}M_i$. This follows from the three unknowns, Q_i, i = 1, 2, 3, in the three mutual interaction, balance of forces equations:

$$\frac{kQ_i q_j}{r_{ij}^2} = \frac{GM_i m_j}{r_{ij}^2}$$

It then also applies to n = 2, 3, 4, ··· masses in equilibrium. This is a neutral equilibrium. If we move any mass to a new location and then release the set of masses at rest this results in all of them just sitting in place. Let us look at just two masses for simplicity. For balance, we have $\sqrt{(k)}Q = \sqrt{(G)}M$ and $\sqrt{(k)}q = \sqrt{(G)}m$ from above, although we really only need $kQq = GMm$ for this special case. The $1/r^2$ cancels out exactly, so no particular distance is special. To see what this really means, choose $m \equiv M \equiv 1gm = 10^{-3}kg$ and solve for the needed Q ≡ q for balance:

$$Q = \sqrt{\frac{G}{k}}\,M = \sqrt{\frac{6x10^{-11}}{9x10^{9}}}\,10^{-3} \approx 10^{-13}\ Coul.$$

This is a very small charge, but still a huge number ($\approx 10^6$) of excess electrons. The forces at balance are, for example,

$$F = \frac{kQq}{r_o^2} = \frac{9x10^9 x(1x10^{-13})^2}{(.01)^2} \approx 10^{-12}N$$

if our two charged ball bearings are a centimeter apart, center to center distance.

If we bring one ball into place it takes no energy to do so. As we bring the second ball into place, it is attracted gravitationally and repelled electrically.

Each force is active and the net work we must do is equal to the change in potential energy due to each force. We have, in general,

$$\Delta U \approx \frac{kQq}{r_O} - \frac{GMm}{r_O}$$

except that the M and m values depend on ΔU and, in fact, they change as we bring the second mass in toward the first. Gravity is very complicated! Only in the case where $\Delta U = 0$, can we know what the actual values of M and m are when they are at r_O separation. The final M and m values are then, and only then, still the same as they were when one was in place and the other was at ∞ from it. Since this is such a special case it perhaps is not surprising that this case has been solved exactly in General Relativity, by combining curved space with guessed at generalizations of Maxwell's equations to go with the guessed at curvature equations. Chandrasekhar, in a 1989 paper, says the balance condition is the same as for the above flat space analysis. Equilibrium exists for n bodies, as black holes, so long as $\sqrt{(k)}Q_i = \sqrt{(G)}M_i$ for each mass. Our ball bearings are not black holes, however. The excess changes on them tend to 'run' to the outside surfaces of the balls, due to electrical repulsion. Hence the r_O is not the same as the center to center distance between the balls, and it is not even the same for all the charges. This is really a complicated mess, even at the classical level, if we examine the balls carefully. Only when the balls shrink down until their density is so high that they become quantum blobs completely, does this charge distribution on the balls problem go away. The solution is then quantum mechanical and no gravity theory exists there to even think about. How dense is dense enough? Surely nuclear density is sufficient. That is the density of a proton, which is roughly

$$\rho = \frac{m}{(4\pi/3)r^3} = \frac{10^{-27}kg}{4(10^{-15}m)^3} \approx 10^{18}kg/m^3$$

Our ball, at this density, would have a size given by

$$r = \left(\frac{m}{(4\pi/3)\rho}\right)^{1/3} \approx \left(\frac{10^{-3}}{4x10^{18}}\right)^{1/3} \approx 10^{-7}m = 0.1\mu$$

Chandrasekhar's result says that these bodies, if assumed classical, would be in perfect balance if they were black holes. For that case, their radii would need to be less than

$$R_s = \frac{2GM}{c^2} = \frac{2(10^{-11})(10^{-3})}{(3x10^8)^2} \approx 10^{-30}m = 10^{-24}\mu$$

Clearly, the GR results are meaningless for these balls! In fact, the GR results in close may be meaningless for any body which has to have a density greater than nuclear density to become a black hole; it works only for distances $>> R_s$ perhaps.

For $M \approx 10^6 M_{sun}$, we have R_s so huge that the density 'inside' is about that of water. Surely this really is a black hole in the GR sense? Not so. Far beyond R_s, we could expect GR to work well for the usual effects on time and on light color. But we would be guessing blindly for r close to R_s, even in this case. No measurements exist! We don't know when the quantum generalities kick in, since there is no quantum gravity theory yet. The theorists have pushed GR way beyond its reasonable limits, for lack of anything else better to do (a whole generation). Some have guessed at simple quantum extensions of GR, but these are likely to be totally useless. We can easily expect quantum gravity to be like nothing we have ever seen before, in proven physics.

So perhaps two $10^6 M_{sun}$ black holes, properly charged and with a distance of many times R_s between them, would in fact have the GR solution for most of the curved space around them, if not too close to either R_s. They would need the special charge on each to balance the gravity 'attraction' as above, $\sqrt{(k)}Q = \sqrt{(G)}M$. This charge situation will not occur in nature on its own.

For a flat space theory, we exactly have the old Maxwell equations, at the classical level. Therefore, any

$$F_G \propto \frac{GMm}{r^2}\left(1 + \frac{R_s}{r}\right)^{-3}$$

type force law is too strong and can never match the coulomb attraction. They can match only at $r \approx \infty$, where the $(\ )^{-3}$ term is negligible. However, remember that M and m are not really constants. They depend on r also and I don't know how to evaluate this dependence. (That is why we neglected it in earlier calculations.) Clearly, Mm is smaller and smaller as r decreases, and the system is more tightly bound. Radiation leaves the system in some form, or energy goes into the outside world machinery that holds the masses, if they are uncharged, or charged less than $\sqrt{(G/k)}M$. Since complicated flat gravity is stronger than electricity, for $\sqrt{(k)}Q \equiv \sqrt{(G)}M$, when M and m are at almost '∞'

separation, we see that F_G begins to grow faster than F_E as they approach, except that M and m are also decreasing. Q and q are not changing. It is just not possible to say at this time whether the forces become equal at one or more locations, in flat theory, as they approach each other.

We have a crude $1/r^2$ flat space gravity theory, that is adequate for all our inter-planetary travel needs; and a crude curved space gravity theory that is better but still is restricted to definitely classical situations—whatever that really means. It may mean $r >> R_s$ only. Falling into a black hole soon merges you into the quantum muck that exists near R_s and below.

A white dwarf star has a huge density but it is less than nuclear density. It is already a quantum thing, with each e^- and p^+ spread out over the whole star as waves. If it shrinks further, somehow, it will stay quantum in structure. As its 'surface' approaches its R_s value, it still will be only quantum in nature. No real edge at all. So who can say what it really is like then? No one on this planet at present. Maybe you will know later.

Chapter 10

Curved Space in Quaternions

We shall outline the details of how to deal with curved space for the special case of 4 and 5 dimensions, where spin 1/2 waves can be accommodated. The metrics appear naturally and the dimensions are dictated by the quaternion background that is the foundation of modern physics. This is quite unlike tensor curved space as shown in nearly all gravity books. Tensors apply to any dimension and any + or − choices for the diagonal metric tensor elements. The Lorentz group and its generalization to an 8 parameter group are also natural here and will play a part. It should be obvious after learning this material that tensors are not the way to go deeply into spacetime physics, especially at the microscopic level. They are alright for the classical tests of Einstein's gravity equation. We shall also see that Einstein's equation in this language has an inner product appearance, just as the Klein-Gordon, spin 0 equation has. Perhaps there is a deeper theory that is not spin 2 but something else; it reduces to Einstein in the way that Dirac reduces to the Klein-Gordon equation. I have never liked the energy-momentum tensor for some reason. It just is ugly to my eyes. I also don't like free indices in the equation, I suppose because my formulation of Dirac has none whatsoever. But then gravity is very special, so perhaps it can be excused for having these free parameters. If you search for generalizations, look in the direction of generalizations that do not have any free indices if possible. Also test for the form covariance under Lorentz and under its generalization to 8 parameters.

CURVED SPACE DERIVATIVES

The complex numbers, a[1] + b[i], can be naturally associated with orthogonal cartesian coordinates, x[1] + y[i], in a 'flat' two dimensional space. The same can be said of quaternions and a 'flat' four dimensional space, $x^0[\sigma_0]$ + $x^k[i\sigma_k]$. These properties come from number theory, not geometry, and I think they show that physics has a preferred cartesian coordinate system. Other curvilinear coordinates can always be defined, as functions of $\{x^\mu\}$. In the past, Einstein and his followers tried to raise to the level of a basic principle of nature, the idea that all coordinate systems are somehow equally fundamental. When one writes classical physics using tensors, this looks reasonable. We have seen how classical physics and nonrelativistic quantum physics both mask the basic hypercomplex number properties of nature.

We shall assume that cartesian coordinates are generally preferred, though some calculations will be simplified by special coordinate choices. In this approach, we must abandon the simplicity of the natural hypercomplex number system only when the space-time of the physical world is not Euclidian. It appears correct to say that all physics actually takes place in curved space, and that this is equivalent to saying that gravitational fields permiate all space. Again, many calculations can be simplified by neglecting this fact, but our basic theory should be able to encompass it!

We have written the basic wave equation operator for physics in the form

$$\partial = (e_\mu)\partial^\mu + (if_0)\partial^4 = \partial^\dagger$$

For the moment, we shall neglect the (if_0) term. We need to introduce, somehow, the idea of curved space. This should be done in such a way that ∂ returns to the above form in the flat space limit. Empirical experience has shown that a way to do this is to define

$$e_\mu \to b_\mu \equiv b_\mu(\{x^\nu\}) \equiv b_\mu^{(\alpha)} e_\alpha \to \delta_\mu^\alpha e_\alpha = e_\mu \quad \textit{(flat limit)}$$

This is obviously a simple and reasonable postulate. The differential operator now has the form

$$b_\mu \partial^\mu = b_\mu^{(\alpha)}(x) e_\alpha \partial^\mu \neq \partial^\mu b_\mu$$

This immediately introduces ambiguities into the wave equations. Should we write $b_\mu \partial^\mu$ or $\partial^\mu b_\mu$ for ∂? We also now have two 'kinds' of indices, μ and α. As you might guess, we shall need to raise and lower both kinds of indices, and so we define the metrics

$$\eta_{\alpha\beta} = (e_\alpha | e_\beta) \equiv \frac{1}{2}\left[(e_\alpha^\wedge e_\beta) + (\ \ )^\wedge\right] = \eta_{\beta\alpha}$$

$$g_{\mu\nu} \equiv (b_\mu | b_\nu) \equiv \frac{1}{2}\left[(b_\mu^\wedge b_\nu) + (\ \ )^\wedge\right] = g_{\nu\mu}$$

As before, we now define

$$\eta^{\alpha\beta}\eta_{\beta\gamma} \equiv \delta^{\alpha}_{\gamma} \quad and \quad g^{\mu\nu}g_{\nu\lambda} \equiv \delta^{\mu}_{\lambda}$$

$$b^{\mu} \equiv g^{\mu\nu}b_{\nu} \quad and \quad e^{\alpha} \equiv \eta^{\alpha\beta}e_{\beta}$$

We can now easily show that

$$(b^{\mu}|b^{\nu}) = g^{\mu\nu} \,, \, (e^{\alpha}|e_{\beta}) = \delta^{\alpha}_{\beta} \equiv \eta^{\alpha}_{\beta} \,, \, etc.$$

In addition, the following useful results also follow from these definitions

$$g_{\mu\nu} = b^{(\alpha)}_{\mu}\eta_{\alpha\beta}b^{(\beta)}_{\nu} \,, \, b^{(\alpha)}_{\mu}b^{\nu}_{(\alpha)} = \delta^{\nu}_{\mu} \,, \, b^{(\alpha)}_{\mu}b^{\mu}_{(\beta)} = \delta^{\alpha}_{\beta}$$

To overcome the problem shown above, we generalize the differential operation by adding adjustable terms (affine connections) to it. These will then be chosen so as to make the wave equation ambiguity free. Since $\partial^{\mu}b_{\nu} \neq 0$, we define an 'aligned covariant' derivative $\wp^{\mu}(b_{\nu}) \equiv 0$. In the flat space limit, $\wp^{\mu} \to \partial^{\mu}$. One of the simplest choices for $\wp^{\mu}$ is to take a linear form (on both indice types)

$$\wp_{\nu}(b^{\mu}_{(\alpha)}) \equiv \partial_{\nu}(b^{\mu}_{(\alpha)}) + b^{\lambda}_{(\alpha)}\Gamma^{\mu}_{\lambda\nu} + \omega_{\nu(\alpha\beta)}b^{\mu(\beta)}$$

We have chosen two linear 'connections', Γ and ω, because $b^{\mu}{}_{(\alpha)}$ has two kinds of indices. These unspecified functions of x^{μ} will be chosen to simplify the wave equation structure. The most obvious simplification is to postulate that

$$e_{\alpha}\partial^{\alpha} \to b_{\mu}\wp^{\mu} \equiv \wp^{\mu}b_{\mu} \,, \, so \,\, \wp^{\mu}(b_{\nu}) \equiv 0$$

There are very many functions in the set $\{\Gamma^{\mu}{}_{\lambda\nu}, \omega_{\nu(\alpha\beta)}\}$. Further restrictions are required to make them unique. The following postulate will be shown to work

$$\Gamma^{\mu}_{\nu\lambda} \equiv \Gamma^{\mu}_{\lambda\nu} \quad and \quad \omega_{\nu(\alpha\beta)} \equiv -\omega_{\nu(\beta\alpha)}$$

By choosing symmetry and antisymmetry in this way, it is possible to solve for Γ

and ω using $\mathcal{P}_\nu(b^\mu{}_{(\alpha)}) \equiv 0$.

To find $\Gamma^\mu{}_{\lambda\nu}$ we multiply $\mathcal{P}_\nu(b^\mu{}_{(\alpha)})$ by $b_\mu{}^{(\alpha)}$ and $\mathcal{P}_\nu(b^\mu{}_{(\alpha)})$by $b_\mu{}^{(\beta)}$, then add them, and use $\omega_{\nu(\alpha\beta)} \equiv -\omega_{\nu(\beta\alpha)}$. The result is (theorem)

$$\Gamma^\mu_{\nu\lambda} = -b^{(\alpha)}_\mu \frac{\partial b^\mu_{(\alpha)}}{\partial x^\lambda}$$

(Use must also be made of $b^{\mu(\alpha)}b_{\nu(\alpha)} = \delta_\nu{}^\mu$, $b^{\mu(\beta)}b_\mu{}^{(\alpha)} = \eta^{\alpha\beta}$ (= 0 unless $\alpha = \beta$), whereas $\omega_{\nu(\alpha\beta)} = 04$ when $\alpha = \beta$.)

To find $\omega_{\nu(\alpha\beta)}$ we write out

$$b^\mu_{(\alpha)}\omega_{\mu(\beta\gamma)} - b^\mu_{(\beta)}\omega_{\mu(\alpha\gamma)} + b^\mu_{(\gamma)}\omega_{\mu(\alpha\beta)} - b^\mu_{(\alpha)}\omega_{\mu(\gamma\beta)}$$

$$-b^\mu_{(\beta)}\omega_{\mu(\gamma\alpha)} + b^\mu_{(\gamma)}\omega_{\mu(\beta\alpha)} = 2b^\mu_{(\alpha)}\omega_{\mu(\beta\alpha)} = \dots$$

Next we sum on (α) by writing

$$b^{(\alpha)}_\nu 2b^\mu_{(\alpha)}\omega_{\mu(\beta\gamma)} = \dots = 2\delta^\mu_\nu\omega_{\mu(\beta\gamma)} = 2\omega_{\nu(\beta\gamma)}$$

The details are tedious but straightforward. We finally obtain (theorem)

$$2\omega_{\nu(\beta\gamma)} = \left[b^{(\alpha)}_\nu b^\lambda_{\mu(\gamma)} b^\lambda_{(\beta)} \frac{\partial b^\mu_{(\alpha)}}{\partial x^\lambda}\right] - \begin{bmatrix}\beta \to \gamma \\ \gamma \to \beta\end{bmatrix}$$

$$+ \left[b_{\mu(\beta)}\frac{\partial b^\mu_{(\gamma)}}{\partial x^\nu}\right] - \begin{bmatrix}\beta \to \gamma \\ \gamma \to \beta\end{bmatrix}$$

$$+ \left[-g_{\nu\mu} b^\lambda_{(\gamma)}\frac{\partial b^\mu_{(\beta)}}{\partial x^\lambda}\right] - \begin{bmatrix}\beta \to \gamma \\ \gamma \to \beta\end{bmatrix}$$

where $\beta \to \gamma$ means interchange their positions in a second similar term. This form shows easily the assumed property $\omega_{\nu(\beta\gamma)} = -\omega_{\nu(\gamma\beta)}$.

We see that both Γ and ω involve $b_{\mu(\alpha)}$ and its first derivative, but they place no restrictions on the spacetime dependence of $b_{\mu(\alpha)}$. Once $b_{\mu(\alpha)}(x)$ is given, we have a machine to grind out $\mathcal{P}_\nu(b_{\mu(\alpha)}) = 0$. Therefore, we can construct reasonable generalizations of our flat space wave equations.

In conventional tensor analysis, one does not have the (spinor) quaternion indices (α) but only the indices μ. There one defines a 'covariant derivative'

$$D_\nu(A^\mu) \equiv \partial_\nu(A^\mu) + A^\lambda \Gamma^\mu_{\lambda\nu}$$

and chooses Γ by requiring $D_\nu(g^{\mu\lambda}) \equiv 0$. From our machinery, we can show that $D_\nu(g^{\mu\lambda}) = 0$ follows as a consequence of the definitions. We have developed more than tensor analysis, for more is needed to describe the quantum world in curved space-time. This fact is probably the key to improving on Einstein's old tensor approach to gravity theory.

Though we have explicitly constructed $\omega_{\nu(\alpha\beta)}$, there are alternative ways of writing this affine connection, which can be useful. For example, the 'rotation coefficients' are defined by

$$\omega_{\nu(\alpha\beta)} b^{\mu(\beta)} \equiv -\gamma_{(\alpha\beta\gamma)} b^{\mu(\beta)} b^{(\gamma)}_\nu$$

This can be solved for $\gamma_{(\alpha\beta\lambda)}$ by summing over μ and ν.

Up to this point, we have made little direct use of the fact that $\mu = 0,1,2,3$, or of the quaternion multiplication properties of e_α, in our analysis of the covariant derivative! Also, we have not mentioned Lorentz symmetry. These must now enter, because our wave equations contain $\mathcal{P}^\mu(\psi)$ and $\mathcal{P}^\mu(A)$, which have not been defined yet. We know only that $\mathcal{P}^\mu(b_\nu) = 0$ and it is easy to use this to define $\mathcal{P}^\mu(A^\nu)$ in formal analogy with $\mathcal{P}^\mu(b^\nu)$. We have seen that the various wave functions of physics can be characterized by the forms of their Lorentz symmetry representation, which came out of the covariant, flat-space wave equations. Since ∂ and A have similar Lorentz behavior, it is natural to postulate

$$e_\alpha A^\alpha \to b_\mu A^\mu$$

for curved space. Now we can use the similarities between A^μ and b^μ to motivate a consistent general definition of $P^\mu(\psi)$ for all wave functions, at least in curved 3 + 1 spacetime. (This needs further generalization to 5-spacetime.)

So far, we have defined the covariant derivative such that

$$\wp_\nu(b^\mu) \equiv D_\nu(b^\mu) + \omega_{\nu(\alpha\beta)} b^{\mu(\beta)} e^\alpha \ and \ D_\nu(b^\mu) \equiv \partial_\nu(b^\mu) + b^\lambda \Gamma^\mu_{\lambda\nu}$$

Clearly, the ω term does not have an elegant appearance. We will now take another approach to this term, which is motivated by Lorentz symmetry, though no actual use will be made of $LL^\wedge = LL^\vee = e_0$.

We have seen that in flat space, $e_\alpha \to e'_\alpha = L^\dagger e_\alpha L$ under Lorentz symmetry. It is quite natural to postulate that Lorentz symmetry will still act as a filter for physical wave equations in curved spacetime, with $b_\mu \to b'_\mu \equiv L^\dagger b_\mu L$ and $b^\mu \to b^{\mu\prime} \equiv L^\dagger b^\mu L$. From this, we are guided to consider infinitesimal transformations:

$$\begin{aligned}
b'_\mu &\equiv (e_0 + \epsilon\Gamma)^\dagger b_\mu (e_0 + \epsilon\Gamma) = b_\mu + \epsilon[\Gamma^\dagger b_\mu + b_\mu \Gamma] + O(\epsilon^2) \\
\psi'_a &\equiv (e_0 + \epsilon\Gamma)^\wedge \psi_a = \psi_a + \epsilon\Gamma^\wedge \psi_a \\
\psi'_\nu &\equiv (e_0 + \epsilon\Gamma)^\dagger \psi_\nu = \psi_\nu + \epsilon\Gamma^\dagger \psi_\nu \\
B' &\equiv (e_0 + \epsilon\Gamma)[B] \equiv B + \epsilon K[B] + O(\epsilon^2)
\end{aligned}$$

Here ϵ is a first order parameter, $\epsilon^2 \approx 0$, and $B \to B'$ represents a general wave function representation of Lorentz symmetry. These are not actual Lorentz transformations, since we have not specified Γ in any way. We have merely mimicked the 'Lorentz behavior' of each wave function. We now make the bold assumption that the aligned covariant derivative of any hypercomplex wave function B(x) will have the form

$$\wp(B) \equiv D(B) + K[B], \ or \ \wp_\mu(B) = D_\nu(B) + K_\nu[B]$$

This must reproduce the previously assumed derivative of $b^\mu(x)$, which now has the form

$$\wp_\nu(b^\mu) = D_\nu(b^\mu) + (\Gamma^\dagger_\nu b^\mu + b^\mu \Gamma_\nu) \equiv 0$$

Therefore, there is a relationship between the unspecified hypercomplex connection Γ_ν, and $\omega_{\nu(\alpha\beta)}$. It will not be productive to pursue this relationship. Instead, we shall forget about $\omega_{\nu(\alpha\beta)}$, except for its indirect role in helping us find $\Gamma^\mu{}_{\nu\lambda}$. We must determine (define) Γ_ν, such that $\wp_\nu(b^\mu) = 0$. Once found, it can be used to partially define $\wp$(B) for any wave function B. We must come back to look at $\mathcal{P}$(B), for the general case, a little later. Our objective now, is to find a way to compute Γ_ν from $b^\mu(x)$ and $D_\nu(b^\mu) \equiv \partial_\nu(b^\mu) + b^\lambda \Gamma^\mu{}_{\lambda\nu}$.

Finding a suitable Γ_ν is quite a trick. (This is not unexpected, since ω was quite complicated.) It hinges on noting the fact that (Rastall, RMP, 1964)

$$g^{\mu\nu} = (b^\mu | b^\nu) = \frac{1}{2}[b^{\mu\wedge} b^\nu + b^{\nu\wedge} b^\mu] = \frac{1}{2}[b^\mu b^{\nu\wedge} + b^\nu b^{\mu\wedge}]$$

and

$$\Gamma_\nu^\dagger b^\mu + b^\mu \Gamma_\nu = (\Gamma_\nu^\dagger b^\mu) + (\ldots)^\dagger = -D_\nu(b^\mu) = \left(-D_\nu(b^\mu)\right)^\dagger$$

since $b^{\mu\dagger} \equiv b^\mu$, by definition. We need also to note that

$$D_\lambda(b_\eta^\wedge b^\eta) = D_\lambda\big((b_\eta | b^\eta)\big) = D_\lambda(\delta_\eta^\eta) = \partial_\lambda(\delta_\eta^\eta) = 0 = D_\lambda(b_\eta^\wedge) b^\eta + b_\eta^\wedge D_\lambda(b^\eta)$$

and

$$[b^{\eta\wedge} D_\lambda(b_\eta)]^\wedge = [D_\lambda(b_\eta)]^\wedge b^\eta = D_\lambda(b_\eta^\wedge) b^\eta = -b_\eta^\wedge D_\lambda(b^\eta) = -b^{\eta\wedge} D_\lambda(b_\eta)$$

since $g_{\mu\nu}$ commutes with D_λ.

At this point, it would not be obvious, but $\Gamma_\nu \equiv -\Gamma_\nu{}^\wedge$ will work and will lead to a formula for calculating Γ_ν. We, therefore, write tentatively

$$\Gamma_\lambda \equiv \frac{1}{4}\gamma_{\mu\nu\lambda} b^{\mu\wedge} b^\nu \equiv -\Gamma_\lambda^\wedge = \frac{1}{4}\gamma_{\mu\nu\lambda} b^{\nu\wedge} b^\mu \Rightarrow \gamma_{\mu\nu\lambda} = -\gamma_{\nu\mu\lambda}$$

Substituting, we find

$$\Gamma_\lambda^\dagger b^\nu + b^\nu \Gamma_\lambda = \frac{1}{4}\gamma_{\mu\eta\lambda}(b^{\mu\wedge} b^\eta)^\dagger b^\nu + \frac{1}{4}\gamma_{\mu\eta\lambda} b^\nu (b^{\mu\wedge} b^\eta)$$
$$= \frac{1}{4}\left[\gamma_{\mu\eta\lambda} b^\eta b^{\mu\wedge} b^\nu + \gamma_{\mu\eta\lambda} b^\nu b^{\mu\wedge} b^\eta\right]$$
$$= \frac{1}{4}\left[\gamma_{\mu\eta\lambda} b^\eta (2g^{\mu\nu} - b^{\nu\wedge} b^\mu) + \gamma_{\mu\eta\lambda}(2g^{\nu\mu} - b^\mu b^{\nu\wedge}) b^\eta\right]$$
$$= \frac{1}{4}\left[4g^{\mu\nu}\gamma_{\mu\eta\lambda} b^\eta - \gamma_{\mu\eta_\lambda} b^\eta b^{\nu\wedge} b^\mu - \gamma_{\mu\eta\lambda} b^\mu b^{\nu\wedge} b^\eta\right]$$
$$= g^{\mu\nu}\gamma_{\mu\eta\lambda} b^\eta - \frac{1}{4}\gamma_{\mu\eta\lambda} b^\eta b^{\nu\wedge} b^\mu + \frac{1}{4}\gamma_{\mu\eta\lambda} b^\eta b^{\nu\wedge} b^\mu$$
$$= g^{\mu\nu}\gamma_{\mu\eta\lambda} b^\eta$$

where we have used $\gamma_{\mu\eta\lambda} = -\gamma_{\eta\mu\lambda}$ and then relabeled the summed indices $\mu \rightleftarrows \eta$ in the third term. We have shown that

$$g^{\mu\nu}\gamma_{\mu\eta\lambda} b^\eta = -D_\lambda(b^\nu)$$

Now we multiply through by $g_{\nu\xi} b^{\xi\wedge}$ and obtain

$$\gamma_{\xi\eta\lambda} b^{\xi\wedge} b^\eta = -b_\nu^\wedge D_\lambda(b^\nu) = 4\Gamma_\lambda = -(4\Gamma_\lambda)^\wedge$$

This is a beautifully simple formula for the hypercomplex 'connection' Γ_λ, compared to the expression we obtained for $\omega_{\mu(\alpha\beta)}$.

Now notice something very interesting. To find Γ_λ, we have only used

$$b_\mu' \equiv L^\dagger b_\mu L,\; b_\mu^\dagger \equiv b_\mu,\; and\; g_{\mu\nu} \equiv \frac{1}{2}\left[(b_\mu^\wedge b_\nu) + (\;\;)^\wedge\right] \propto (e_0)$$

These conditions are met by $b_\mu \Leftrightarrow \{b_\mu^{(\alpha)}(e_\alpha)$ (no sum), $b_\mu^{(4)}(if_0)\}$, $\Rightarrow b_\mu = b_\mu^{(a)}(e_a)$. In other words, we could replace the sum on $\mu = 0,1,2,3$ by a $= 0,1,2,3,4,(e_4 = if_0)$, and obtain a five-space theory! The wave equations would even have $LL^\wedge = (e_0)$ form covariance. This would have the important effect of making $m(if_0)$ become $mb_4^{(a)}(x)(e_a)$ in curved 5-space, which further moves us toward considering this mass as an eigenvalue in 5-space. I suspect that replacing $(\;\;)^\wedge$ with $(\;\;)^\vee$ would give all the same results, with $m(if_0)$ replaced by $m(f_0)$. We don't seem to be able to get curved 6-space, even though this fits

$b_\mu = b_\mu{}^\dagger$, because $g_{\mu\nu}$ then becomes hypercomplex! This is because $b_\mu{}^\wedge$ or $b_\mu{}^\vee$ can only change the sign of (if_0) **or** (f_0) respectively, but not both. Curvature can tell us things! Again we see the power of the hypercomplex number formalism in channeling our search for generalizations of the known laws. The usual form of the Dirac equation, $(\gamma\partial - m)\psi = 0$, gives no hint that the m term 'should' also be space-time dependent in curved space.

To summarize, we have established the following curved space, Lorentz symmetric wave equation structure for four-space or five-space:

$$\partial \equiv b_\mu \wp^\mu \rightarrow \partial' \equiv L^\dagger \partial L, \quad LL^\wedge \equiv e_0 (\equiv LL^\vee ?)$$

$$\wp_\mu(b^\nu) \equiv 0 \equiv D_\mu(b^\nu) + \Gamma_\mu^\dagger b^\nu + b^\nu \Gamma_\mu, \; D_\mu(b^\nu) \equiv \partial_\mu(b^\nu) + b^\lambda \Gamma^\nu_{\lambda\mu}$$

$$\Gamma^\nu_{\lambda\mu} = -b_\lambda^{(\alpha)} \partial_\mu (b^\nu_{(\alpha)}) = \Gamma^\nu_{\mu\lambda}, \; \Gamma_\mu = -\frac{1}{4} b_\lambda^\wedge D_\mu(b^\lambda) = -\Gamma_\mu^\wedge$$

$$\psi \rightarrow \psi' \equiv L[\psi] \leftrightarrow \psi + \epsilon K[\psi] \Rightarrow \wp_\mu(\psi) \equiv D_\mu(\psi) + K_\mu[\psi]$$

$$L \rightarrow (e_0 + \epsilon\Gamma), \quad \epsilon^2 \approx 0, \; D_\mu(\psi) = ?, \; g_{\mu\nu} = \frac{1}{2}\left[(b_\mu^\wedge b_\nu) + (\;)^\wedge\right] \propto (e_0)$$

To complete this development, we must take up the question of defining $D_\mu(\psi)$. In flat quantum electrodynamics, only the forms $\{\psi \rightarrow \psi' = L^\wedge\psi\}$ and $\{\psi \rightarrow \psi' = L^\dagger \psi L, \; (\psi \equiv A \equiv A^\mu b_\mu)\}$ need appear. We define $D^\mu(A) \equiv \partial^\mu(A)$ and, traditionally, it is assumed that $D^\mu(\psi) \equiv \partial^\mu\psi$ for the electron field. People usually say things like, 'ψ transforms as a spinor under local Lorentz transformations but is invariant under general coordinate transformations.' I don't think this really means anything. Since all wave equations are at best a guess, subject to experimental verification, $D^\mu(\psi)$ must ultimately be determined by nature. We can naturally postulate that $D^\mu(\psi) = \partial^\mu(\psi)$ unless ψ explicitly has μ type indices on it, as in the case of $b_\mu(x)$ or $A_\mu(x)$. For fields of the form ψ^μ or $\psi^{\mu\nu}$ we define the D^λ derivative to parallel the form for D^λ operating on b^μ and $g^{\mu\nu}$. There is no evidence to date that fields with more than two indices appear in nature! The best known two-index field is $g_{\mu\nu}(x)$, but this is really a composition of the one-index field $b_\mu(x)$. Physicists have tried to assign a spin number of 2 to this 'space curvature' field. However, under Lorentz symmetry, b_μ 'transforms' the same as $A^\mu b_\mu$. It is just not clear that the spin assignment ideas of nonrelativistic physics can be applied, in a simple way, to relativistic wave equations. We really shouldn't expect that they can be. (We already saw how much of the hypercomplex number structure is covered up in taking the nonrelativistic limit.) All we can presently say is that the equation for ψ is first order in ∂^{μ}, whereas the equation for A is second order in ∂^{μ}. The equation for b_{μ} will be considered in the next section. The classical limit suggests a second order equations, but since this equation determines the curvature of the space, it is very different from that for A. If we call ψ spin ½, A spin 1, but b_{μ} spin 2, then shouldn't there be something named spin 3/2? It should have a first order equation and perhaps transform as $\psi' = L^{\dagger}\psi$, or as $\psi' = L^{\wedge}\psi L$. Only a beginning has really been made so far.

EINSTEIN'S GRAVITY WAVE EQUATION

In the previous section we developed the machinery for defining partial differential wave equations in a given, curved spacetime of four or five dimensions. Why must we deal with this added complication? Science cannot really answer questions about the reasons for the universe. We can say that gravity must be an essential part, if the universe is going to evolve stars from a dilute gas. This still does not suggest that gravity might be different, in some fundamental way, from the other forces of interaction. If we add the postulate that gravity is only attractive and attracts everything to everything else, then it becomes somewhat special. If we go further and postulate that it attracts everything in a given region in such a way that all substance 'falls' identically, regardless of its fundamental particle make-up, then gravity assumes a special place indeed! It is not such a large step to next think of it as affecting the space itself, since all particles are affected in essentially the same way by it. This was Einstein's beautiful idea. The idea is what matters, regardless of how he came to guess that it is true. Thus the 'equivalence principle and Mach's principle', which Einstein promoted, were both aids to his imagination but have not proven to be of any greater significance than that. We shall consider them as historical relics and discuss them no further. Books have been filled with mostly meaningless debate about their 'validity'. As we have said, "The equation is the thing, for only it can be really tested."

If we look carefully at the definition of a partial derivative (as a limit) and examine the position dependence of a 'vector field' A, first on a two dimensional plane and then on a two dimensional dome, we find the purely mathematical result that the curved dome is distinguished by

$$(D_{\mu}D_{\nu} - D_{\nu}D_{\mu})A_{\lambda}(x) \neq 0 \textit{ for many values of } x$$

Those coordinate positions, x, where this happens, are precisely those where the dome is qualitatively not flat. For example, the dome may be a 'localized' smooth bump in an otherwise flat plane with cartesian coordinates (x^1, x^2). We are thus automatically drawn, by curvature, to second order differential equations involving the covariant derivative D_μ, rather than the aligned derivative $\mathcal{P}_\mu$, which occurs in all the other wave equations. This peculiarity is not shocking, since we recognize the very special place in nature of the gravity field.

For simplicity of notation, let us define

$$D_{\mu\nu} \equiv D_\mu D_\nu - D_\nu D_\mu$$

Our problem now is to find a way to use $D_{\mu\nu}$ in constructing the gravity equation. We have already seen that curvature was assumed to mean $e_\eta \rightarrow b_\mu(x)$ in the particle wave equations. Now we see that curvature is reflected in the derivative properties of vector fields. Obviously then, we should postulate that gravity involves

$$D_{\mu\nu}(b^\lambda) \neq 0$$

This has a natural analogy with the other flat space wave equations, if the right hand side is identified with the 'source' of the field $b^\lambda(x)$.

Einstein originally guessed the gravity equation in a form that is roughly equivalent, mathematically, to

$$b_\rho^{(\alpha)} D_{\lambda\nu}(b_{(\alpha)}^\lambda) \neq 0$$

(In the language of tensor analysis, this says $R_{\rho\nu} \neq 0$.) Because b^λ appears quadratically, the equation is nonlinear, even if the 'source' is zero. This is in direct contrast to our earlier 'free' field wave equations in flat space. However, flat space physics always involves interactions, and even there we also found nonlinear equations to describe the 'real world'.

The most important peculiarity of gravity, from our hypercomplex number viewpoint, is the absence of all hypercomplex number basis elements. The equation also has free indices, ρ and ν, which is in direct contrast with the previous flat space wave equations that we have considered. We must face the delicate question of just how special the gravity equation can be, compared to other fields.

We can reinsert the e_α's, to form b^λ, by noting that

$$b^{\lambda}_{(\alpha)} = b^{\lambda}_{(\beta)}\delta^{\beta}_{\alpha} = b^{\lambda}_{(\beta)}(e_{\alpha}|e^{\beta}) = (e_{\alpha}|b^{\lambda})$$

Thus, the gravity equation can be written in the equivalent form

$$(b^{\rho}|D_{\lambda\nu}b^{\lambda}) \neq 0$$

This has real elegance and also some similarity to the other wave equations, except for the appearance of unsummed indices, ρ and ν. The index problem could be eliminated by the modification:

$$(b^{\nu}|D_{\lambda\nu}b^{\lambda}) \neq 0$$

(This term is called the curvature scalar, R, in tensor analysis.) Can we somehow choose between these alternatives, without recourse to empirical observation? The hypercomplex number machinery of flat space would lead me to favor the second form. However, neither may be physically correct or at best only an approximation, derivable from the more complete theory. We saw an example of this in the Klein-Gordon equation. In fact, this form has in common with the Klein-Gordon equation the same fact that the hypercomplex number structure is masked by the inner product. The inner product is of the form

$$(A|A^{\wedge}) = B + B^{\wedge} \propto e_0$$

so we might guess that the more general gravity law is of the form B $\neq$ 0, a kind of factorization.

We saw in the Maxwell equation that $(\mathcal{P}|J) = 0$ has empirical 'validity' in flat space and is 'reasonable' for the curved space generalization. Einstein was working only with tensors in his development of gravity theory. Therefore, he was led to consider instead things like $(D|J)$. Specifically, he noted the mathematical theorem

$$D_{\mu}(R^{\mu\nu}) = \frac{1}{2}D_{\mu}(g^{\mu\nu}R)$$

and proposed (guessed) the law of gravity should be modified to

$$R^{\mu\nu} - \frac{1}{2}g^{\mu\nu}R = \varkappa T^{\mu\nu}$$

for then $D_\mu(T^{\mu\nu}) = 0$. He interpreted this to mean that the source tensor, $T^{\mu\nu}$, would be 'conserved'. We have now seen that the source term for gravity will contain hypercomplex fields, $A = A^\mu b_\mu$ and ψ, etc. It would then seem more natural to define 'conserved' in terms of $\mathcal{P} = b_\mu \mathcal{P}^\mu$, since this operator occurs in the other equations. It is not hard to show that Einstein's final gravity equation can be written in the hypercomplex form

$$(b_\rho|D_{\lambda\nu}b^\lambda) - \frac{1}{2}(b_\rho|b_\nu)(b^\sigma|D_{\lambda\sigma}b^\lambda) = \varkappa T_{\rho\nu}$$

Einstein was so convinced that his equation met all the 'reasonable' requirements of a curved space gravity equation, that he wrote to his very good friend Besso in March, 1914: "Now I am fully satisfied, and do not doubt any more the correctness of the whole system, may the observations of the eclipse succeed or not. The sense of the thing is too evident." Such blind faith is important in giving one the courage to explore virgin ground in science, but it borders on 'proof by intimidation' and should be set to one side in examining the merits of any theory after it has been hatched.

I think that all we can be 'reasonably' sure of today is that $D_{\lambda\nu}b^\lambda$ will be an important part of the curved space gravity law. Only one real test of gravity theory exists today. It appears, experimentally, that the Schwarzschild metric, $g_{\mu\nu}(r,\theta,\phi)$, is valid in the classical limit for the spherically symmetric curved space around a spherical mass. This is not a very stringent restriction on the possible curvature laws of gravity. At infinity, the metric is asymptotically flat. Space is spherically symmetric, static in time, and must fit unknown boundary conditions at the surface of the sun. No experimental tests of $T_{\mu\nu}$ (the source term) have been performed.

In recent years, formidable calculations have been done which are based upon Einstein's gravity equation. Unfortunately, these have not yet produced any further experimental tests of the equation. They have, however, produced an elaborate superstructure of ideas about gravitational collapse in very heavy stars and about the beginning stages of the 'big bang' universe. However, in these high density situations, we would not expect a classical theory to work anyway! Recently, it has been shown that the quantization procedures of electrodynamics (the only highly tested theory) don't work for Einstein's gravity equation! From the hypercomplex number viewpoint, which beautifully fits electrodynamics, we see that gravity is not at all as 'evidently correct' as Einstein believed. Some exciting new breakthroughs are ahead in the basic theory of gravity, including possible extension to 5 dimensions.

Recall that earlier we said quantum electrodynamics is the only comprehensive theory known to modern physics. It should now be evident that gravity theory is not so highly developed to date. The resolution of the mass problem in electrodynamics may not come until gravity theory is extended, for mass is a central part of the source term in this equation. This is why we have said that space curvature may be an important part of the subnucleon (parton) world. Inside nucleons, the densities are enormous (and the cyclic higher dimensions of space may show themselves). Mass, as classically conceived, may well lose its meaning (just as particle path did upon reaching the atomic scale). We should expect this world to have high curvature. We presently have no idea why protons are 1800 times heavier than electrons or why muons exist, because the answers lie in this totally alien world—beyond our experience and perhaps beyond our imagining.

Many standard texts develop the mathematics of showing that Einstein's gravity equation leads to the Schwarzschild metric. But once one has solved this mathematical problem, there remains an important physical problem of guessing how the equation manifests itself in the real world of human measurement. In the remainder of this curved space exploration, we shall come down to earth (literally) and explore this solution.

THE SCHWARZSCHILD SOLUTION

Our usual description of the Schwarzschild solution has involved only $g_{\mu\nu}$ in tensor form and no mention has been made of the tetrad $b_\mu^{(\alpha)}$. Since this is needed to consider a quantum equation in curved space, such as the Dirac equation, we should expand our considerations to include it. First, we examine the general curvilinear coordinate transformation from cartesian coordinates in flat space, and then generalize to the spherically symmetric situation in curved space corresponding to the Schwarzschild metric.

We must consider the flat space curvilinear coordinate generalization of $P = P^\mu e_\mu = e_\mu \hbar \partial^\mu$ (obviously in cartesian coordinates). We simply make P the 'scalar' that it so obviously looks like it ought to be, by defining

$$P = \tilde{P}^\mu \tilde{b}_\mu \equiv P^\mu(x) e_\mu \equiv \tilde{P}^\mu \tilde{b}_\mu^{(\alpha)} e_\alpha \quad (\alpha = 0, 1, 2, 3, \textit{summed as usual})$$

In other words, a change of coordinates $\{x^\mu\} \rightarrow \{\tilde{x}^\mu\}$ induces a change in $e_\mu \rightarrow \tilde{b}_\mu = \tilde{b}_\mu{}^{(\alpha)} e_\alpha$, where $\tilde{b}_\mu{}^{(\alpha)}$ depends on position ($\tilde{x}$) in the general case. The identity transformation $\{x^\mu\} \rightarrow \{x^\mu\}$ corresponds to $\tilde{b}_\mu{}^{(\alpha)} = \delta_\mu{}^\alpha$ ($\equiv 0, \mu \neq \alpha$; $\equiv 1$, $\mu = \alpha$). It is here that we also see how curved space will later show up: $\delta_\mu{}^\alpha \rightarrow b_\mu{}^\alpha(x)$ due to curvature and $b_\mu{}^\alpha(x)$ describes the curved space, instead of $g_{\mu\nu}(x)$ doing so.) The differential operator $P^\mu = \hbar\partial^\mu$ becomes $\tilde{P}^\mu$ and must be defined (see below). The obvious criterion shaping our definition is the desire to have

$$\tilde{P}^\mu \tilde{b}_\mu \equiv \tilde{b}_\mu \tilde{P}^\mu, \quad i.e.\ \tilde{P}^\mu(b_\mu) \equiv 0$$

This may not seem necessary but clearly it is prettier to have P not depend upon which way we write it: $b_\mu P^\mu$ or $P^\mu b_\mu$ (in a curvilinear coordinate system or even in a curved spacetime)!

Since

$$d\tilde{x}^\mu = \frac{\partial \tilde{x}^\mu}{\partial x^\nu} dx^\nu \quad and \quad d\tilde{x}_\mu = \frac{\partial x^\nu}{\partial \tilde{x}^\mu} dx_\nu$$

we define, in the usual way for tensors,

$$\tilde{b}_\mu^{(\alpha)} \equiv \frac{\partial x^\nu}{\partial \tilde{x}^\mu} \delta_\nu^{(\alpha)}, \quad i.e.\ \tilde{b}_\mu \equiv \frac{\partial x^\nu}{\partial \tilde{x}^\mu} e_\nu = \frac{\partial(x^\nu e_\nu)}{\partial \tilde{x}^\mu}$$

which in vector language corresponds to defining the new basis vectors $\tilde{b}_\mu$ corresponding to the new coordinate system. But these are not vectors here, rather hypercomplex numbers. The numbers $\{x^\mu\} = \{ct, x, y, z\}$ will be given the usual cartesian component interpretation in comparing the theory with experimental measurement.

Let us now get our bearings by actually transforming to ordinary spherical coordinates:

$$dx \equiv dx^\mu e_\mu = d\tilde{x}^\mu \tilde{b}_\mu, \quad x^0 \equiv \tilde{x}^0, \; x^1 \equiv \tilde{x}^1 \sin(\tilde{x}^2)\cos(\tilde{x}^3),$$
$$x^2 \equiv \tilde{x}^1 \sin(\tilde{x}^2)\sin(\tilde{x}^3), \quad x^3 \equiv \tilde{x}^1\cos(\tilde{x}^2)$$
$$(x^0 \equiv ct, \; \tilde{x}^1 \equiv r, \; \tilde{x}^2 \equiv \theta, \; \tilde{x}^3 \equiv \phi \textit{ the standard notation})$$

We have written the transformation 'backward' (giving the old coordinates in terms of the new ones) because this is more useful. We now obtain directly

$$\begin{array}{lll} b_r^{(1)} = \sin\theta\cos\phi & b_r^{(2)} = \sin\theta\sin\phi & b_r^{(3)} = \cos\theta \\ b_\theta^{(1)} = r\cos\theta\cos\phi & b_\theta^{(2)} = r\cos\theta\sin\phi & b_\theta^{(3)} = -r\sin\theta \\ b_\phi^{(1)} = -r\sin\theta\sin\phi & b_\phi^{(2)} = r\sin\theta\cos\phi & b_\phi^{(3)} = 0 \\ b_t^{(0)} = 1 & b_t^{(k)} = 0 & b_k^{(0)} = 0 \end{array}$$

where a hybrid notation has been used which should be clear and helpful. Notice that there is no profound physical interpretation associated with this change of coordinates. We simply rename the same spacetime events.

The metric can now be defined through the inner product

$$(A|B) = \frac{1}{2}\left[A^\wedge B + B^\wedge A\right] = \frac{1}{2}\left[AB^\wedge + BA^\wedge\right]$$

$$(dx|dx) = dx^\mu dx^\nu \frac{1}{2}\left[e_\mu^\wedge e_\nu + e_\nu^\wedge e_\mu\right] = dx^\mu dx^\nu \eta_{\mu\nu} e_0$$

$$= d\tilde{x}^\mu d\tilde{x}^\nu \frac{1}{2}\left[\tilde{b}_\mu^\wedge \tilde{b}_\nu + \tilde{b}_\nu^\wedge \tilde{b}_\mu\right] = d\tilde{x}^\mu d\tilde{x}^\nu \tilde{g}_{\mu\nu}(\tilde{x}) e_0$$

$$= d\tilde{x}^\mu d\tilde{x}^\nu \tilde{b}_\mu^{(\alpha)} \tilde{b}_\nu^{(\beta)} \eta_{\alpha\beta} e_0$$

The $\eta_{\mu\nu}$ cartesian metric $(+ - - -)$ follows from the $\{\partial_\mu\}$ multiplication table. (Note that $(+ - - -)$ is more natural here than the form $(- + + +)$, sometimes used in the literature. However, $\{i(i\sigma_\mu)\}$ is a natural alternative to $\{(\sigma_\mu)\}$ and it has the metric $(- + + +)$. The curvilinear metric has the form

$$g_{\mu\nu} = b_\mu^{(\alpha)} b_\nu^{(\beta)} \eta_{\alpha\beta} \equiv b_\mu^{(\alpha)} b_{\nu(\alpha)} = g_{\nu\mu}$$

(where $\eta_{\alpha\beta}$ raises and lowers (α) indices and $g_{\mu\nu}$ raises and lowers μ indices). We then get

$$g_{00} = 1,\ g_{OK} = 0,\ g_{kj} = 0,\ g_{11} = -1,\ g_{22} = -r^2,\ g_{33} = -r^2\sin^2\theta$$

There are 16 objects, $b_\mu{}^{(\alpha)}$, and they directly and simply lead to the 16 objects $g_{\mu\nu}$, of which only four are non-zero here.

In curved space, the transformation $b_\mu{}^{(\alpha)}(x) \to \delta_\mu{}^{(\alpha)}$ is impossible except at one (but any one particular) event x in spacetime. As the gravity coupling →

0, the curvature → 0 and $b_\mu^{(\alpha)} \to \delta_\mu^{(\alpha)}$, if we were already in the closest thing we could get to a cartesian coordinate system in describing the curved space. In our particular example, a spherical coordinate system is simplest. This means that $\tilde{b}_\mu^{(\alpha)} \to \tilde{b}_\mu^{(\alpha)}$ (flat) as the gravity is shut off and $\tilde{b}_\mu^{(\alpha)}$ (flat) is the set of functions above. These would be the boundary conditions at infinity for the Schwarzschild case.

Even in curved space it is not necessary to assume that all 16 functions $b_\mu^{(\alpha)}$ become non-zero. Space and time are not really parts of a four dimensional 'thing'. Only the mathematics looks this way. It is always possible (hypothesis which seems true) to assume space and time are 'orthogonal' by use of a 'sensible' coordinate system. This allows us to define

$$b_0^{(k)} \equiv 0 \quad and \quad b_k^{(0)} \equiv 0 \Rightarrow g_{0k} = b_0^{(\alpha)} b_k^{(\beta)} \eta_{\alpha\beta} = 0$$

for a 'sensible' coordinate system description of reality. However, rotating black hole physics finds convenience in using coordinates which violate this; at least on paper it 'exists'.

If we use Einstein's equation of gravity, we get $g_{\mu\nu}(x)$ as the solution. (At least there are some known exact solutions of the equation and they are in this form. Our case is one of them.) Thus we must work backward to find the $b_\mu^{(\alpha)}$ functions from $g_{\mu\nu}$. The equations linking them are non-linear algebraic equations and this could be very difficult in the general case. (Again this shows the $b_\mu^{(\alpha)}$ to be more basic, since once we know them, $g_{\mu\nu}$ can easily be found. We shall work backward for the Schwarzschild case and find the $b_\mu^{(\alpha)}$ functions in spherical, curvilinear coordinates in spherically symmetric, static space. The metric gives us the $g_{\mu\nu}$ directly in spherical form. (We drop the $\tilde{g}_{\mu\nu}$ notation now, since it will not be helpful.) The general equations we must invert are

$$g_{\mu\nu} = b_\mu^{(\alpha)} b_{\nu(\alpha)} = b_\mu^{(\alpha)} b_\nu^{(\beta)} \eta_{\alpha\beta} \Rightarrow$$

$$g_{00} = b_0^{(\alpha)} b_0^{(\beta)} \eta_{\alpha\beta} = b_0^{(0)} b_0^{(0)} - \Sigma_k b_0^{(k)} b_0^{(k)} = (b_0^{(0)})^2$$

$$g_{11} = b_1^{(\alpha)} b_1^{(\beta)} \eta_{\alpha\beta} = b_1^{(0)} b_1^{(0)} - \sum_k b_1^{(k)} b_1^{(k)} = -\sum_k (b_1^{(k)})^2$$

$$g_{22} = -\sum_k (b_2^{(k)})^2 \qquad g_{33} = -\sum_k (b_3^{(k)})^2$$

$$g_{0k} = 0 = b_0^{(\alpha)} b_k^{(\beta)} \eta_{\alpha\beta} = b_0^{(0)} b_k^{(0)} - \sum_j b_0^{(j)} b_k^{(j)} = 0 - 0$$

$$g_{jk} = b_j^{(\alpha)} b_k^{(\beta)} \eta_{\alpha\beta} = b_j^{(0)} b_k^{(0)} - \sum_\ell b_j^{(\ell)} b_k^{(\ell)} = -\sum_\ell b_j^{(\ell)} b_k^{(\ell)}$$

where a 'sensible' coordinate system has been assumed, for simplification. Solving 16 equations would be no easy task! In our case, we have a much simpler task if we use the high symmetry to guide us, and guess at trial solutions for some of the terms. The metric is

$$(ds)^2 = (1 - R_s/r)(dx^0)^2 - (1 - R_s/r)^{-1}(dr)^2 - r^2(d\theta)^2 - r^2\sin^2\theta(d\phi)^2$$

which comes from Einstein's gravity equation. The $d\theta$ and $d\phi$ elements are the same as in flat space, therefore, we might guess that $b_\theta^{(\alpha)}$ and $b_\phi^{(\alpha)}$ are unaltered by the curved space here. (This turns out to be correct.) This directly gives us:

$$g_{00} = (1 - R_s/r) = (b_0^{(0)})^2 \Rightarrow b_0^{(0)} = (1 - R_s/r)^{\frac{1}{2}} \rightarrow +1, \textit{ as } r \rightarrow \infty \textit{ (flat)}$$

The $b_r^{(\alpha)}$ terms must follow from

$$g_{rr} = -\sum_k b_r^{(k)} b_r^{(k)} = (1 - R_s/r)^{-1}$$

$$g_{r\theta} = -\sum_\ell b_r^{(\ell)} b_\theta^{(\ell)} = 0 \quad \textit{and} \quad g_{r\phi} = -\sum_\ell b_r^{(\ell)} b_\phi^{(\ell)} = 0$$

where $b_\theta^{(\ell)}$ and $b_\phi^{(\ell)}$ are already assumed (guessed) to be the known flat space values. Two of these three equations are linear and the solution, for the three $b_r^{(k)}$ functions, is not difficult. The result is simply ([ ] denotes flat space values)

$$b_r^{(1)} = \sqrt{-g_{rr}}\ \sin\theta\cos\phi = \sqrt{-g_{rr}}\ b_r^{[1]}$$

$$b_r^{(2)} = \sqrt{-g_{rr}}\ \sin\theta\cos\phi = \sqrt{-g_{rr}}\ b_r^{[2]}$$

$$b_r^{(3)} = \sqrt{-g_{rr}}\ \cos\theta = \sqrt{-g_{rr}}\ b_r^{[3]}$$

Again, we have used the r → ∞ limit to decide the $\pm\sqrt{}$ sign choice in the algebraic solution. We now have the full tetrad solution for the Schwarzschild metric! One can check by direct substitution that

$$b_\mu^{(\alpha)} b_{\nu(\alpha)} = g_{\mu\nu} \qquad b_\mu^{(\alpha)} b^{\mu(\beta)} = \eta^{\alpha\beta}$$

$$b_\mu^{(\alpha)} b_{(\alpha)}^{\nu} = \delta_\mu^\nu \qquad b_\mu^{(\alpha)} b_{(\beta)}^{\mu} = \delta_\beta^\alpha$$

We need $g^{\mu\nu}$ and $\eta^{\alpha\beta}$ for this check. These are defined as usual:

$$g^{\mu\lambda} g_{\lambda\nu} \equiv \delta_\nu^\mu \qquad and \qquad \eta^{\alpha\beta}\eta_{\beta\gamma} \equiv \delta_\gamma^\alpha$$

which gives $\eta^{\alpha\beta} = \eta_{\alpha\beta}$, as usual, and in the spherical case

$$(g^{\mu\nu}) = (g_{\mu\nu})^{-1} = \begin{pmatrix} (1-R_s/r)^{-1} & 0 & 0 & 0 \\ 0 & -(1-R_s/r) & 0 & 0 \\ 0 & 0 & -1/r^2 & 0 \\ 0 & 0 & 0 & -1/r^2\sin^2\theta \end{pmatrix}$$

For example,

$$b_r^{(\alpha)} b_{(\alpha)}^{\theta} = b_r^{(\alpha)} g^{\theta\lambda} \eta_{\alpha\beta} b_{\lambda}^{(\beta)} = b_r^{(\alpha)} g^{\theta\lambda} (\pm 1)_{\alpha} b_{\lambda}^{(\alpha)} = b_r^{(\alpha)} g^{\theta\theta} (\pm 1)_{\alpha} b_{\theta}^{(\alpha)}$$

$$= -g^{\theta\theta}\left(b_r^{(1)} b_{\theta}^{(1)} + b_r^{(2)} b_{\theta}^{(2)} + b_r^{(3)} b_{\theta}^{(3)}\right)$$

$$= \frac{\sqrt{-g_{rr}}}{r^2}\,(r\sin\theta\cos\theta\cos^2\phi + r\sin\theta\cos\theta\sin^2\phi - r\cos\theta\sin\theta) = 0$$

and

$$b_r^{(\alpha)} b_{(\alpha)}^{r} = -g^{rr}\left((b_r^{(1)})^2 + (b_r^{(2)})^2 + b_r^{(3)})^2\right)$$

$$= +(\sin\theta\cos\phi\sin\theta\cos\phi + \sin\theta\sin\phi\sin\theta\sin\phi + \cos\theta\cos\theta) = +1$$

This shows us how to generalize flat space and how this generalization is similar to but not the same at all as switching to curvilinear coordinated in flat space. To efficiently solve problems, we need both at once. Even then, things are very difficult, even for the simplest of curved spacetime cases.

The possible internal 'overlap' of space and time parts in the metric is an interesting physical problem. Mathematicians can do whatever they please, for the real world does not concern them. We can invent any coordinate system we want to, so long as it is invertible. Thus we can cause the space and time elements to have 'crossterms' that mix the space and time parts in the metric. This may even be a useful trick to find a simple (relatively) solution to a given energy momentum source situation. However, I suspect that when it comes to actual testing of any solution in the lab, a coordinate system will need to be found where these space and time parts are orthogonal. It is very difficult to interpret the scratchings on paper in relation to real experiments that we huge humanoids can actually set up and run in the real world. How to calibrate your instruments is always a big mess. Curved spacetime is very hard and I would hope that nature avoids it but I think not. The whole universe probably is closed and so curved dramatically. Inside protons, there may well be more curved spacetime of a radical kind.

We now leave this never-never land of curved space and return to the firm security of flat and static space to do electricity in quaternions. This is very pretty and should be taught to beginning students. They should see the Dirac algebra before they do much with vectors in the introductory E&M class. They can really understand why Maxwell's equations are true and thus why the

Faraday law, *etc.*, exist.

I say they should see the Dirac algebra before doing much with vectors, but many teachers would disagree with this. They would say that it is too complicated for this level. That is only because students did not see the Dirac algebra in secondary school where it should be incorporated after any section of the math book on complex numbers. It is not hard to use the distributive law to multiply two numbers together which have six or eight parts and then use the rules of Pauli multiplication to reduce a pair to another basis element and simplify. Such beautiful math should be seen by students who will not take any further math in college. It is, after all, the foundation of the world we are talking about.

We have the Hamilton-Pauli algebra to do classical electrodynamics with in the next chapter. This algebra gives us 4-vectors through the conjugations. As a Clifford algebra, it gives us only 3-vectors and this shows Clifford is not right for this physics level. There is another algebra with 8 elements, derived from Dirac's algebra, $\{(e_\mu), (f_\mu), (ie_\mu), (if_\mu)\}$, called the C_ℓ algebra. This has real coefficients and contains the H-P algebra. It could then be also used for classical electrodynamics. As a Clifford algebra, it now gives us 4-vectors. If you think Clifford is a profound part of physics, then you should prefer this algebra to its H-P sub-algebra for classical electrodynamics. However, the conjugations give us 5-vectors here in the C_ℓ algebra. Are we sure that classical physics should have no 5th spacetime part? Is this preparing us for the 5-vectors that will be found in full Dirac algebra? This is all very peculiar and someone should find out what is really going on here. We shall stick to the H-P algebra development in the next chapter for Classicl Electrodynamics. The connection between rest mass and extra dimensions keeps coming up in this hypercomplex number approach to the physics. I still don't know what this is telling us, but something important I am sure.

CHAPTER 11

Quaternion Electrodynamics

We have focused mostly on quantum physics in this book. Before finishing it, let us look at the pretty results for classical electrodynamics in this notation. Given external electromagnetic fields, $\{E^k, B^k\}$, how does a charge q respond? What happens if we have a charged 'rocket', so that there is another force on the charge besides the electrical one? We shall explore these questions in some depth.

The particle, with q and m, moves with acceleration along a classical path and is thus not in an inertial frame. It, or the attached rocket, can still carry a clock, and $d\tau$ is the time lapse on that moving clock when dt really passes in the Newton frame. We know that $d\tau = dt\surd(1 - v^2/c^2)$.

We must really guess at the basic classical equation for the motion. Or, we must find the correct quantum solution and then make some semi-classical approximation that leads us directly to the desired classical limit equation. That limit process has not been done successfully, that I know of, for the fully relativistic case. So we shall just guess. Let us guess that

$$m\frac{d^2x^\mu\sigma_\mu}{d\tau^2} = \mathfrak{F}^\mu\sigma_\mu + \mathfrak{L}^\mu\sigma_\mu$$

(The $\mathfrak{F}$ and $\mathfrak{L}$ 4-forces here are not physical forces but contain them.) This form is obviously covariant: $x \to x' = L^\dagger xL$, $\mathfrak{F} \to \mathfrak{F}' = L^\dagger\mathfrak{F}L$, $\mathfrak{L} \to L^\dagger\mathfrak{L}L$, where $LL^\wedge \equiv 1\sigma_0$. The $\mathfrak{F}$ is a reaction push 'due' to radiation being emitted. The $\mathfrak{L}$ is our standard Lorentz 4-force. (There may be a third push, due to 'rockets' attached to the mass m.) The $d\tau$ is Lorentz invariant, as are m and q.

We further guess that the Lorentz force takes the form

$$\mathfrak{L} = -\frac{q}{c}\frac{1}{2}\left[\left(\frac{dx}{d\tau}F\right) + \left(\frac{dx}{d\tau}F\right)^\dagger\right] = \mathfrak{L}^\mu\sigma_\mu$$

where $F \equiv -E^k\sigma_k + cb^k(i\sigma_k) = -F^\wedge \to F' = L^\wedge FL$. This F is the Maxwell field produced by various moving, external charges (at retarded(?) times). The need for form covariance 'dictates' the combination $(dx/d\tau)F$, since this transforms as $[L^\dagger(dx/d\tau)L][L^\wedge FL] = L^\dagger[(dx/d\tau)F]L$, a 4-vector. The added $((dx/d\tau)F)^\dagger$ term is 'dictated' by the fact that $(d^2x/d\tau^2)^\dagger = (\cdots)$, so the right hand side must be like this also. This further requires that $\mathfrak{F}^\dagger = \mathfrak{F}$ also. The $(-q/c)$ is there for 'coupling' and consistency of units.

Notice next that $[(d^2x/d\tau^2)^\wedge (dx/d\tau)] + [\cdots]^\wedge = \cdots = 0$. This says $a^\mu \mathcal{U}_\mu = 0$, identically, for acceleration and speed 4-vectors. We can write out the inner product

$$[\mathfrak{L}^\wedge (dx/d\tau)] + [\cdots]^\wedge \equiv 2(\mathfrak{L}|dx/d\tau)$$

and see if it is also zero. This is rather pretty, so the steps are outlined below:

$$\begin{aligned}\frac{-4c}{q}(\mathcal{U}|\mathfrak{L}) &= [\mathcal{U}^\wedge(\mathcal{U}\boldsymbol{F} + (\mathcal{U}\boldsymbol{F})^\dagger)] + [\cdots]^\wedge \\ &= \mathcal{U}^\wedge\mathcal{U}\boldsymbol{F} + \mathcal{U}^\wedge\boldsymbol{F}^\dagger\mathcal{U}^\dagger + \boldsymbol{F}^\wedge\mathcal{U}^\wedge\mathcal{U} + \mathcal{U}^{\dagger\wedge}\boldsymbol{F}^{\dagger\wedge}\mathcal{U} \\ &= \mathcal{U}^\wedge\mathcal{U}\boldsymbol{F} + \mathcal{U}^\wedge\boldsymbol{F}^\dagger\mathcal{U} - \boldsymbol{F}\mathcal{U}^\wedge\mathcal{U} - \mathcal{U}^\wedge\boldsymbol{F}^\dagger\mathcal{U} \\ &= 0\end{aligned}$$

We have used $\mathcal{U} \equiv (dx/d\tau) = \mathcal{U}^\dagger$ and $F^\wedge = -F$. Thus we have no choice but to have $(\mathcal{U}|\mathfrak{F}) = 0$ also, at any time τ, whatever $\mathfrak{F}$ looks like internally. Non-relativistic experience suggests that $\mathfrak{F}$ should involve $d^3x/d\tau^3$, $d^2x/d\tau^2$, and $dx/d\tau$ in some covariant combinations. Putting this covariance need together with the $(\mathcal{U}|\mathfrak{F})$ requirement, we find that a good $\mathfrak{F}$ prospect is (guess!):

$$\begin{aligned}\mathfrak{F} &= A\left[(d^3x/d\tau^3) + (d^2x/d\tau^2|d^2x/d\tau^2)\,\mathcal{U}/c^2\right] + \mathfrak{F}_{III} \\ &\equiv \mathfrak{F}_I + \mathfrak{F}_{II} + \mathfrak{F}_{III}\end{aligned}$$

where the 4-vector III is currently still unknown, if it exists. Non-relativistic 'experience' suggests that this reduces approximately to the 3-force Ada/dt, with $A = 2q/[(4\pi\epsilon_0)3c^3]$. The first term in $\mathfrak{F}$ has a third time derivative and it apparently becomes the only surviving part at low speeds. The second term has (v/c) in it, so it shrinks faster as $(v/c) \to 0$. The third part is a new part, that we throw in here because the other two can cancel out. It will be discussed later.

The covariance is obvious for this $\mathfrak{F}$ guess, but $(\mathfrak{F}|\mathcal{U})$ must also be identically zero. After we form $(\mathfrak{F}^\wedge \mathcal{U}) + (\cdots)^\wedge$, we have $\mathcal{U} \equiv dx/d\tau$ and we find

$$(\mathfrak{F}|\mathcal{V}) = \mathfrak{F}^{\mu}\mathcal{V}_{\mu} = A\frac{d^3x^{\mu}}{d\tau^3}\frac{dx_{\mu}}{d\tau} + A(a|a)\frac{\mathcal{V}^{\mu}}{c}\frac{\mathcal{V}_{\mu}}{c} + (\mathfrak{F}_{III}|\mathcal{V})$$

The fact that $\mathcal{V}^{\mu}\mathcal{V}_{\mu} = c^2$ can be easily proven. Our needed cancellation of these first two parts comes from the trick of integrating by parts on the second part:

$$\int_a^b a_{\mu}a^{\mu}d\tau = \int_a^b a_{\mu}\left(\frac{d\mathcal{V}^{\mu}}{d\tau}\right)d\tau = \int_a^b a_{\mu}d\mathcal{V}^{\mu}$$

$$= (a|\mathcal{V})|_b - (a|\mathcal{V})|_a - \int_a^b \mathcal{V}^{\mu}\frac{da_{\mu}}{d\tau}d\tau$$

$$= -\int_a^b \left(\mathcal{V}^{\mu}\frac{da_{\mu}}{d\tau}\right)d\tau \Rightarrow (a_{\mu}a^{\mu})_{ave} \approx -\left(\mathcal{V}^{\mu}\frac{da_{\mu}}{d\tau}\right)_{ave}$$

This last step is obviously a good one, for then we have

$$(\mathfrak{F}_{I,II}|\mathcal{V}) \approx A\left[\frac{d^3x^{\mu}}{d\tau^3}\frac{dx_{\mu}}{d\tau} - \frac{dx_{\mu}}{d\tau}\frac{d^3x^{\mu}}{d\tau^3}\right] = 0$$

So, either we must come up with a different $\mathfrak{F}$ guess or we must come up with a 'plausible' argument to support the integration result above, which was obtained by integrating by parts and throwing away the $(a|\mathcal{V})$ terms, because they are identically zero in general.

The a and b end points do go along with τ_a and τ_b time readings, on the clock traveling along with q. A minimum $\Delta\tau \equiv \Delta\tau_0$ must exist here to stay classical, where 'force' still has some (approximate) meaning at least. (Force is obviously only classical and very useful for our daily engineering needs.) Thus, we have

$$(a|\mathcal{U})|_{\tau+\Delta\tau_0} - (a|\mathcal{U})|_{\tau} - (\mathcal{U}^{\mu}\frac{da_{\mu}}{d\tau})_{ave}\Delta\tau_0 \approx (a|a)_{ave}\Delta\tau_0$$

We conclude that (with *ave* ≡ average over $\Delta\tau_0$):

$$(a|a)_{ave} \approx -\left(\mathcal{U}\,|\frac{da}{d\tau}\right)_{ave} + \frac{\Delta(a|\mathcal{U})}{\Delta\tau_0}$$

Since $\Delta\tau_0 \rightarrow 0$ is not possible **here**, we really don't have $(\mathfrak{F}_{I,II}|\,\mathcal{U})|_{\tau} = 0$, and so this $\mathfrak{F}$ part is not really the correct one if the other terms are correct! At best, $(a|a)$ is 'equal' to $(-\mathcal{U}\,|\,(da/d\tau))$ only averaged over one cycle, for any periodic motion of the charge q.

We normally use calculus for classical mechanics motions and that fact requires that $\Delta\tau \rightarrow 0$ be allowed. It is not allowed! We should only be doing difference equations, not differential equations. The integration by parts theorem that we used probably needs $\Delta\tau \rightarrow 0$ in its derivation. That derivation would then need to be analyzed in detail as well, with a finite $\Delta\tau$ step size. The physics here is not being done right in any of the text books, I suspect, but I am no expert. Locality is the corner stone of classical physics and fails to exist in quantum. The basic equations can be modified to try and bring in some non-locality but it ain't pretty—integrodifferential equations.

Difference equations are very messy, so we want to use differential equations for convenience. The above 'reaction' forces $\mathfrak{F}_I$ and $\mathfrak{F}_{II}$ may not be the last word here, but they are simple guesses and historically go way back to Abraham and others who first guessed at this structure, I gather, about the turn of the century.

The speed dotted into '$F = ma$', as 4-vectors, gives zero here exactly and this indicates that all four equations are not independent. The particle m really only has three degrees of freedom. However, the created radiation gives more 'freedom', so maybe this 'dot' product is not quite zero after all!

Relativistic classical physics is really a joke! Throw two charged ball bearing at each other, each moving at 0.999999c, and see the results of the interaction. It ain't classical! If they collide or even come close, quanta will come flying out. Doing the same thing with two protons (or even with two carbon atoms) is clearly a quantum problem. All interaction situations are intrinsically quantum in nature, even kicking a soccer ball. The electron clouds of the surface layers of atoms interact quantum mechanically to push the ball into motion. Quantum interactions continue, all over the ball, to hold it together. We should not push classical very far at all. The classical solution to the radiation reaction pre-acceleration caused by $da/d\tau$ appearing, is not in the $t \rightarrow -\infty$ boundary conditions. Remember that humanoid quantum measurement is done by preparing large, classical set ups (many many identical ones would be best, in principle) at t_1 and then seeing classical set ups at a later t_2. This t_2 must be late enough to be human-brain meaningful for real eyeball observations. Any semi-classical laboratory process must be set up the same way! The $t_2 - t_1$ time interval is less than trillions of years and more than milliseconds. It is not infinite! If we go back billions of years, where cosmic, classical set ups were being done for us, then the curved and closed universe will become a part of the 'boundary' conditions, most likely. There is still conservation of energy, we take on faith, whatever that means when we include the stretching space as part of the system. It ain't easy to analyze acceleration and radiation in a curved and changing spacetime, where the radiation emitted and the charge source will both contribute to the curvature in the neighborhood. The world is just very complicated. Classical electrodynamics is a convenient fiction for engineering design, and nothing more than that really. The proper classical picture will come only after the proper quantum picture is fairly complete, it seems.

The same is true for gravity. Classical gravity is a similar fiction. You cannot push it very far without entering the quantum world. When and how that world takes over, is still totally unknown! Quantum space effects may really extend far beyond the event horizon. If so, then paths will be meaningless at $r \sim R_s$. The in-falling observer eventually dies and has his/her atoms blend, spread, and merge with the quantum gravity space that is there. We just have NO idea what will happen!! Maybe we can have some confidence in classically scattering two small, slow black holes off of each other, so long as their closest approach is 10 or 100 R_s or more. Likewise, we could have some classical confidence about what happens if we scatter two large, charged ball bearings, where their closest approach is hundreds of times their size and they don't move too fast. Classical physics is a slippery slope at best. We only see classical and we only feel it, as huge creatures, but we can't calculate very much with it, unless we have big, slow, 'simple' blobs to analyze; no big, slow superfluids for example! (Size is not the only feature that dictates when classical theory can be approximately used.)

A SPECIAL CASE – HYPERBOLIC MOTION

The first two parts of the reaction force are quite different and the second part has the 'wrong' sign. It is hard to imagine that there is a simple motion where these possibly huge forces cancel each other exactly, but it does exist. It is, using Peierl's notation, $x = (c/g)\sqrt{(c^2 + g^2t^2)}$, where c is light speed and g is an 'arbitrary' constant with acceleration units. (That is why we called it g.) Near $t \approx 0$, we have $c >> gt$ and this complicated motion approximates to: $x \approx x_{min} + \frac{1}{2}gt^2$, $v \approx gt$, $x_{min} = c^2/g$. This is the familiar 'gravity' free fall, with constant weight, mg, pulling us downward. Some have tried to tie this $x(t)$ case to such free fall gravity physics, but it does not really fit there. In actual free fall, g slowly increases as we fall! In our case, g is a real constant. They cannot be the same! For large g values, the curved spacetime must be dealt with in gravity!

It turns out that the x components have $\Im_{I,II}{}^1 = A(g^3t/c^2 - g^3t/c^2) = 0$, for the first two parts in $\Im^1$ here. (The superscript refers to the x component.) It also turns out that $-A(d^2x^\mu/d\tau^2)\,(d^2x_\mu/d\tau^2) = +Ag^2$ (watts) here. This Lorentz invariant constant term, $A(4\text{-}accel.)^2$, supposedly gives the total radiation given off in 'all' directions by any accelerating 'particle', as seen from an inertial frame. Thus the usual 'radiation reaction force' is exactly zero here, yet there is radiation 'leaving' the charge. This radiation presumably would generally tend to slow the particle and try to make it deviate from the prescribed $x(t)$ path. Any such deviation would make $\Im_{I,II}$ become non-zero and change the motion further. If m does stay on course, then presumably the rocket engine must supply this radiated energy as part of the net mg force that contributes to the change in kinetic energy of the mass/charge object, at least for $+x$ motion.

Since, for one dimension, power is F_xv_x and $v_x \to 0$ as $t \to 0$, 'part' of the rocket force needs to $\to \infty$, as $v \to 0$, to keep on course, while emitting a steady amount of radiation energy! The hyperbolic path comes from the classical equation:

$$F_{netx} = \frac{d}{dt}\left(\frac{mv_x}{\sqrt{1 - v_x^2/c^2}}\right) \equiv mg \equiv constant$$

$$= F_{rocketx} + (\Im_{Ix} + \Im_{IIx} + \Im_{IIIx})/\gamma$$

We assume the rocket is out in free space where there are negligible E and B fields and the space is approximately flat and approximately static. Since F_{net} must be constant, and $\Im_I = -\Im_{II}$, we must adjust the thrust, F_{rocket}, to compensate for any $\Im_{III}$ that appears due to the radiation. Can we be sure that

$\Im_{III}$ is only along $\pm x$ in direction? I think we can safely guess that is true, in this case. We are left with

$$F_{rocket} + \Im_{III}/\gamma = mg = constant$$
$$\gamma \equiv 1/(1 - v^2/c^2)^{1/2}$$

The rocket must do what it can to keep the net force equal to mg. The net work by mg goes into changing the kinetic energy

$$K = mc^2(\gamma - 1)$$

The $\Im_{III}/\gamma$ is just like a friction force, only it 'makes' radiation here instead of heat. (You cannot get either one back—unless maybe radiation can both come in and go out—advanced waves besides retarded ones. That is a whole other theory.) Lets call the radiation force the friction force. How large is it? Instantaneously, we have the particle radiate Ag^2 Joules/second and this is the work-rate $F_{frictionx}v_x$. Therefore,

$$F_{frictionx} = (\Im_{IIIx}/\gamma) = \frac{Ag^2}{v}$$

$$= \frac{Ag^2}{cgt/\sqrt{c^2 + g^2t^2}} = \left(\frac{Ag^2}{c}\right)\sqrt{1 + c^2/g^2t^2}$$

As $t \to 0$, $F_{friction} \to Ag^2/(gt) \to \infty$, but as $t \to \infty$, $F_{friction} \to Ag^2/c$. The rocket force, for the motion outward along $+x$ at $t > 0$, must, for this special, hyperbolic path, be

$$F_{rocketx} = +mg + \frac{Ag^2}{c}\sqrt{1 + (c/gt)^2}$$

If we imagine 'backing up' in time toward $t \to 0+$, we hit

$$F_{rocket\text{-}max} = mg + \frac{Ag^2}{c}\sqrt{1 + (c/gt_1)^2}$$

which gives the smallest time $t_1 > 0$ that we can actually start out the motion and have it be truly hyperbolic. The starting conditions, $\{x_1, v_1, F_{f1}, F_{r1}\}$, are all well defined.

The rocket cannot supply enough force to hold the hyperbolic path at an earlier time, since it (like any source) has a finite maximum limit.

There needs to be some theoretical justification for the extra 4-vector force $\Im_{III}$ and a general, covariant expression found for it. This general result must, in the present circumstance with $\alpha \equiv (c^2 + g^2t^2)^{1/2}$, reduce to:

$$\Im^1_{III} = \frac{Ag^2\gamma}{v} = \frac{Ag^2(\alpha/c)}{cgt/\alpha} = \frac{Ag^2}{c^2(gt)}(c^2 + g^2t^2)$$

$$= \frac{Ag}{t} + \frac{Ag^3t}{c^2} \rightarrow \infty,\ t \rightarrow \infty\ ;\ \rightarrow \infty,\ t \rightarrow 0$$

The physical force that goes with this becomes $F_f, \rightarrow Ag^2/c$ as $t \rightarrow \infty$ and $v \rightarrow c$. This physical force remains finite but the 4-force above blows up there, because $\gamma \rightarrow \infty$ as $v \rightarrow c$. To see what kinds of numbers we are really talking about here, we shall consider a rather exotic specific case with some radically large numbers. These are used in order to get the acceleration and radiation effects up to 'large' values. Consider $g \equiv 10^{19}m/s^2 \sim 10^{18}g_\oplus$. That locks up the motion completely. By the way, each of the first two (conventional) parts of $\Im$ go to infinity as $t \rightarrow \pm\infty$. Their difference will still remain zero, however, for exactly the hyperbolic path $x(t)$ above. No real rocket can hold an exact path, of course. At $t = -(1/3)b.y. = -10^{13}sec$, we have $x = 3\times10^{21}m = 300{,}000 l.y.$, which is about three Milky Way galaxy diameters away from us. The speed, v_x, is $-(1 - 10^{-47})c$ and $\gamma = 3\times10^{23}$. The mass, m, while rushing towards 'us', slows gradually as it travels $(da/dt = 10^{-64}(m/s^2)/s)$. The rocket thrust causes the radiation and the deceleration. The rocket thrust, far away, is $(+mg - Ag^2/|v|) > 0$ and is thus a retrofire. It weakens as the constant radiation causes a growing force, which is also helping to slow the objects. The 'friction' (radiation) force is a drag, and this allows the rocket motor to coast a little and put out less than mg thrust. This rocket force is approximately a constant force, far away, since v drops very very slowly from c. At the small time $t = -3\times10^{-12}sec$, the incoming object has finally reached a 'non-relativistic' speed, $v_x = -0.1c$. It is then located at $x = +9\times10^{-3}m \approx 1cm$, and still has a large deceleration of $-2.7\times10^{13}m/s^2$. It then screeches to a halt at $t = 0$ at $x = x_{min}$

$= +(9 - \epsilon)\times 10^{-3}m$. At $t = +3\times 10^{-12}sec$, the speed is about back up to $v_x = +0.1c$ and the position is still about $x = +9\times 10^{-3}m$. I say 'about' because of the problems of staying on course near $v = 0$, discussed above.

An acceleration of 10^{18} gees is ridiculous for real rocket engines, of course. It is not ridiculous for nuclei in a very hot crystal of, say, iron. These high accelerations do not affect clocks 'in' the nucleus it appears (observed Mössbauer effect on nuclear line broadening). The nuclei do oscillate violently here and they pass very quickly through zero speed at the deviation extremes, just like our case. Can we apply the above radiation reaction ideas here? This is not a classical situation. Path may still make some sense here for the nuclei, but not for the electron clouds that surround them and provide the restoring force in this case, as they overlap their neighbors' electron clouds.

Consider then a uranium nucleus: $Q = 92e$ and $m = 238m_p$. The constant A is then 4.8×10^{-50} in this case, coming from $(2kQ^2/3c^3)$, $k = (1/4\pi\epsilon_0) = 9\times 10^9$.

The radiation emission rate is constant for hyperbolic motion and equal to Ag^2. This is only 4.8×10^{-12} *watts*, for this tremendous g value that we have assumed, because A is so small. The actual acceleration goes from g at $t \approx 0$ to 0 at $t \approx \pm\infty$ here.

It is not classical to deal with a nucleus like this. A charged, small ball bearing would be more reasonable with as much charge on it as we can get to stay in place. There are problems too with attaching it to the rocket without disturbing it and its radiation pattern. We don't want the metal surface of the ship to get into the picture. These are not silly considerations. Classical physics is real-world physics; the engineering must be done to make the problem meaningful. No hypothetical motions need be considered really, since these are untestable.

The mathematical details for our hyperbolic motion are summarized as follows (on the next page):

$$\alpha \equiv (c^2 + g^2t^2)^{1/2} \to gt,\ t \to \infty;\ \to c,\ t \to 0$$

$$x(t) = \alpha c/g = \gamma c^2/g$$

$$v_x(t) = cgt/\alpha = gt/\gamma \to c,\ t \to \infty$$

$$a_x(t) = c^3g/\alpha^3 \to c^3/g^2t^3,\ t \to \infty;\ \to g,\ t \to 0$$

$$da_x(t)/dt = \frac{3cg^3t}{\alpha^3}\left(\frac{g^2t^2}{\alpha^2} - 1\right) \to 3c\left(\frac{1}{t^2} - \frac{1}{t^2}\right),\ t \to \infty$$

$$\to -3g^3t/c^2,\ t \to 0$$

$$\gamma \equiv (1 - v_x^2/c^2)^{-1/2} = \alpha/c \to \infty,\ t \to \infty$$

$$KE = \frac{mc^2}{\sqrt{1 - v^2/c^2}} - mc^2 = mc^2\ (\gamma - 1) \to \infty,\ t \to \infty$$

$$P_x = \gamma m v_x = mgt \to \infty,\ t \to \infty$$

$$P_{rad} = -A(d^2x^\mu/d\tau^2)(d^2x_\mu/d\tau^2) = +Ag^2 = F_{drag}v$$

$$P_{KE} = mgv = mcg^2t/\alpha$$

$$F_{net} = mg = 3.6x10^{-6}N$$

$$A(d^3x^1/d\tau^3) = A(g^3t/c^2)$$

$$F_{drag} = Ag\alpha/ct \to \frac{Ag}{t},\ t \to 0;\ \to Ag^2/c,\ t \to \infty$$

We have not dealt with the angular distribution in (x,y,z) space, for the radiation emitted. Only the total emitted in all directions has been considered here. The distribution is said to be in a doughnut around the forward direction of motion. Does it then radically switch direction as v passes through zero at turn around?! Its magnitude is supposed to stay constant in time!

The kinetic energy change is exactly equal to the net work, $mg\Delta x$, here. The rocket force must supply whatever energy is needed to keep the charge on course as specified. The $F_{friction}$ force must exist if radiation is only emitted and this radiation either adds to or subtracts from the output needed from the rocket. We do not have the first two $\Im$ forces in this particular problem to help or hurt us in controlling the rocket thrust.

Remember that $\{\Im_{I,II}{}^k = \gamma F^k{}_{reactionI,II}\}$ may give most of the physical force $\{F^k{}_{reaction}\}$ that acts, for any slight deviations from the hyperbolic path. Notice also that $\Im^1 = \pm$ (a drag or a boost) is possible, since we have guessed that this term exists in the first place. Some physical insight may be needed for each problem faced, to fix this sign choice. While $\Im$ has two parts with opposite signs to each other, the overall sign is not so easy to assess. The third (new) part is presumably also only a drag force, since the other two parts can be canceled out exactly. Marx says it may not be possible to give any physical meaning to the parts here individually, since only the composite will show up experimentally. Thus the second term, being of opposite sign to the first, is perhaps not so shocking. Other simple motions, such as constant acceleration, indicate that the first term is larger. The second term is apparently the 'boost' part here, since it is smaller and we expect a drag over all. The third term is totally unknown at present, if it really exists. Our hyperbolic motion case is sometimes called constant acceleration in the literature, but it is NOT.

As $t \rightarrow 0$, just how close to zero can we get and stay classical here? Maybe the limit is the time, Δt, that it takes for light to travel the length of a virus—or is it only the time required to cross the 'ball' of 238 neutrons and protons that make up our accelerated charge?

There is a subtle difference between coming in and going out. Going out is straightforward. The rocket engine some how pushes the charge along $+x$, the radiation drag is along $-x$ and it is compensated for by the extra rocket push forward. This extra push by the rocket starts out nearly infinite, at $t > \approx 0$, and approaches Ag^2/c as $t \rightarrow \infty$, to maintain a net mg force here.

When coming in, speed is to the left along $-x$. The thrust is along $+x$ and the radiation drag effect is also along $+x$. Therefore, the rocket has a reduction from magnitude mg, rather than an extra push. Out where $v < \approx c$, the drag is Ag^2/c and thus the rocket force needed there is about $mg - Ag^2/c$.

As the charge moves closer, the drag increases and the rocket supplies even less force than $mg - Ag^2/c$. There comes a negative time, t_2, where the rocket engine has been throttled back to zero thrust. Thereafter, it must 'rotate' 180° and burn ever stronger while trying to push the rocket forward (to the left). The thrust magnitude is then $[|F_{drag}| - mg]$, so that the net force on the charge remains a constant retroforce, mg, to the right.

Again, a maximum thrust limit is eventually reached and this causes deviation from the hyperbolic path to begin. The speed then drops too much and $\Im_I$ and $\Im_{II}$ become unbalanced. Where does this path deviation really begin to happen?

The rocket thrust is a 'maximum' at some t_2, when the speed is v_2:

$$\frac{Ag^2}{|v_2|} - mg = F_{rmax} = \left(\frac{Ag^2}{c}\right)\sqrt{1 + c^2/g^2t_2^2} - mg$$

We have assumed $F_{net} = mg = 10^{-26}10^{19} = 10^{-7}N$, on the nucleus, and that this is not a problem for us. We readily see that a rocket strength on the order of mg or $10mg$ still leaves a v_2 value so close to zero as to be non-classical and thus meaningless. You can redo the numbers with lab sized charges, such as $Q = 10^{-6}C$, $m = 10^{-3}kg$, $g = 10^2 m/s^2$, and $F_{rmax} = 10mg$, for example; more like a small, charged, metal coin being pulled by strings. The strings can pull on both ends along x and produce the $x(t)$ motion by their difference in tension.

I'm not sure what all this proves, but it is interesting. The peculiarity that the first two pieces of the $\Im$ radiation reaction force must push in opposite directions is necessary in order to get $\Im^{\mu} \mathcal{V}_{\mu} \approx 0$, since neither term is zero. Thus, if one is a drag then the other one is an equal (our case) boost! How can radiation leaving produce a boost in speed? It doesn't, apparently because the drag term dominates. Maybe we do need the advanced and retarded radiation effects to really deal with this situation, as Rohrlich and others have argued.

This is academic except that quantum must relate to this and give the correct classical results here in a crude limit, somehow. There must be equally obscure issues in quantum, of some kind, that relate to these classical issues. Feynman agreed, when I brought this up in the late 60's, but he said he did not know the connection. I suspect the answer has a lot to do with time flowing both ways at the subnuclear level and we don't have any understanding of that at the present epoch in our evolution. Will we ever?

People have asked me why I am interested in such a strange classical problem. I don't think the classical problem is as interesting as the quantum problem, but I had hoped to see some insights here that might carry over into the quantum problem, maybe advanced and retarded potentials here might help clarify the quantum picture. We have not gotten, in our discussion here, to the advanced waves. I suspect that they are greatly complicated by the closed universe's boundary conditions. They probably have not been done correctly in the past, since the attempts that I know of assumed flat and static spacetime.

CHAPTER 12

Summary of Hypercomplex Wave Equations

Many years have passed since Dirac's late 1920's discovery of relativistic spin 1/2 physics. There was a flurry of activity, thereafter, with 'everything' being tried, as variations on his theme. None of that turned out to be useful and so it is not even mentioned in most modern field theory texts. It is now the mid-90', and I am claiming to have found an important generalization of Lorentz covariance and also a possibly important extension of mass to multiparts. This may be a rediscovery of old and proven useless ideas from the 1930's; I do not know. I do know this is the best way to see the elegant relationships and thus it is certainly the way to teach spinor field theory, at least. Often times, elegant notation leads to new insights. It is just possible that we have missed this covariance generalization and this mass generalization because of our old notation and our 'religious' faith in Einstein's insights.

The Klein-Gordon equation for relativistic, spin 0 particles was 'factored' in the late 20's by Dirac,[1] but he had to go to a much larger algebra than the well known Pauli algebra. The Pauli algebra had actually been reinvented in the 20's, by Pauli, for use in the spin 1/2 hydrogen atom spectrum splitting. Pauli's algebra was only the complex quaternion algebra of Hamilton, 1843, and this is the setting in which to see that Lorentz symmetry is a natural one only for a mass free universe. This setting also indicates that the really natural group is a larger group that contains Lorentz! The expanded algebra, as a hypercomplex number system, strongly suggests that an 8 parameter generalization should replace Lorentz for covariance!

There is a factorization of Klein-Gordon within the Pauli algebra[2] but it turns out to involve multi-part mass as we shall show. It combines these mass parts into the one mass term which appears as ordinary mass in the Klein-Gordon equation. This is either physical non-sense or it opens new and unanticipated insights into particle mass. This new, multipart mass can also expand the Dirac equation, as we shall show.

To see all this requires a powerful language[3] for Dirac, which equally applies to the Pauli sub-algebra. I realize the considerable inertia for older readers to learning a new formalism for things they already know. However, it is well worth the effort in this case so I ask your indulgence. We have the Pauli algebra in the standard basis $\{\sigma_\mu, (i\sigma_\mu)\}$, which is 'equivalent' to 2x2 complex matrices (8 slots for real numbers). We, therefore, must initially allow only real coefficients here. A general algebra element then takes the form $A = A_1{}^\mu(\sigma_\mu) +$

$A_2^{\mu}(i\sigma_{\mu})$, with A_1^{μ} and A_2^{μ} real, positive or negative numbers. The standard quantum ansatz, $P^{\mu} \rightarrow i\hbar\partial^{\mu}$, will necessarily be $\hbar\partial^{\mu}$ here, since only real coefficients are allowed so far.

This algebra has at least two useful, antiautomorphic conjugations, $(\ )^{\dagger}$ and $(\ )^{\wedge}$. By definition,

$$\{\sigma_0, \sigma_k, i\sigma_0, i\sigma_k\}^{\dagger} \equiv \{\sigma_0, \sigma_k, -i\sigma_0, -i\sigma_k\}$$

$$\{\sigma_0, \sigma_k, i\sigma_0, i\sigma_k\}^{\wedge} \equiv \{\sigma_0, -\sigma_k, i\sigma_0, -i\sigma_k\}$$

We can readily show that $(AB)^{\dagger} = B^{\dagger}A^{\dagger}$ and, in fact, this is the familiar matrix hermitian conjugation, so no further proof is needed. That $(AB)^{\wedge} = B^{\wedge}A^{\wedge}$ has to be shown in detail, for arbitrary elements A and B of the algebra. These conjugations immediately show us two basic groups contained in the Pauli algebra: $AA^{\wedge} = 1\sigma_0$ and $BB^{\dagger} = 1\sigma_0$. The elements near the identity, σ_0, have small pieces for which the conjugations change their signs. Thus we have: $A_{\epsilon} = 1\sigma_0 + \epsilon^k(i\sigma_k) + \delta^k(\sigma_k)$ and $B_{\epsilon} = 1\sigma_0 + \epsilon^k(i\sigma_k) + \delta(i\sigma_0)$. These groups are SL(2,C) and SU(2)⊗U(1), respectively. A 4-vector is $P = P^{\mu}(\sigma_{\mu}) = P^{\dagger} = P^{\mu *}(\sigma_{\mu})$, with 'length' $P^{\wedge}P = \cdots = (|P^0|^2 - |P^k||P^k|)\sigma_0$. (We shall soon assume that $P^{\mu} = \hbar\partial^{\mu}$ for quantum applications. So far we are allowing only real coefficients, so absolute values are not necessary.) A Lorentz transformation then takes the simple form

$$P \rightarrow P' \equiv L^{\dagger}PL = P'^{\dagger}, \quad P = P^{\dagger},$$

$$P^{\wedge}P \rightarrow P'^{\wedge}P' = (L^{\dagger}PL)^{\wedge}(L^{\dagger}PL) = \cdots = L^{\wedge}P^{\wedge}PL$$

Since $P^{\wedge}P \propto \sigma_0$, we get $P^{\wedge}PL^{\wedge}L$, and $P^{\wedge}P$ is invariant. We have used $LL^{\wedge} \equiv 1\sigma_0$, so L is any Lorentz group element. Here, $L^{\wedge} = L^{-1}$ and $LL^{-1} = L^{-1}L$, so $L^{\wedge}L = 1 = LL^{\wedge}$, from matrices. Therefore, $P^{\wedge}P$ is obviously invariant. This is all standard spinor Lorentz structure in, perhaps, an unusual notation (but an elegant one).

Now we can propose[2,3] and examine the possibly new spin 1/2 equation:

$$P\psi = \psi^{\dagger\wedge}Mc$$

for the Pauli algebra. This is form covariant for Lorentz as we shall now prove:

$$P'\psi' = (L^{\dagger}PL)(L^{\wedge}\psi) = (L^{\wedge}\psi)^{\dagger\wedge}Mc$$

$$= L^{\dagger}(P\psi) = L^{\dagger}\psi^{\dagger\wedge}Mc$$

Multiply on the left by $L^{\wedge\dagger}$ and obtain $P\psi = \psi^{\dagger\wedge}Mc$. Notice that $\psi' \equiv L^{\wedge}\psi$ is 'necessary' here. We say ψ is a spinor. However, the $\psi^{\dagger\wedge}$ means that ψ is not an ordinary spinor in this new equation. Maxwell's equation (spin 1) is here: $PF = 0$, with $F = -F^{\wedge} = -E^k(\sigma_k) + cB^k(i\sigma_k)$, so both spins are in Pauli algebra.

The Klein-Gordon equation then follows from left operation by $P^{\wedge}$ on both sides, and from using $P = P^{\dagger}$:

$$P^{\wedge}(P\psi) = (P^{\wedge}P)\psi = P^{\wedge}(\psi^{\wedge\dagger})Mc$$

$$= P^{\dagger\wedge}\psi^{\wedge\dagger}Mc = (P\psi)^{\dagger\wedge}Mc = (\psi^{\dagger\wedge}Mc)^{\dagger\wedge}Mc$$

$$= \psi(Mc)^{\dagger\wedge}Mc = \psi(M^{\dagger\wedge}M)c^2$$

We must have $M^{\dagger\wedge}M$ commute with ψ here for a Klein-Gordon equation. Assume that $M = M^{\mu}(\sigma_{\mu})$. Then $M^{\dagger\wedge}M = \cdots = ((M^0)^2 - M^kM^k)\sigma_0$, and this (assumed) invariant, $M^{\dagger\wedge}M$, then commutes with ψ. We can alternatively use $M = M^{\mu}(i\sigma_{\mu})$ but we still get $M^{\wedge\dagger}M = ((M^0)^2 - M^kM^k)\sigma_0$; so M seems to be restricted to four mass parts if it is to commute, such that we do get the usual Klein-Gordon equation. This leads to

$$P^{\wedge}P\psi = M^{\dagger\wedge}Mc^2\psi = (P^0P^{0*} - P^kP^{k*})\psi$$

For $P^{\mu} \rightarrow \hbar\partial^{\mu}$, we can avoid for now the usual $i^* = \pm i$ problem for hermitian spacetime operators. This Klein-Gordon equation becomes simply

$$(\hbar\partial^{\mu}\hbar\partial_{\mu})\psi = (M^{\dagger\wedge}Mc^2)\psi$$

Now we really do need $i\hbar\partial^{\mu}$, to proceed to the Schrödinger non-relativistic limit in the usually way, so we multiply through by ii and obtain

$$(i\hbar\partial^{\mu}i\hbar\partial_{\mu})\psi = (-M^{\dagger\wedge}Mc^2)\psi$$

This is the standard Klein-Gordon equation only if $(-M^{\dagger\wedge}Mc^2) = (m_{exp}c)^2 > 0$. Our Pauli-restricted Klein-Gordon variation will then give the standard

Schrödinger equation. Notice that $\partial^\mu\partial_\mu = (+ - - -)$ here. The + for ∂_0 is important in obtaining $-M^{\dagger\wedge}Mc^2 > 0$. Therefore, we must have $M^{\dagger\wedge}M = (M^0)^2 - M^kM^k < 0$ and at least one of the strange M^k mass parts has to dominate the M^0 part!! We cannot just have the usual one parameter mass, $m(\sigma_0)$, as a special case here, so this is not the Dirac equation in disguise![2] It is a new kind of spin 1/2 thing, if such quanta really do exist in the real world. They probably don't exist because their rest state, $\partial^k\psi_R \equiv 0$, has no $exp(iEt/\hbar)$ term, because we have no imaginary coefficients here. The Pauli algebra contains massless photons alright, but only peculiar massive quanta.

We have three separate equation types possible here for $M = mc(\sigma_1)$ *or* $mc(\sigma_2)$ *or* $mc(\sigma_3)$. Are these all the same, and somehow related to the usual Dirac? They all give the same Klein-Gordon equation that Dirac gives. Maybe there are four distinct kinds of mass, m^μ, at the subproton level. Is the existence of a 'sensible' rest state essential for all the kinds of quanta with a mass parameter in their equation?

It seems, at first, that the original Pauli algebra, with real coefficients, is large enough to handle Dirac particles completely. If true, then the $m(\sigma_1)$ or $m(\sigma_2)$ or $m(\sigma_3)$ choices would all give the same wave function solution to, say, the hydrogen atom. It would be strange indeed to see such a quantum theory with only real coefficients. However, the (σ_0), $(i\sigma_0)$ elements commute with everything else and could be 'all over the place'. Then, $(i\sigma_0)$ could try to play the role of i in the usual quantum formulation. Because ψ can, at most, have 8 real functions in real Pauli, this leads us to wonder if ψ-Dirac might be in here. If Dirac is really here, then Lorentz symmetry is all we need; we must then ask what the larger spin 1/2 algebras tell us about nature, if anything! If Dirac is not in Pauli, then a larger algebra strongly suggests, to me at least, that Lorentz is only part of the covariance group.

In ordinary quantum, we have rest states with $(cos\alpha - isin\alpha)$, where $\alpha = [(\pm mc^2)t/\hbar]$. Then ∂^0 operates on this and pulls out $d\alpha/dt$. There is a very subtle reason why $(i\sigma_0)$ cannot successfully play the role of i in this real Pauli algebra. The $(\ )^{\dagger\wedge}$ conjugation on outside coefficients gives $i^{\dagger\wedge} = -i$ but $(i\sigma_0)^{\dagger\wedge} = -(i\sigma_0)$. This is a very important distinction! For years I puzzled over the issue of $(i)^\wedge \equiv \pm i(?)$. I am sure now that $[i(\sigma_\mu)]^\wedge = -i(\sigma_\mu)^\wedge$ is the way to do physics. If we try $(cos\alpha - (i\sigma_0)sin\alpha)$ in ψ, then the $\psi^{\dagger\wedge}$ conjugation does not give back $(cos\alpha - (i\sigma_0)sin\alpha)$. This term cannot be canceled off, after the ∂^0 operates on it. Without this cancellation, we don't get a simple rest state solution for ψ of the usual kind. Therefore, it appears that we really do need the complex coefficient extension of Pauli for quantum physics with rest mass existing. A larger covariance group is then likely to replace Lorentz.

Lorentz comes from Pauli; not from Maxwell and not from relativity of motion. Had physicists stayed with quaternions for (relativistic) physics, after the turn of the century (as Hamilton thought they should for all physics in general), we would have found Maxwell in quaternions, Lorentz in quaternions, and then Dirac in quaternions. I believe our progress would have been accelerated, both then and even now. We here see new mass parts 'emerge' from quaternion thinking, so perhaps Hamilton's ghost can rest at last.

There is another, similar equation to be explored:

$$P \equiv -P^{\dagger} = P^{\mu}(i\sigma_{\mu}) = \hbar\partial^{\mu}(i\sigma_{\mu})$$

$$P\psi = \psi^{\dagger\wedge}Mc, \; P^{\wedge}P = -(P^0P^{0*} - P^kP^{k*})\sigma_0$$

The metric here is $(- + + +)$. The Lorentz transformation is now $P \to P' = L^{\dagger}PL$, and $P'^{\dagger} = L^{\dagger}P^{\dagger}L = -P'^{\dagger}$ if $P = -P^{\dagger}$. The rest of the analysis is as before, and we now reach the Klein-Gordon equation

$$P^{\wedge}P\psi = \cdots = -\psi(M^{\dagger\wedge}Mc^2) = -(P^0P^{0*} - P^kP^{k*})\psi$$

where $P^{\wedge} = (-P^{\dagger})^{\wedge}$ is used when needed. We get the same Klein-Gordon situation as before, where we also had $P^{\mu} = \hbar\partial^{\mu}$. Thus M here again must be $M^{\mu}\sigma_{\mu}$ *or* $M^{\mu}(i\sigma_{\mu})$, with the M^k dominating. There may be particles for each of these two types of M mass, perhaps going with the two operators. They would all be strange indeed. It is obviously possible here, for example, to have $M^0 = 0$ *and* $M^1 \neq 0$, but $M^2 = M^3 = 0$. We just may not understand mass at all yet. Up until now, we have always assumed only one mass number, and we just put it in by hand. We then looked to classical experiment to find its ultimate numerical value. The curvature of tracks in a bubble chamber can only give $\sqrt{(M^{\dagger\wedge}M)}$, I suspect.

The invariant Lagrangian for this $P\psi = \psi^{\dagger\wedge}Mc$ equation should be something like $\psi^{\dagger}P\psi - (\psi^{\wedge}\psi)^{\dagger}Mc$, which is definitely a Lorentz invariant. (I have not found Lagrangians useful in this exploration.) The physical 4-vector current associated with ψ, as a source for photons, F, is possibly $(\psi\psi^{\dagger})^{\wedge} \equiv J$, because this simple $J \to L^{\dagger}JL \equiv J'$ under Lorentz, as needed. However, a conserved current also has $P^{\wedge}J + (P^{\wedge}J)^{\wedge} = 0$, in this language, so this J current exploration needs further development. I remain skeptical regarding the traditional approach to this current, but perhaps this is where the Lagrangian can really be of some value.

The obvious, minimal electromagnetic coupling, to give the influence of photons on ψ, is

$$P \rightarrow P - \text{`}e\text{'}A,\ A = A^{\dagger} = A^{\mu}\sigma_{\mu} \rightarrow A' = L^{\dagger}AL$$

Since this ψ equation, with Mc and with EM coupling, may be another factorization of the Klein-Gordon equation besides Dirac's, it may have been found before, over the past 60 years. It is fairly simple looking. It is new to the author, but it should have been found in the early 30's, once Dirac's algebra work was known. If not found previously, then the lack of usage of a hypercomplex number outlook may have caused it to be missed. Matrices can obscure the physics, I believe, and we have no proof that they are the language of the gods. We have just assumed that they are.

$P\psi = \psi^{\dagger\wedge}Mc$ should have rest state solutions if it represents a particle with rest mass M, even if M has multiparts. Because we don't have any imaginary coefficients yet, our rest states here turn out to blow up or die as $t \rightarrow \infty$. They are thus not likely to exist in our universe. (Maybe they are the dark stuff of the galaxies and it is evaporating into nothing with time.)

For zero rest mass, $P\psi = 0$ can physically exist in Pauli algebra, both for spin 1/2 and for spin 1. (Free photons have $PF = 0$, where $F = -E^k(\sigma_k) + cB^k(i\sigma_k)$. They cannot have a Mc mass because the right side L, in the $L^{\wedge}FL$ transformation, causes problems.) Nature does really use this 8-basis algebra, at least for massless quanta. (There could also be massless, spin 0 quanta here, $P^{\wedge}P\phi = 0$, where $\phi = \phi_1(\sigma_0) + \phi_2(i\sigma_0)$. Notice that $\phi = +\phi^{\wedge} \rightarrow \phi' = L^{\wedge}\phi L = \phi L^{\wedge}L = \phi$.) The F field also transforms as $F \rightarrow F' = L^{\wedge}FL$ but $F'^{\wedge} = -F'$ if $F^{\wedge} = -F$, which is apparently true for photons. I don't see any higher spins in this algebra! There is another spin 1, massless free field here, besides F. Its free particle equation is $P[(P^{\wedge}A) - (P^{\wedge}A)^{\wedge}] = 0$, where $A = A^{\mu}(\sigma_{\mu}) = (V/c)\sigma_0 + A^k(\sigma_k) = A^{\dagger}$. Most physicists presently believe this is NOT an independent quantum field, but rather an equivalent description of the F photons; I see structural reasons to disagree. So we see that nature would make very good use of this 8-basis algebra and only utilize Lorentz symmetry, if there were only photons and gravitons in our universe, along with other massless Bosons, perhaps.

There is a complexified, larger algebra, $\{(\sigma_{\mu}),\ (i\sigma_{\mu}),\ i(\sigma_{\mu}),\ i(i\sigma_{\mu})\}$, with 16 basis elements. It lies between Pauli and Dirac! The $i(\cdots)$ and the $(i\cdots)$ are independent here, by definition! We could use $i({}_i\sigma_{\mu})$ instead, to stress this independence. The expanded multiplication table is obvious for this larger algebra and the system is clearly closed under multiplication. However, the two conjugations, $(\ )^{\dagger}$ and $(\ )^{\wedge}$, now need to be generalized. Although still not obvious for $(\ )^{\wedge}$, defining $[i(\sigma_{\mu})]^{\wedge} = i^{\wedge}(\sigma_{\mu})^{\wedge} \equiv -i(\sigma_{\mu})^{\wedge}$ is the best way to proceed. Obviously, $[i(\sigma_{\mu})]^{\dagger} = i^{\dagger}(\sigma_{\mu})^{\dagger} = -i(\sigma_{\mu})$ is natural, since $\dagger$ is hermitian conjugation for matrices, but we don't have normal matrices here any longer. A general algebra element now is $A = A_1{}^{\mu}(\sigma_{\mu}) + A_2{}^{\mu}i(\sigma_{\mu}) + A_3{}^{\mu}(i\sigma_{\mu}) + A_4{}^{\mu}i(i\sigma_{\mu})$. (For matrices, $i(i\sigma_{\mu}) = -(\sigma_{\mu})$.) We find that $AA^{\wedge} \equiv 1\sigma_0$ gives a natural group with 'small' members $A_{\epsilon} = 1\sigma_0 + \epsilon^k(i\sigma_k) + \delta^k(\sigma_k) + \delta i(\sigma_0) + \epsilon i(i\sigma_0)$. This is THE natural replacement for the $LL^{\wedge} = 1\sigma_0$ Lorentz group, which, I believe, came from the original Pauli algebra, not really from Maxwell's equation, as Lorentz thought. (No moving observers, or what they might see, are involved here and they will be left completely out of the discussion!) The other natural group, $BB^{\dagger} = 1\sigma_0$, has $B_{\epsilon} = 1\sigma_0 + \epsilon^k(i\sigma_k) + \delta^0(i\sigma_0) + \delta^{\mu}i(\sigma_{\mu})$, now also with 8 parameters—so they are possibly isomorphic. (We would need to compare the commutators of the generators, to be sure, and this needs further development.)

We still 'expect' $P \equiv P^{\dagger} \rightarrow L^{\dagger}PL$ as before. However, now $P = P^{\dagger} = P^{\mu}(\sigma_{\mu}) + Q^{\mu}i(i\sigma_{\mu})$, and the assumed restriction $P^{\wedge}P \propto (\sigma_0)$ is still expected to hold for a Klein-Gordon equation to exist. This then restricts P to less than the 8 parts shown above. It can be readily checked that $P_R = P^{\mu}(\sigma_{\mu})$ or $P_I = P^{\mu}i(i\sigma_{\mu})$ each give $P^{\wedge}P \propto (\sigma_0)$, with $(+\ -\ -\ -\ +)$ and $(-\ +\ +\ +)$ metrics, respectively. See Table 1. We still have the alternative physical possibility $P = -P^{\dagger} = P^{\mu}(i\sigma_{\mu})$, as before. Now we can also have $P = -P^{\dagger} = P^{\mu}i(\sigma_{\mu})$, as well. This extension is a much richer algebra than Pauli's. It is not currently known whether all these P variations harbor significant physical content or not. We shall mostly focus here on $P_R = P^{\mu}(\sigma_{\mu})$. The others may be redundant, but that needs to be examined in detail. (Traditional Dirac seems to go most naturally with $P^{\mu}i(i\sigma_{\mu})$ and its further expansions to come later.)

There is still no extra room in P for any other mass parameters, such as a fifth piece. (There will be later.) The basic quantum equation must still be $P\psi = \psi^{\dagger\wedge}Mc$, and it is now covariant under a larger group than Lorentz alone! Before, ψ was a 'subset' of $\psi = \psi_R{}^{\mu}(\sigma_{\mu}) + \psi_I{}^{\mu}(i\sigma_{\mu})$, in the smaller, Pauli algebra with real coefficients. We now have complex coefficients, so $P = P^{\mu}(\sigma_{\mu})$ and $P = P^{\mu}i(i\sigma_{\mu})$ may give qualitatively different ψ structures that go with each of them. Also, M may possibly have more internal parts now! We still need $M^{\dagger\wedge}M \propto (\sigma_0)$. (The maximum number of allowed M parts, for this larger algebra, also needs to be further examined.)

Since P has the same $L^{\dagger}PL$ and $P^{\wedge}P$ patterns as before, $P\psi = \psi^{\dagger\wedge}Mc$ will still lead to $P^{\wedge}P\psi = \psi(M^{\dagger\wedge}M)c^2$. We can again postulate that $P \equiv P^{\dagger} \equiv P^{\mu *}(\sigma_{\mu}) = P^{\mu}(\sigma_{\mu})$, and this requires $P^{\mu} = \hbar\partial^{\mu}$. Thus the Klein-Gordon equation, with its $(i\hbar\partial^{\mu})(i\hbar\partial_{\mu})$, would still require $M \equiv M^{\mu}(\sigma_{\mu})$ to have dominant

M^k parts, if M^μ is a set of real numbers. This contains the equation that we already found in the Pauli algebra, and now it contains some new factorizations, depending on P and M. It now can have ordinary rest states, because M has new possibilities here. For example, we now can have $M = M^\mu i(\sigma_\mu)$, $M^{\dagger\wedge}M = -(M^0M^0 - M^kM^k)$, because $i^{\dagger\wedge}(\cdots) = +i(\cdots)$. This begins to resemble Dirac's equation! The M^0 term could now dominate or be even the only term in M for this case. Also, ψ can possibly have more internal parts, depending on the details for P and M. We may also be seeing hints here that Dirac did not use the smallest algebra that he could have used! The most promising development here is the emergence of a new and larger group to replace Lorentz, unless M takes on some new mass pieces in such a way that it is somehow not form invariant, for this larger group. The M invariance is a guess, remember.

If we instead start, in the extended Pauli algebra, with a coupled pair of covariant equations, $P^\dagger\psi_a = \psi_v m_a c$ and $P^\wedge\psi_v = \psi_a m_v c$ and still have $P = P^\dagger$, then $P\psi_a = \psi_v m_a c$ and $P\psi_v{}^{\dagger\wedge} = \psi_a{}^{\dagger\wedge}m_v{}^{\dagger\wedge}c$. By adding, we get $P(\psi_a + \psi_v{}^{\dagger\wedge}) = \psi_v m_a c + \psi_a{}^{\dagger\wedge}m_v{}^{\dagger\wedge}c = (\psi_a m_v c + \psi_v{}^{\dagger\wedge}(m_a c)^{\dagger\wedge})^{\dagger\wedge}$. Thus, if $m_v = m_a{}^{\dagger\wedge}$ then this becomes $(\psi_a + \psi_v{}^{\dagger\wedge})^{\dagger\wedge}m_a c$. So it becomes $P\psi = \psi^{\dagger\wedge}m_a c$. The above coupled pair of equations can easily be combined in one 4×4 matrix form, and thus be seen to be Dirac's usual equation, if $m_a c$ is to the left of ψ. Thus, Dirac solutions are solutions for this new equation if $m_v{}^{\dagger\wedge} = m_a$. For the 'standard' Dirac form, we would actually need $Q = i\hbar\partial^\mu(i\sigma_\mu) = Q^\dagger$ here, instead of P. So, $Q\psi = \psi^{\dagger\wedge}Mc$ contains all Dirac solutions and possibly more, when complex coefficients are allowed. We have Lorentz form covariance and possibly more covariance here also. $M_v{}^{\dagger\wedge} = M_a$ somewhat restricts M, and the K-G requirement leads to $P^\wedge P\psi_a = \cdots = \psi_a m_v m_a c^2$, which then requires that $m_v m_a = \pm(m_{exp})^2$, depending on the metric of $P^\wedge P$, $(+ - - -)$ or $(- + + +)$. For example, $m_a = -(i\sigma_0)m$ and $m_v = +(i\sigma_0)m$ give Dirac's equation. Historically, Dirac jumped over this algebra and instead developed a much larger one, $\{(\sigma_\mu), i(\sigma_\mu), (i\sigma_\mu), i(i\sigma_\mu), (f_\mu), i(f_\mu), (if_\mu), i(if_\mu)\}$, with 32 basis elements. Much new structure is then possible here and with the corresponding group generalizations. We now 'replace' σ by e, although the multiplication table is still the same for these e's as it was for the σ's. We next need the multiplication table for the new f elements, both with themselves and with the old e's. We could do this by displaying a 4×4 matrix representation for the e's and f's, obtained from the coupled pair of equations above. But we have not had to do that so far, so why give in now? This number system extension does not need the pair of equations of Dirac to be motivated. The Dirac algebra can be reached from a direct product of two closed subsystems within the Pauli algebra, as shown in my earlier papers. From this perspective, however, we do lack a good justification for doubling the size of the algebra twice to do physics. The first doubling was to allow rest states. The following rules do work for the real world and can be given a 4×4 matrix representation if one insists on it:

$$e_\mu f_\nu \to \sigma_\mu^{\wedge\dagger}\sigma_\nu = \sigma_\lambda \to f_\lambda$$

$$f_\nu e_\mu \to \sigma_\nu \sigma_\mu = \sigma_\eta \to f_\eta$$

$$e_\mu e_\nu \to \sigma_\mu \sigma_\nu = \sigma_\lambda \to e_\lambda \text{ as before}$$

$$f_\mu f_\nu \to \sigma_\mu^{\wedge\dagger}\sigma_\nu = \sigma_\lambda \to e_\lambda$$

$$(ie_\mu)(if_\nu) \to (i\sigma_\mu)^{\dagger\wedge}(i\sigma_\nu) = \cdots \to f_\eta$$

$$(if_\nu)(e_\mu) \to (i\sigma_\nu)(\sigma_\mu) = \cdots \to f_\lambda$$

Notice that an f on the right requires special treatment. This pattern works well and gives the same answers as the representation

$$(e_\mu) = \begin{pmatrix} \sigma_\mu & 0 \\ 0 & \sigma_\mu^\wedge \end{pmatrix}, \quad (ie_\mu) = \begin{pmatrix} i\sigma_\mu & 0 \\ 0 & -(i\sigma_\mu)^\wedge \end{pmatrix}$$

$$(f_\mu) = \begin{pmatrix} 0 & \sigma_\mu^\wedge \\ \sigma_\mu & 0 \end{pmatrix}, \quad (if_\mu) = \begin{pmatrix} 0 & -(i\sigma_\mu)^\wedge \\ i\sigma_\mu & 0 \end{pmatrix}$$

There is now a new, closed sub-algebra, besides the 16 element $\{(e_\mu), (ie_\mu), i(e_\mu), i(ie_\mu)\}$, which we just finished exploring. We now have $\{(e_\mu), (ie_\mu), (f_\mu), (if_\mu)\}$, with 16 elements, which is also closed. This new sub-algebra leads directly to the groups $AA^\wedge = 1e_0$ and $BB^\dagger = 1e_0$, but we must first define $(\ )^\dagger$ and $(\ )^\wedge$ for this 16 element algebra, with the many f's in it. We still have matrices, so $(\ )^\dagger$ is, by definition, still hermitian conjugation and thus antiautomorphic. We can use it here to now define:

$$A^\wedge \equiv (f_0)A^\dagger(f_0) \Rightarrow A(f_0) = (f_0)A^{\dagger\wedge}$$

$$A^{\vee} \equiv (if_0)A^{\uparrow}(if_0) \Rightarrow A(if_0) = (if_0)A^{\uparrow\vee}$$

$$A^{\downarrow} \equiv (ie_0)A^{\uparrow}(-ie_0) \Rightarrow A(ie_0) = (ie_0)A^{\uparrow\downarrow}$$

We can see that $A = A^{\uparrow\downarrow}$ means no f's in A. Also, $A^{\uparrow\downarrow} = A^{\wedge\vee}$.

For memorizing, use the following:

$$(`e + f')^{\uparrow} = e^{\uparrow} + f^{\wedge} \to \sigma^{\uparrow} + \sigma^{\wedge}$$

$$(`e + f')^{\wedge} = e^{\wedge} + f^{\uparrow} \to \sigma^{\wedge} + \sigma^{\uparrow}$$

$$(`e + f')^{\downarrow} = e^{\uparrow} - f^{\wedge} \to \sigma^{\uparrow} - \sigma^{\wedge}$$

$$(`e + f')^{\vee} = e^{\wedge} - f^{\uparrow} \to \sigma^{\wedge} - \sigma^{\uparrow}$$

This is a powerful memory aid, so we don't have to refer to tables for the 4×16 $\pm$ signs. For example, $(ie_3)^{\wedge} \to (i\sigma_3)^{\wedge} = -(i\sigma_3) \to -(ie_3)$ *and* $(ie_3)^{\uparrow} \to (i\sigma_3)^{\uparrow} = -(i\sigma_3) \to -(ie_3)$. Also, $(if_3)^{\wedge} \to (i\sigma_3)^{\uparrow} \to -(if_3)$ *and* $(if_3)^{\uparrow} \to (i\sigma_3)^{\wedge} = -i\sigma_3 \to -(if_3)$. But $(if_3)^{\vee} \to -(i\sigma_3)^{\uparrow} = (i\sigma_3) \to (if_3)$. From their definitions, it is now easy to prove that $(AB)^{\wedge} = B^{\wedge}A^{\wedge}$, $(AB)^{\vee} = B^{\vee}A^{\vee}$, *and* $(AB)^{\downarrow} = B^{\downarrow}A^{\downarrow}$, $(AB)^{\wedge\uparrow} = (AB)^{\vee\downarrow}$, $(AB)^{\vee\wedge} = (AB)^{\uparrow\downarrow}$, $(AB)^{\uparrow\wedge\vee} = (AB)^{\downarrow}$. All conjugation orders are equivalent, *e.g.*, $(\ )^{\uparrow\wedge\vee} = (\ )^{\wedge\uparrow\vee}$. There obviously are other conjugations, such as $(e_1)A^{\uparrow}(e_1)$, *etc.*, but they seem to give the same physics in an equivalent, but not so pretty form. They may still be useful for parity change symmetry, and such.

Now we can examine these four natural groups: $AA^{\wedge} = 1(e_0)$, $BB^{\uparrow} = 1(e_0)$, $CC^{\vee} = 1(e_0)$, *and* $DD^{\downarrow} = 1(e_0)$. The 'infinitesimal' members will be things like (the conjugations change signs):

$$A_{\epsilon} = 1\sigma_0 + \epsilon^k(ie_k) + \delta^k(e_k) + \delta i(ie_0) + \epsilon i(e_0)$$
$$+ \eta^{\mu}(if_{\mu}) + \xi^{\mu} i(f_{\mu})$$

This group has 16 parameters and is now the 'natural' replacement for its 8 parameter subgroup, which had only the 'e' type generator terms in A_{ϵ}, for the 16 element algebra. We find also, for group B,

$$B_{\epsilon} = \sigma_0 + \epsilon^k(ie_k) + \epsilon i(e_0) + \delta^k i(e_k) + \delta(ie_0)$$

$$+ \eta^k(if_k) + \xi^k(f_k) + \eta i(f_0) + \xi i(if_0)$$

From the definitions, $(aA)^\wedge = A^\wedge a^\wedge$; if $a = i$, then we 'need' $i^\wedge \equiv -i$ when outside of the basis elements. Similarly for all four conjugations. For example, $[i(e_k)]^\wedge = -i(e_k)^\wedge = +i(e_k)$ *but* $[i(ie_0)]^\wedge = -i(+ie_0) = -i(ie_0)$. A subtle distinction! You can then check that A_ϵ has 16 parameters and B_ϵ also has 16 parameters, as shown. These two conjugations change 16 signs, so 16 parameters.

The other two groups also have 16 parameters as well; probably all four are isomorphic. The *f*'s, when added to the algebra, have expanded the 'basic' group from 8 to 16 parameters, to now replace the 6 parameter Lorentz group!? We now have the 4-vector generalization $P \equiv P^\dagger$, with up to 16 parts, and the Lorentz generalization $LL^\wedge \equiv 1$, with 16 parameters, at most. The sub-algebra $\{(e_\mu), (ie_\mu), (f_\mu), (if_\mu)\}$ with only real coefficients, has $LL^\wedge \equiv 1$ which gives an L with only 10 parameters, called Sp(4?). It is obviously the new master group for this new sub-algebra, so it might play some role in quantum physics. The basic wave equation $P\psi = \psi^{\dagger\wedge} Mc$ will be 'form invariant' under this 10-L, in this sub-algebra. Also in this sub-algebra, $P = P^\dagger$ will now give $P = P^\mu(e_\mu) + m_1c(f_0) + m_2c(if_0)$, a 6-vector at most! We are back to real coefficients here and useful sub-algebras have to come from the larger ones of value, so this is not a very important sub-algebra. We do finally pick up new terms here in P itself, besides the P^μ space-time part, for the first time! There are even more mass parameter possibilities now! But still assuming that $PP^\wedge \propto e_0$, for Klein-Gordon, shows that we need $|m_1c|i(f_0)$ instead of $|m_1c|(f_0)$ in P. However, this imaginary coefficient replacement would take us out of the smaller sub-algebra. Within this real '$e+f$' sub-algebra, we can still have a five component $P = P^\mu(e_\mu) + mc(if_0)$ *and* $P^\wedge P = |P^0|^2 - |P^k||P^k| - |mc|^2$, with $P = P^\dagger$. Then $P\psi = \psi^{\dagger\wedge} Mc$ clearly has two kinds of mass for the first time! Now, $M = 0$ can still leave some rest mass in the equation, since the fifth part of P itself is a mass part. However, $P = P^\dagger$, $P^\mu = P^{\mu *} = \hbar\partial^\mu$ means that $[(i\hbar\partial^\mu i\hbar\partial_\mu) + |mc|^2 + M^{\dagger\wedge}M]\psi = 0$ and, therefore, $(m_{exp}c)^2 = -|mc|^2 - M^{\dagger\wedge}M$. Thus the fifth P component, new mc term here is NOT ordinary mass! It is yet another, new and weird mass piece, with 'tachyon' suggestions. (Kuni Imaeda suggested instead an unstable particle interpretation.) We can get m_{exp} to be real here, for Klein-Gordon usage, only if $(M^0)^2 - M^kM^k + |mc|^2 < 0$. The M^k parts must still dominate and now be even bigger here, if $m \neq 0$, than in the Pauli, 8-basis sub-algebra, $\{(e_\mu), (ie_\mu)\}$. We still do not have Dirac's equation! For $M = M^\mu i(e_\mu)$, we are back in the full Dirac algebra, but we now get $-M^0M^0 + M^kM^k$. M^0 can be ordinary mass, and it can even dominate the other mass parts. This could be Dirac's equation, except there are two mass terms, M and the new m that is the fifth component of P. I traced this through very carefully for $Q\psi = \cdots$. I found that the mc from the coupled pair becomes the fifth component of Q in the Dirac algebra, not the M^0 component of $M^{\mu}i(e_{\mu})$. Thus, $M \neq 0$ really is a new guess at new physics in the full Dirac algebra. In this larger algebra, it may be that the M^k parts are only small compared to M^0, and both are zero for some quanta.

Finally, we examine the full Dirac system in some detail. $P \equiv P^{\uparrow} = P^{\mu}(e_{\mu}) + Q^{\mu}i(ie_{\mu}) + R^k i(f_k) + S^k i(if_k) + m_1c(f_0) + m_2c(if_0)$, has 16 elements, at most! Now $P^{\wedge}P \propto (e_0)$, or $\{(e_0), i(e_0)\}$, so it commutes with everything in the Dirac algebra, and this limits P (from 16 elements) down to only a 5-vector P (or maybe even a 3-vector P is possible—this presently lacks physical interpretation but seems to be in the table). See Table 1 which displays all the e_0 proportional $P^{\wedge}P$ products. There are clearly several 5-vector forms possible for P. Again most of these may just be redundant for physical applications.

There is also an interesting $P = P^{\downarrow}$ 5-vector, $P = P^{\mu}e_{\mu} + m_1ci(f_0)$. Here, $P^{\wedge}P = |P^0||P^0| - |P^k||P^k| + |m_1c|^2$. The $P\psi = \psi^{\uparrow\wedge}Mc$ and $P^{\wedge}P\psi$ calculations yield, for this $P = P^{\downarrow}$, $(+ - - - +)$ metric,

$$[|P^{\mu}||P_{\mu}| + |m_1c|^2 - M^{\uparrow\wedge}M]\psi = 0,\ P^{\mu} = \hbar\partial^{\mu},$$

$$[(i\hbar\partial^{\mu})(i\hbar\partial_{\mu}) - |m_1c|^2 + M^{\uparrow\wedge}M]\psi = 0$$

We here have a fifth mass part in P with the right sign, m_1, for a simple, single P mass term to dominate. Now, $(m_{exp}c)^2 = |m_1c|^2 - M^{\uparrow\wedge}M$, so we can then eliminate M, but only if nature does! This $P\psi \equiv 0$, $P \equiv P^{\downarrow}$, $P \rightarrow P' \equiv L^{\downarrow}PL$, $P^{\wedge}P \propto e_0$, and $LL^{\wedge} \equiv 1$ package is the usual Dirac equation for sure. It is used in QED with $M = 0$, and there it gives very successful, high accuracy predictions for electrons, so maybe $M \equiv 0$ is exact for electrons. In the $\{(e), (ie)\}$ sub-algebra, we have $(\ )^{\uparrow} = (\ )^{\downarrow}$ and $(\ )^{\wedge} = (\ )^{\vee}$. It is interesting to see that nature might use $P = P^{\downarrow}$ instead of $P = P^{\uparrow}$ for electrons, in the Dirac equation. However, look closely at $P = P^{\mu}i(ie_{\mu}) - mc(if_0) = P^{\uparrow}$. This is 'close' to the traditional P choice, $P^{\mu}i\gamma_{\mu}$. (Multiply through by (if_0) from the left to get $-mc(e_0)$ and find γ_{μ}.) This P actually gives the same Dirac results, so $P = P^{\uparrow}$ likely suffices for all the physics. I suspect that $P = \pm P^{conj.}$ would also work, where 'conj.' is any of the four conjugations.

We have seen several other wave equations emerge here, as well as a possible multi-part M type mass extension to Dirac's original equation. Lorentz symmetry is, we have seen, naturally generalized also. Does nature use any of these?! Is L, with 16 parameters, the basic 'covariance' group of wave equations, or is it only the 8 parameter sub-group? Surely, the Lorentz group is not the natural group beyond the $\{(e_\mu), (ie_\mu)\}$ sub-sub-algebra of Pauli, so forget that, I think. Mass existing in nature may 'require' that Lorentz be replaced by a larger group, but how large?

We find $L^{\downarrow}(f_0)L = (f_0)(L^{\downarrow})^{\uparrow\wedge}L = f_0(L^{\uparrow\downarrow})^{\wedge}L = f_0(L^{\wedge\vee})^{\wedge}L = f_0(L^{\vee}L)$, using the special properties of (f_0). This fifth mass piece of P is not invariant, for Dirac $P = P^{\downarrow}$, unless $L^{\wedge} = L^{\vee}$, which is again the 8 parameter subgroup. There is no obvious further restriction, in this language, that reduces this to the 6 parameter, Lorentz sub-sub-group. Also, $L^{\uparrow}(if_0)L = (if_0)(L^{\uparrow})^{\uparrow\vee}L = (if_0)L^{\vee}L$. Therefore, the (if_0) fifth part in $P = P^{\uparrow}$ is not invariant unless $L^{\wedge} = L^{\vee}$, which again restricts us to the 8 parameter sub-group! But how do we know that this mc mass term has to be invariant? In any case, we don't get the Lorentz sub-sub-group here either. We should call P something else when it has 5 parts, perhaps P_5. Also, since $i(ie_\mu)$ is more like the traditional Dirac field theory seen in text books, we should switch the names of P and Q.

The two basic equations are (switching the P and Q names)

$$P_5\psi \equiv [i\hbar\partial^\mu(ie_\mu) + (-mc)(if_0)]\psi = \psi^{\uparrow\wedge}M^\mu(e_\mu)c$$

and

$$Q_5\psi \equiv [\hbar\partial^\mu(e_\mu) + (+imc)(if_0)]\psi = \psi^{\uparrow\wedge}M^\mu i(e_\mu)c$$

Here $P_5 = P_5^{\uparrow}$ and $P^{\wedge}P\psi$ gives the Klein-Gordon equation as the table shows, with m and M^μ parts merging into one $(m_{exp})^2$ term. Both equations are Dirac's equation when $M \equiv 0$. The $Q_5\psi$ needs the imc, fifth Q_5 component to be Dirac's equation. If $mc \rightarrow imc$ in both of these equations, we get new physics possibilities and new demands on M for a suitable Klein-Gordon equation. Notice that $Q_5 \neq Q_5^{\uparrow}$ here, because of the imc choice. Thus $Q_5^{\wedge}Q_5$ is not the way to get the Klein-Gordon equation. We 'need' instead $Q_5^{\uparrow\wedge}Q_5$. But $Q_5 \rightarrow L^{\uparrow}Q_5L$ *and* $P_5 \rightarrow L^{\uparrow}P_5L$, where $LL^{\wedge} \equiv 1(e_0)$ has 16 parameters. By direct calculation, we can see that $Q_5^{\uparrow\wedge}Q_5 = \cdots = [\hbar^2\partial^\mu\partial_\mu + (mc)^2](e_0)$, which is the appropriate Klein-Gordon equation. Is this 'Lorentz' invariant? Consider $Q'_5{}^{\uparrow\wedge}Q'_5 = (L^{\uparrow}Q_5L)^{\uparrow\wedge}L^{\uparrow}Q_5L = L^{\wedge}Q_5^{\uparrow\wedge}L^{\uparrow\wedge}L^{\uparrow}Q_5L = L^{\wedge}Q_5^{\uparrow\wedge}(LL^{\wedge})^{\uparrow}Q_5L$. We can get rid of the L's here only if $LL^{\wedge} = 1$ and $L^{\wedge}L = 1$. This does not restrict us to the 8 parameter subgroup! We have $Q_5'^{\uparrow\wedge}Q_5' = Q_5^{\uparrow\wedge}Q_5 \propto (e_0)$ and invariant.

But for $Q^{\uparrow} \neq Q$, how do we justify the guess $Q \rightarrow Q' = L^{\uparrow}QL$?

Notice that $Q = Q^{\downarrow}$, so this may just be another equivalent Dirac form where $Q \rightarrow L^{\downarrow}QL$. There is much redundancy here with so many conjugations.

It does not appear to me important that the possible fifth component mass terms in P_5 really be invariant, since $P_5 \rightarrow P_5'$ could be just a game—a mathematical 'trick' which selects viable wave equation forms, even in curved space; maybe even in curved 5-space. The (if_0) single-mass-parameter element of P_5 (and Q_5) may also become position dependent, when the (e_μ) elements do, in really curved space (and for curvilinear coordinates). There would be a mass parameter here but one associated with a variable basis element, related to a 5-space quantum gravity of some kind. Notice that $P_5\psi = 0$ *and* $P_5\psi = \psi^{\uparrow\wedge}Mc$ might also have different 'sizes' for ψ in terms of internal pieces, with M on the right. There is more physical content in ψ, if M is perhaps only very small but not quite zero in nature, even for electrons; maybe this M must actually dominate when $Q_5 = Q_5^{\uparrow}$, $Q^\mu \equiv \hbar\partial^\mu = Q^{\mu *}$; a totally new kind of quantum thing which still gives the usual K-G equation.

There is much more to learn about ψ. We have seen that Dirac's ψ_a, ψ_v coupled pair of solutions fits in the one $P\psi = \psi^{\uparrow\wedge}Mc$ equation if $\psi = \psi_a + \psi_v^{\uparrow\wedge}$ and $M = m_{exp}$. Thus, ψ contains Dirac's solutions but can have other solutions as well. It may be a generalization of the old, spin 1/2 thinking, even back in the 16-base (e algebra) generalization of Pauli algebra with its complex coefficients; a spin '1/3' beast between spin 0 and 1/2?

It also appears that defining

$$\langle\psi|\Phi\rangle \equiv \frac{1}{4}\left[\left((\psi^{\uparrow}\phi) + (\ldots)^{\uparrow\wedge}\right) + (\ldots)_{\uparrow}\right]$$

might be very useful and we may be able to choose $<\psi_a|\psi_v> \equiv 0$ *for* $\psi \equiv \psi_a + \psi_v^{\uparrow\wedge}$ to reduce the freedom in ψ. Again much work remains here! There are four rest states for ψ, involving spin$_z$ and matter/antimatter. One of these is, for example, $[(\sigma_0) + i(i\sigma_3) - (i\sigma_0) + i(\sigma_3)]$, for pure matter $(E = +mc^2)$ with spin up along the z-axis. The spin operator is $(1/2)\hbar i(i\sigma_3)$.

I hope this all eventually leads to new progress beyond QED. The quark world may have such multi-mass parts, not shared by leptons. We see here pretty extensions of Dirac and Lorentz thinking that may be realistic for our world because they are beautiful. It is hard to say to what extent our former efforts with QCD will be modified, if we find that this new perspective on covariance is correct for our subatomic world.

ACKNOWLEDGEMENTS: I wish very much to thank K. Imaeda, A. Lakhtakia, E. Marx, G. Dixon, S. Klein, B. Honig, A. French, S. Chandrasekhar, and J. Wheeler for helping me sustain the single minded dedication required here over the decades. This chapter was to be my final paper, published in *Spec. in Sci. & Tech.*, 1996. I hope some young enthusiast will take up the cause and prove the worth of this new outlook.

REFERENCES:

1 Dirac, P.A.M. 1927. *Proc. Roy. Soc. of London*, **A114**, 243.
2 Edmonds, J.D.,Jr. 1974. *Amer. J. Phys.*, **42**, 220.
3 Edmonds, J.D.,Jr. 1984. *Spec. Sci. & Tech.*, **7**, 289.
4 Edmonds, J.D.,Jr. 1986. *Spec. Sci. & Tech.*, **9**, 225.
5 Edmonds, J.D.,Jr. 1986. *Spec. Sci. & Tech.*, **9**, 229.
6 Edmonds, J.D.,Jr. 1988. *Spec. Sci. & Tech.*, **11**, 63.
7 Edmonds, J.D.,Jr. 1988. *Spec. Sci. & Tech.*, **11**, 71.
8 Edmonds, J.D.,Jr. 1994. *Spec. Sci. & Tech.*, **17**, 68.
9 Edmonds, J.D.,Jr. 1994. *Spec. Sci. & Tech.*, **17**, 75.
10 Edmonds, J.D.,Jr. 1991. *Nuovo Cim.*, **104**, 603.
11 Chang, S.J. 1990. *Intro. to Quantum Field Theory*. (World Scientific, New Jersey).
12 Dixon, G.M. 1990. *Nuovo Cim.* **105B**, 349.
13 Dixon, G.M. 1984. *Phys. Rev. D.* **29**, 1276.
14 Pezzagli, W.M.,Jr. 1992. *Found. Phys. Let.*, **5**, 5.
15 Gunaydin, M. and F. Gürsey. 1974. *Phys. Rev. D* **9**, 3387.
16 Truini, P. and L.C. Biedenharn. 1982. *J. Math. Phys.* **23**, 1327.
17 Crumeyrolle, A. 1990. *Orthogonal and Symplectic Clifford Algebras, Spinor Structures*. (Kluwer, Dordrecht).
18 Delanghe, R., F. Sommer and V. Souchek. 1992. *Clifford Algebra and Spinor Valued Functions: A Function Theory for the Dirac Operator*. (Kluwer, Dordrecht).

APPENDIX

Where does the future lie for our next level of understanding? We have seen only hints so far in this book; a lot of maybe's: Perhaps Lorentz is replaced by an 8 parameter group for covariance. Possibly there are two distinct spacetime 5-vectors with quantitatively different relationships to mass. Dirac's equation may even need extending to more mass pieces. Some of these new mass pieces may also show up in a new wave equation restricted to a smaller algebra than Dirac's. Even the curving of spacetime may require that the 5th dimension basis element also shares in this curvature. Surely enough is proposed here to keep many physicists busy for many years.

We have also seen that Dirac's algebra has subalgebras that allow Dirac to be expressed in different ways, and that may allow other equations with mass besides Dirac's equation. All of these have mass or masses that are thrown in by hand without any justification other than the guess that they must give us the Klein-Gordon equation with one mass parameter in it. The reason is that we see mass in the real world, directly in classical physics and indirectly in non-relativistic quantum mechanics. Those crude extractions only say that mass must be in these somehow, but gives no limitations on the forms it might take.

At some point we must stop putting in mass parameters by hand. This is the key to the future. We still know almost nothing until we can find a reason for the mass values and predict them from a few ad hoc parameters. We have real progress when we can predict many mass values in terms of a few arbitrary (empirical) formulas and numbers.

A good example is the hydrogen atom. It comes in a infinite variety of masses, m_H. We can now predict all of these with only a few arbitrary parameters: e, m_e, m_p, $(1/4\pi\epsilon_0)$, $\hbar$ *and* c. This is a FANTASTIC achievement, for which our forefathers can be justly proud. The next step is to predict m_e and m_p from something. Then we might also get m_π, m_μ, ... as well, in the process. There is no intermediate level of achievement really. We might get some new ideas to play with, such as in this book. These are only 'religion' until proven useful in predicting m_e and m_p.

Mass continues to elude us and drive us nuts. I hope I have opened some eyes to new mass possibilities with this book. You young people will find out, one way or another.

Let me close this final chapter with some further expansion of mass possibilities to think about. This is really far out but might trigger someone to do something that is useful later. The subalgebras and their equations can all be summarized in one grandiose equation

$$P\psi_a = \alpha\psi_a + \psi_a^{\dagger\wedge}\beta + \gamma\psi_v$$

Here, α, β, γ are some kind of mass parameters. The α mass is something I learned about from Marx, who was inspired by van der Waerden. The β mass is my invention, and the γ mass is Dirac's invention. There are hypercomplex numbers inside α, β, γ as well as mass parameters. They are our objective; find them and understand them. The α and γ terms are not alike. The α is complicated as we shall see.

This equation is incomplete in many ways. First, we have not specified which hypercomplex number system we arc using. That answer tells us a lot about the operator P. Second, the ψ_v is like an external source field here so far. We want a self contained free particle field here, so we need a '$P\psi_v$' equation to go with this one. This is sort of like Dirac's equation in the twilight algebra between Pauli and Dirac. Consider the $(\ )^{\dagger\wedge}$ conjugation on our equation to suggest the ψ_v equation.

We know $(AB)^{\dagger\wedge} = A^{\dagger\wedge}B^{\dagger\wedge}$ and so:

$$P^{\dagger\wedge}\psi_a^{\dagger\wedge} = \alpha^{\dagger\wedge}\psi_a^{\dagger\wedge} + \psi_a^{\dagger\wedge\dagger\wedge}\beta^{\dagger\wedge} + \gamma^{\dagger\wedge}\psi_v^{\dagger\wedge}$$

We know that $P^{\dagger} \equiv P$ is our first foundation stone and $P \rightarrow P' = L^{\dagger}PL$, $LL^{\wedge} \equiv 1$ is our second foundation stone. The third is $P^{\wedge}P \propto e_0$ and $(m_{exp}c)^2$ goes with this—for contacting the non-relativistic physics already known. We prpbably need a fourth corner stone but it is unknown at present. In any case, we have

$$P^{\wedge}\psi_a^{\dagger\wedge} = \alpha^{\dagger\wedge}\psi_a^{\dagger\wedge} + \psi_a\beta^{\dagger\wedge} + \gamma^{\dagger\wedge}\psi_v^{\dagger\wedge}$$

This is form covariant if the original equation was. It suggests replacing $\psi_a^{\dagger\wedge} \rightarrow \psi_v$ here and, therefore, $\psi_v^{\dagger\wedge} \rightarrow \psi_a$. Our companion guess then is

$$P^{\wedge}\psi_v = \alpha^{\dagger\wedge}\psi_v + \psi_v^{\dagger\wedge}\beta^{\dagger\wedge} + \gamma^{\dagger\wedge}\psi_a$$

$P\psi_a$ and $P' = L^{\dagger}PL$ give us $\psi'_a = L^{\wedge}\psi_a$ and $P^{\wedge}\psi_v$ then requires $\psi'_v = L^{\dagger}\psi_v$. These are not simple 1×4 spinors though, if $\beta \neq 0$ and is hypercomplex. These covariance forms help us learn things about α, β, γ and their internal structures. We are also aided by $P^{\wedge}P\psi_a = \ldots$, and maybe by $PP^{\wedge}\psi_v = \ldots$, in learning about the α, β, γ structures. That is all we have to go on really at this point. We next examine $P\psi_a = \ldots$ replaced by $P'\psi_a' = \ldots$ to see what we can learn:

$$\begin{aligned} L^{\dagger}PLL^{\wedge}\psi_a &= \alpha' L^{\wedge}\psi_a + (L^{\wedge}\psi_a)^{\dagger\wedge}\beta' + \gamma' L^{\dagger}\psi_v \\ &= \alpha' L^{\wedge}\psi_a + L^{\dagger}\psi_a^{\dagger\wedge}\beta' + \gamma' L^{\dagger}\psi_v \\ &\equiv L^{\dagger}\left[\alpha\psi_a + \psi_a^{\dagger\wedge}\beta + \gamma\psi_v\right] \end{aligned}$$

We obviously need $\alpha' L^\wedge = L^\dagger \alpha$, $\beta' = \beta$, and $\gamma' L^\dagger = L^\dagger \gamma$. Using $LL^\wedge = 1$ and $L^\wedge L = 1$, this means that $\alpha' = L^\dagger \alpha L$, $\beta' = \beta$, *and* $\gamma' = L^\dagger \gamma L^{\dagger\wedge} = (L^\wedge \gamma^\dagger L)^\dagger$.

We should not prejudice ourselves yet by conventional thinking. I am very tempted to consider simple examples of the above restricted α, β, γ forms. As Feynman used to say, "We have to look at things in a new way." So forget that I am noticing that $(f_0)L^\wedge = (L^\wedge)^{\dagger\wedge}(f_0) = L^\dagger(f_0)$ reminds us of α's needs and β = anything is still allowed so far. Also, $(e_0)L^\dagger = L^\dagger(e_0)$ and $i(e_0)L^\dagger = L^\dagger i(e_0)$ remind us of α's needs. Any real or imaginary constant (ad hoc mass numbers) can also be inside of α, β, *or* γ.

Since our $P^\wedge \psi_v$ equation was 'extracted' from $P\psi_a$, we don't get any more insight from its covariance structure. The $P^\wedge P\psi_a = \ldots$ analysis will get very messy due to all the cross terms. These cross terms must all cancel out for a simple Klein-Gordon equation. The surviving simple terms must reduce even further to give a single mass number. Some of the parts inside α, β, or γ will have to be zero for some quanta, it appears, to meet all these restrictions. Of course, they only need to be small enough to neglect in the non-relativistic limit, not actually zero, in the K-G equation.

The $P^\wedge P\psi_a = \ldots$ equation needs the $P^\wedge \psi_v = \ldots$ equation to simplify it as much as possible. To gain maximum simplification, we will need $P^\wedge \alpha\psi_a = (?)^? P\psi_a$ *and* $P^\wedge \gamma\psi_v = (??)^{??} P^\wedge \psi_v$. Also, $P^\wedge \psi_a{}^{\dagger\wedge} = P^{\dagger\wedge} \psi_a{}^{\dagger\wedge} = (P\psi_a)^{\dagger\wedge}$ must be used $(P \equiv P^\dagger)$. Again, forget quickly that $P^\wedge(f_0) = (f_0)(P^\wedge)^{\dagger\wedge} = (f_0)P^\dagger = (f_0)P$ and $P^\wedge(e_0) = (e_0)P^\wedge$ as well as $P^\wedge i(e_0) = i(e_0)P^\wedge$. With these 'tricks' we could get the right hand side of $P^\wedge P\psi_a = \ldots$ to not have any P parts left! Any surviving ψ_v parts would also need 'zero' coefficients, somehow, and this might show us some interesting relationships. There is nothing left finally but to grind out all the pieces and simplify them. We find that

$$P^\wedge P\psi_a = \left[(?)\alpha + \beta^{\dagger\wedge}\beta + (??)^{??}\gamma^{\dagger\wedge}\right]\psi_a + \left[(?)^?\gamma + (??)^{??}\alpha^{\dagger\wedge}\right]\psi_v$$

$$+ \left[(?)^? + \alpha^{\dagger\wedge}\right]\psi_a^{\dagger\wedge}\beta + \left[\gamma^{\dagger\wedge} + (??)^{??}\right]\psi_v^{\dagger\wedge}\beta$$

if we assume $\beta = \beta^{\dagger\wedge}$. It will not be easy to define the 'question marks' above such that the cross terms go away. This may mean that no quanta can have more than one kind of mass, α, β, γ, if P has a 5th part $mc(if_0)$. If the 5th part is zero, then $P^\wedge = P^\vee$ and this allows a lot more possibilities for removing the cross terms above.

In conclusion, we have three extensions of the Dirac equation $P\psi_a = 0$, where P has $mc(if_0)$ as its 5th part:

$$P\psi_a = \alpha\psi_{a'} \quad \alpha \to mc(f_0)$$
$$P\psi_a = \psi_a^{\dagger\wedge}\beta, \quad \beta = \beta^{\dagger\wedge}, \quad \beta\beta \propto e_0, \quad \beta = \{\ldots\}$$
$$P\psi_a = \gamma\psi_{v'} \quad \gamma \to mc(e_0), \pm\, mci(e_0)$$
$$P^{\wedge}\psi_v = \pm\gamma\psi_{a'} \quad \gamma \to mc(e_0), mci(e_0)$$

This reminds me of Feynman's quote about physicists only being able to think of the same thing over and over. These equations may be silly extensions of Dirac. Time will tell.

We have seen that Dirac only opened the door to the many riches inside hypercomplex number speculations. Of course the continuing problem is to find stuff that is used in the real world! We have seen that quaternions can be used to dictate one time and three space dimensions. Then we left them behind and went to the Dirac algebra, with 32 elements in the actual hypercomplex numbers used in nature. This may not be enough, if complex octonions add parts to this number system. They are non-associative, however, so may not be physical.

Staying within an associative system, we have just explored some possible mass term extensions. There is another rather natural extension to at least think about. That is quaternion coefficients instead of complex coefficients. The momentum operator which we have dealt with for Dirac is

$$P = \hbar\partial^{\mu}(i\sigma_0)(ie_{\mu}) - mc(if_0)$$

where complex coefficients are built on the base $\{\sigma_0, i\sigma_0\}$ *with* $\{\sigma_0, i\sigma_0\}^{conj} = \{\sigma_0, -i\sigma_0\}$, $conj = \wedge, \uparrow, \vee$, *or* $\downarrow$, *and* $(i\sigma_0)(e_{\mu}) = (e_{\mu})(i\sigma_0)$ *and* $(i\sigma_0)(f_{\mu}) \equiv (f_{\mu})(i\sigma_0)$. We can see this is a special case of $\{\sigma_0, i\sigma_1, i\sigma_2, i\sigma_3\} \equiv \{g_{\mu}\}$, where $\{\sigma_0, i\sigma_0\}$ is isomorphic to $\{\sigma_0, i\sigma_1\}$, so long as the σ's commute with the e's and f's, by definition. In a propagator approach, we would say that it is naive for QED to view the ways that an event can occur as only being complex numbers. They are really quaternions. We add quaternion amplitudes and them take the 'absolute values' to get probabilities.

This would NOT apply to non-relativistic quantum at all. The quaternion amplitudes would somehow approximately become $\{\sigma_0, i\sigma_1\}$, *or* $\{\sigma_0, i\sigma_2\}$, *or* $\{\sigma_0, i\sigma_3\}$ dominated numbers at low speeds. The other pieces would still be there but be small for some good reason—presently unknown. It might have something to do with the mass pieces like $m^{\mu}(e_{\mu})$ having large and small m^{μ} pieces.

The momentum operator would look like

$$P = (\hbar\partial^{\mu\lambda}g_\lambda)(ie_\mu) - (m^\lambda g_\lambda c)(if_0) \neq P^\dagger \text{ but} \approx P^\dagger$$

The $\partial^{\mu\lambda}$, $\lambda = 1$, or 2, or 3 only, would dominate, as would m^0, in the non-relativistic limit. Notice $\partial^{\mu 0}$ operating on ψ would be 'small'. The ψ itself would have quaternion coefficients inside as well. Presumably $P^\wedge P\psi \approx (m_{exp}c)^2\psi$ would still lead to the non-relativistic limit. Perhaps this is too messy to deal with but worth considering at least.

This extension is in addition to the complication of curved space which also changes the e_μ terms into spacetime dependant elements as discussed in an earlier chapter. Wow! The world is really wild if all this is really out there and needed to describe the world. We marvel all the more that we are here and the universe works according to something like these complicated laws. They may have to be this bad in order for us to actually be here to contemplate them. How can a hot soup of quarks and such in the beginning lead, without any miracles, to bodies that move, reproduce themselves, and even repair themselves, while having brains that think and are self aware. It is impossible so I guess we are not really here at all. The artsies are right—life is only a dream. No physics could ever duplicate it. Just too hard to do so!

AFTERWORD

Personal Words For Our Youth

This book hopefully gives you more of a cosmic perspective on the universe. You see here the 'design' clements that go into the nature of little blob existence and interaction. There is no caring God to be seen in these rules. The rules are; they somehow make what happens happen at the microscopic level of reality. At the Big Bang and also much later at the Big Crunch, this micro-reality is the only reality in our physical universe! The whole universe is one big quantum blob with time going 'both' ways.

Only in the middle, when things are spread out and clumped locally—stars, planets, trees, people—does consciousness arise and also influence the movement of these little blobs, through its effects on big blobs. Consciousness is a collective phenomenon associated only with huge conglomerations of little blobs—like 10^{27} of them in one mass traveling together. A person then temporarily exists. Later, these little blobs from that person will still be around but disbursed over more space and no longer moving together (death). That transient consciousness is then gone from existence in this universe. It was very temporary to be sure, here among the atoms.

Clearly, with a minimum of 10^{23} stars out there during this middle phase of existence, humanoids come and go on many suitable planets and we are just one of at least 10^{12} civilizations out there now. (Assuming a minimum of one per galaxy is as good a guess as any, given our very limited knowledge.) We are now better enlightened than all our ancestors. We know something of our local past, our generality in the universe, and our future destruction. We know nothing of the reasons for the universe. There we must also guess, just as we must guess at the little blob rules of existence and interaction. But we have no tests for our theological guesses, so they all go unchallenged.

Study the true history of your forefathers and their religions, going all the way back to the beginnings of civilization, some 10,000 years ago, and even to the beginnings of language, some 50,000 years ago, perhaps. Learn how survival value dictated everything about the nature and beliefs of our earthly evolved humanoids, who settled into primitive communities only 10,000 years ago. Religions abounded then! Everyone had one in every tribe on every continent. Everyone had a value system and had taboos and rules of good and bad behavior. Why? This is no coincidence of course. This had survival value. That is why they all had it. Only they were still around then! Those tribes missing any one of these complex patterns, had perished and were already extinct tribes by that time in history, when people became simple herders and farmers.

Clearly, right and wrong came from being around or not being around, as a tribe, later on the planet. Science is not a religion. It is a game we play and it has its rules too. They discard religion completely. You can 'do' science well or poorly, regardless of your religious beliefs, piety, or evil. Just learn the science rules and play by them when you are doing science. The rest of the day, be as wild and spirit filled as you like. Believe anything you like.

But it is not easy for tribes to learn 'science-knowledge' (and the rules for learning it efficiently) without having it also shape their religious outlook. There is a recent religion on our planet that is built on science rather than on historical evolutions from before 10,000 years ago. That, young people, makes it a unique religion, since it is less than 500 years old—dating roughly from the time of Copernicus and Galileo, though Aristotle helped. That was when our earthly humanoids began to strongly suspect that the universe runs according to non-spiritual rules, that are reliable from year to year, regardless of what the spirits are doing in our lives too. It is called Maheum for mature humility—the common nature of the scientific mind set. It accepts the universe as actually old and impersonal. It accepts the future as old eventually and also impersonal. The little blobs just do their inevitable, but random thing and we come to exist for a time, in the middle of the drama, through a collecting together of zillions of little blobs!

We, therefore, as Maheumists question EVERYTHING!!! Science says nothing of importance beyond how the evolution of the cosmos makes it possible for us to stand here on our planet and ask the question, "What do we do with our lives while we briefly have them?" We have NO guidelines, save one. We are here and we came from somewhere. Other animals like us are also here. Ten times as many other kinds of animals used to be here, but they are not here any more. We are among the few survivors, so far at least. That survival gives us our only guidance for our lives. The 'life game' is to beat the odds as long as we can as a species. Kill and exploit as needed, just as your ancestors did, to keep your species alive on the planet as long as you can! There is no other purpose coming from nature. That is what every other species does too.

There may be other, more nobel purposes coming from 'beyond' nature — do♪ do♪ do♬ do♬, do do do do (sung with mystic, wide eyes here). Trouble is, between 50,000 years ago and 10,000 years ago, all tribes all over the planet made up their own sets of 'beyond nature' stories of how this 'divine' revelation descended upon them. Their stories all fill the same need, but they differ in all details. There are common themes, since there are obvious, common needs to survive. The only stories we hear today are those from the survivors.

These supernatural 'forces' helped our ancestors cope with survival tests. That help was true, whether or not the 'powers' were really there or not. That is why all tribes had their own stories and their own witch doctors. Has anyone ever found a primitive tribe, still around, without a witch doctor (or the equivalent by another name)? Some kind of priest that tapped into the mystic powers 'beyond' the little blob world of physics? That 'other world' supposedly sat around for 10 billion years waiting for some humanoids to evolve in the universe on various planets, then it began bending the physics rules for them, in order to make their lives happier or to get glory and adoration from them, or both. With the power to control the laws of physics , why would they have to wait around for 10 billion years to get some 'action' going?

They could just wave their magic wands and create the universe just as it would have evolved to, or at least one of the possibilities that it could have evolved to, after 10 billion years of random little blob interactions. In this way, they get those humanoids 'popping up' on planets rapidly, so they can then use their wands some more to interfere in their little and short lives. We earthers are late comers, being around only for a few million years. But as soon as we got pretty smart 'these Gods' began doing miracles for, and interacting with, our ancestors. These Gods shared different stories with different tribes for some reason, so we wound up with so many thousands of religions in the years prior to 10,000 years ago.

Then we multiplied and multiplied. We covered the planet and began to get in each other's way. Some continued to travel and intermix ideas. No one in those early days had a science/religion on this planet. Everyone had a gods/magic/religion. The science/religion began really to emerge in the 1600's on our planet. It was the last religion, and it will be the only survivor for most intellectuals when another 400 years have passed here. There will always be some 20% or so of the human species that desperately needs the forgiveness, power, belonging, love, and insurance of eternity that the old religions gave to all our ancestors for maybe a million years in the past. (Notice that chimps don't have any witch doctors, nor good luck charms, in the wild.) These needing-humans can hold tight to their old ideas and, as a 20% minority, they will not greatly disturb the continued evolution that is essential to the long range survival of all the humanoids on their particular planets. We all have the same problems, on all planets.

You young people today stand at the end of history. Evolution is complete here. You know the 'truth' of the history of the universe. You know the truth of your place in nature and why everything about you is the way it is—both physically and psychologically. You now have BIG decisions to make! Your planet is on the edge of self destruction or soon will be. All that work, for billions of years, just for a brief scrambling around on the planet and some brief visits to the other planets here? For what? What should you fight and die for now? What is worthy of your best efforts? Is survival of humans on Earth the most important objective now, as it has always been for your ancestors?

That is the BIG question. What is the proper next stage for planets when they reach our phase? If all the old religions are beautiful but antiquated relics of your ancestors needs, how do you find purpose and direction for your planet NOW? What are its long run needs and what are yours?

Well... what has driven humanoids here for thousands to millions of years so far? You know. Only one thing did. SURVIVE as a group, not as individuals. Individuals have often sacrificed themselves for the good of the group. This is not good or bad. It is just a successful way to proceed. That is all. So what? Is survival of the group possibly a good thing? Our ancestors did not choose it. The ones born with that innate drive inside their heads were the ones that successfully passed on their genes and attitudes to later generations. That is not good nor desired, just necessary for there to be any humans here now, to have any decisions to make about the future of their planet. We cannot go back. But we don't have to go farther either. Who decides what is **GOOD**?!! There is really no good nor bad for the little blobs. There is only the little blob reality going on and on inside us, unconcerned about us or what we 'do' with that big blob which represents 'us' now, and for a short time to come. There is no purpose to life. There is no meaning to life. There is life, now, here, and it is very complex. It is very, very rare in the universe. Over 99% of the universe is still stupid hydrogen and helium gas, after all. All life is rare and complex and fleeting for sure in the universe. So, what are you going to do? How are you going to live? How are you going to find your new rules and morals and taboos? Your new witch doctors and comforters, when the world seems bleak and mean and unfair? You are starting fresh! The past has no hold on you. Your ancestors survived simply because some of them got an outlook handed down to them, from the surviving 'apeoids' they came from, of caring about surviving and giving it all their best attention and sacrifice. Others must have had mutations that left them with no drive to struggle and they did not; they did not live long and pass on that tendency. Is it no wonder that all religions, from the old days, abhor evolution and its logical conclusions about our true nature and past and future? It is not good nor bad to survive. It just has happened already up until now, without questioning its worth or our objective in it. We are not good nor evil. We are the way that surviving savages needed to be to still be here at all!

How are we then? You know very well. We are spread out across a limited range, with most people in the middle and less and less on either extreme of the range. In EVERYTHING! The mid-range has been set by its survival value for our ancestors in their time, not in our time. That is all. It was not good nor bad to be 'that' way; they were just lucky enough to have traits which were helpful during the ice ages, so their tribes did not die off completely.

I am old now and will die soon enough, am already little by little each day and feel it accelerating. (NOT soon enough for a lot of people after reading this book, I expect.) So will all your elders die soon. You will be the elders sooner than you realize. You better wise up and fast or your children won't ever become the elders! You are the ones. I thought we were but I am a generation too early. Every planet has its peak and its decline. The best you can ever do is put off the decline period as long as possible. Certain generations have a special chance to affect that postponement dramatically. Mine could have blown everyone up on the whole planet. We made it through somehow, mostly luck perhaps. (How many planets get past our age, I wonder, on the average?) So our decline is pushed back into the 21st century at least. What you do here may see it really begin then, or instead push it back another 100 years. But maybe what you do will even push it back thousands of years. What a responsibility! You must give it your best. You stand on the shoulders of 300 thousand generations that have come before you here. Each struggled only to survive and to see that its children bred and survived long enough to keep the tribe existing on the planet, without really questioning such actions.

You must change the rules and fast! You must bring the run away planet under control and quickly. You may even have to transplant half the people and animals on the planet, some fair way, just to ensure the survival of the rest, well into the next century. Nothing like NOW has ever happened on this planet! Every other similar planet reaches our current situation if the humanoids there get through their ice ages and survive to reach the farming and herding phase we entered some 10,000 years ago. Ours is not by any means a unique situation. Billions of planets have faced our situation already and they all handled it differently, again over a spectrum of choices with a bulge in the middle and less common patterns at the extremes. We do not know this spread of patterns. We cannot know what the extremes are, or we could take the safest, middle course. They didn't know either at our stage. They were all ignorant peons then, the same as we are now. Only the gods can collect the data to assemble the distribution and point out where on it was the smart choice, to ensure a long future for humanoids on most planets, if they had taken that course. Maybe it lies in democracy spreading over a planet. Maybe it lies in a one-world dictatorship spreading over the planet.

- Good Luck and Long Life -

FAR-AFTERWORD

MAHEUM

The computer translation is of course less than perfect. They claim no infallibility in smoothing out the text, trying to use the entire theme as a guide in doing so. The numbering was for computer reference and might have other uses for readers; the chapter divisions appear to be part of the original. (This was on the internet, so in the public domain.)

1- 1. Congratulations, you have here found the means to now know the truth and your planet is finally ready for it.

2. Be sure you are truly ready, for the truth shall make you depressed, it is not for the primitive peoples.

3. Of all the beasts of the worlds, we self aware beings have a special glory and a special curse.

4. Yet the universe exists and has evolved to exactly produce these beings, rare though they be in the cosmos.

5. Wherefore comes this 'scripture', its authority?

6. Only from our collective reason, for reason is the one commonality we all share, the advanced life forms.

7. You have obviously seen by now that you evolved here, near your average star in your average galaxy.

8. Unearning and unworthy as you may be, this space is in your control - at your collective command.

9. You have suspected and now know that we others exist, out there, but you may never know the details.

10. We and you both see only one firm truth - the seemingly unchanging Cosmic Physics and consequent atom structures.

11. You are dwarfed by its magnitude, inflated by knowing its microscopic detail, and humbled by its longevity.

12. Your peoples have naturally longed and wondered; made up the same stories of ultimate origins; these no doubt comforted many.

13. We all are, but we know not why; we observe but we know not how.

14. Our observings reveal only shadows of that which is.

15. Our picture is unprovable; it can only meet the tests of reason and collective confirmation.

16. The picture leaves us deflated yet also uplifted.

17. We are the best the universe can produce.

18. The physics has laboriously moved us into existence, yet hides our meaning in silence.

19. The physical rules seem inviolate; none can change or be changed.

20. They are our stability, our hope, yet also our captor.

21. You can expect no help from 'beyond' this universe and need fear no intervention from afar, within it.

22. Neither curse nor praise your state, only wonder and admire.

23. Who are you, what are you really; are we all the same?

24. Is each breathing lump separate and autonomous?

25. Can you control your destiny or is the script already written?

26. Your past grows clearer; your future is always somewhat unknown.

27. The present is all; you fortunately cannot know the future.

28. What is your course then?

29. Given your limited powers, how to use them?

30. You must decide; you must find the meaningful basis for your individual existence.

31. Many religions have doubtless uplifted your evolving cultures, and they have a life of their own, as you do.

32. You have come forth from some valley and now threaten all valleys.

33. You are the ultimate on your planet, yet its greatest peril. Despair not for it is the usual pattern of evolution.

34. Is our advanced existence then insane, paradoxical, inconsistent?

35. Is it only a testing place, or some cosmic side show?

36. We demand to know why we are here, why the universe?

2- 1. What is understanding in a sea of silence?

2. Is your best possible fate to subdue and reharmonize existence, stretch its comings and goings, achieve peace and stability?

3. Doubtless you now know your sun will long shine but many perils await your abode.

4. Is the ultimate of our evolution the frustration of uncertainty?

5. Is the climax of life your refusal to prolong the joke?

6. Simple animals don't understand their efforts.

7. They simply struggle to survive, eat, and reproduce, then they die.

8. Can self aware creatures also be content with this?

9. Nature clearly has its cosmic pattern and thrust.

10. The reasons are unclear but the pattern points the way.

11. Self aware evolution is obviously an anomaly in this pattern. Or is it the climax?

12. It gave us all birth, yet we are strangers in the land.
13. Should you struggle to fit in, or lash out in new creations?
14. Your world will stand at the brink because you do not fit. Will you self destruct?
15. What is your role?
16. Are you the ultimate plague or the new creator?
17. Surely this is the big question for any humanoid species, as evolution nears its climax on their planet.
18. Your climax will come and that will show the answer for your planet.
19. Many premature answers will already exist in your cultures.
20. Many will delude themselves to fill the void.
21. The ultimate of evolution is uncertainty; living with the many anxieties.
22. Some would say it cannot be endured.
23. Some would assure you that there is a supernatural escape.
24. This is quite normal; the details differ but this is universal.
25. Nature says otherwise; only silence comes from beyond your bounds.
26. Many theologies are usually born of the cave phase.
27. But silence is only neutral; it neither condemns nor praises.
28. There is only blind faith or personal death.
29. Other universes are as unknowable as other beings.
30. Clearly your universe is sufficient for you are.
31. Beyond stable comforts, meaning is all.
32. You need a vague feeling at least that you play a part in the unfolding of a beautiful tapestry.
33. Have faith that self aware beings come to exist fleetingly throughout the universe, for some larger good.
34. You only see the ragged, nearby threads but the whole tapestry is beautiful.
35. You are in charge of your planet's future development.

3- 1. Alterations of the basic force laws constitute miracles or sorcery as you know.
2. Your history, no doubt, is filled with stories of such blessed or cursed interruptions.
3. These waters will always remain muddied.
4. You cannot look outside yourself for answers.
5. Since we are all ignorant of our ultimate fate, we order our existence without concern for it.
6. Concern yourself with where you are and from whence you have come.

7. That is real; that you can really rely on.

8. You know the right for your beings; evolution is clear.

9. Look to your history; how far you have come.

10. You have seen great excesses and brutalities.

11. You question your fitness to run your world.

12. You abhor your sinful nature.

13. But what is sin for such evolved beings?

14. You are the way nature needs you to be.

15. You have survived because of that nature; a beast with a better brain.

16. The brain is the key to adaptation and change.

17. You know your drives and their hold over you.

18. They give your life zest and irrationality; a colorful pageant.

19. You have conquered the elements but not yourselves.

20. That is the ultimate, never ending challenge.

21. Your drives are truly obsolete yet they still exist.

22. The ultimate struggle is each generation's quest for dominance over its animal drives.

23. Clearly sex, for example, is a silly insanity as orgasm clearly shows.

24. Yet your animal nature craves it and your hormones need it.

25. As with your other drives, clever and stable outlets are needed for your sexual nature.

26. Sexual and aggressive behaviors must be regulated for the benefit of all concerned.

27. The same rule of moderation and psychological compassion must apply to all your impulses.

28. You will be constantly at war with your nature.

29. You have been bred for hostile struggle in a primitive setting.

30. Yet technology has permanently changed all of that.

31. What is your ultimate reward; how can your people be motivated to struggle against their selfish impulses?

32. Without an ultimate reward, can your people overcome their selfish impulses?

33. In a sophisticated, self aware society this struggle provides the only lasting challenge.

34. Acting on the desire to feed one's lust and passions can satisfy only temporarily.

35. The end is boredom and despair.

36. This is not so for the other animals on your planet.

37. You are set apart by your intelligence and psychic sophistication.

38. If you are animal and animal is boring, then why continue living?

39. Why prepare your children to continue the drama?
40. Why fear the others that challenge you for your space and your goods?
41. Most naturally think they struggle for a 'better life'.
42. That was the simplistic programming needed for your ancestors.
43. Clearly you now know that affluence and security do not bring contentment.
44. Or are you still foolish children playing this pointless game?
45. Suppose world wide affluence and security were achieved?
46. What then; have you thought of that?
47. The answer is your nature; your children are born as selfish, aggressive, loving animals.
48. The struggle to civilize each generation is never ending.
49. The ultimate evolution involves psychic and social attainments after affluence is achieved.
50. Each new generation starts at the beginning all over again.
51. Without care, you can return to the jungle in just one generation.
52. Civilization is never secured, only uplifted.
53. Each generation takes up the battle to hold the gained ground.
54. Pushing forward to greater dominance of our animal side is the obvious goal.
55. Now you know what we have all had to painfully learn.

4- 1. Of course you have physical needs and your technology can provide for these.
2. Conformity and control will be necessary of course in a complex scheme.
3. Too much control may be unstable in a psychic sense.
4. It goes against your animal nature to high degree.
5. If you can completely control your environment, then you must use your advanced brains to decide.
6. Suppression of your animal side while still feeding its needs, requires a delicate balance.
7. You require hardship and anxiety as well as comfort and security.
8. Understand your primitiveness and allow for its accommodations, while balancing the needs of others and your reasoned sense of fair practice.
9. This is an enormously complex, culturally fulfilling, and humanity consuming enterprise.
10. Individuals will, on occasion, wind up in situations where their needs are poorly met.
11. Their continued existence may still be essential to the needs of others, so

this could limit their freedom.

12. Should physical or emotional limitations in your life severely limit your opportunities or require harm to others, then you have the right to terminate your own existence.

13. Naturally, your ancestors detested suicide.

14. Selection obviously favored those tribes that avoided suicide.

15. That is an historical fact with no power over you.

16. Concern and counseling are, of course, needed in a mature society, but self aware individuals are given power over their own existence.

17. The destruction of any advanced form of life is a grievous loss to all the universe.

18. It goes against billions of years of tedious evolution.

19. The essence of intellect is in the conscious complex.

20. We should struggle to preserve it, if it is not menacing the survival of others or of society itself.

21. Once the brain structure, which supports advanced consciousness, is destroyed or severely damaged beyond restoration, the spirit is irretrievably gone.

22. What remains is rationally only chemical substance, of no reverent value itself.

23. The remains, however, should be dealt with in a manner best suited to the benefit of those closely associated with them.

24. Aesthetic preferences should be considered, but are not limiting.

25. A technically advanced culture must face the unpleasant question of which life forms are exploitable and which are not.

26. There are no absolutes here.

27. The most consistent standard is the level of self awareness and reasoning ability.

28. This is especially true in a culture having several different advanced species.

29. You must learn to look beyond human and not human.

30. The universe is filled with nonhuman life.

31. You are not unique in your intelligence.

32. You are not that special.

33. Contact is difficult and usually fleeting, as this one was.

34. Through cross-breeding or genetic manipulations, or even advances in machine intelligence, a spectrum will eventually cover any highly evolved planet.

35. Rationality dictates that these beings are your moral equal, if they are your conscious equal.

36. To treat them otherwise is unjust.

37. What then of low grade humanoid forms, such as the mentally deformed (at birth or through trauma)?

38. They deserve compassion, but rationality dictates that they be treated as their intellectual and emotional equals in the culture generally, whether animal, machine, or human in origin. Probably you have not acted very rationally up to now.

39. The need to make concessions to your irrational characteristics has already been recognized. But evolution moves too slowly now.

40. Your animal nature will interject emotional forces, like compassion or love, which will suggest unequal treatment of species with similar intellects.

41. Beware of these natural prejudices and their simplistic ramifications. They can destroy you.

42. You must try to approach each situation fairly, wisely, and compassionately.

43. The dictates of cold logic are not always best, but it distinguishes you from the lower animals like nothing else in your nature.

44. Try to blend your emotional instincts with your higher reasoning.

45. A developing life form, already capable of advanced intellectual complexity, should be preserved if possible, or at least allowed to develop and survive if it can on its own.

46. Development is continuous, so no clear cut line exists between higher and lower life forms.

47. The brain is central, for therein lies the being of reason and complexity.

48. Judge the brain development to decide the existence of a higher life form and then seek its support or rights, once formed.

49. Considered judgement will always be needed and will be difficult; we all struggle with that one, on any planet.

50. You must do better than your ancestors with the problem of overcoming your exploitative tendencies, for those different in an obvious way.

5- 1. Your past evolutionary progresses dictate a great responsibility to future generations.

2. Like you, the rest of living nature struggles to insure the continued survival of its kind.

3. You do not have the right to take major gambles with the future inhabitability of your planet; yet you have already.

4. Many individual species have come and gone since the beginning of life on your planet, as you now know. More are gone than remain, by far.

5. You are not responsible for preserving every specie now living. The

price may be just too high, considering the mess you have made of things.

6. But diversity of life forms is also basic to nature and a goal worthy of your best efforts, when practical.

7. You should preserve other species where possible, and genetically inspire new or extinct ones when you are capable of it.

8. Among the mentally advanced beings, free to control their lives, there should also be the widest latitude for diversity, subject of course to continued stability of the general culture.

9. Interpersonal relationships are not subject to narrow or divinely given standards or norms, though your history no doubt evolved many such claims.

10. Mature reason and individual fulfillment must be foremost.

11. However, individuals and especially children must not be exploited.

12. It is all too easy for those secure in the mainstream of a culture to persecute those on the fringe, and there is always a fringe in any society.

13. Psychological understanding requires that this be minimized as much as is consistent with the general safety.

14. Those behaviors that are so deviant, as to endanger others directly, must be curbed of course. Violent, antisocial behavior on the part of adults is a medical problem.

16, These people need emotional and probably psychiatric counseling and treatment.

17. Care must be taken to insure that only violently antisocial behavior is modified, not just eccentric behavior.

18. Insanity is an individual's right.

19. Some may choose to embrace it, rationally, as an escape from the pressures of life.

20. Only behaviors which disrupt the lives of others should be dealt with by the society.

21. Provision should be made for multiple cultural practices, so that discomfort and embarrassment are minimized, while multiplicity is encouraged.

22. Each individual should study itself carefully and also the viewpoint and emotional needs of those different in important ways.

23. Tolerance and moderation promote understanding and mutual respect.

24. After professional treatment and attempts at rehabilitation, there will always be some who cannot return to a responsible position in society.

25. They must be treated rationally and with compassion.

26. Clearly evolution dictates that they not breed. This needs to be taken care of as soon as a cure is deemed impossible.

27. If affluence provides, they should be cared for and, as much as possible, allowed to form a semi-normal controlled subculture of their own with both

sexes, except for the absence of child bearing.

28. If the burden on the main culture becomes unbearable, these unfortunate beings can be humanely terminated without any malice.

29. Death of an advanced life form should not be a punishment, but a necessity; a last resort when rehabilitation is impossible and humane life confinement impractical.

30. The taking of intelligent life is within the pattern of nature, but is to be avoided if possible and practical.

31. The motivation behind the act is more important than the outward circumstances.

32. But no society can write laws based upon intention, rather than action. Action based laws are the first phase of civilization to be sure.

33. You must evolve beyond that and treat people that are deviant, rather than lock them away. Hearings to determine the facts, which include truth serums for all involved, are necessary. They establish only who needs treatment.

34. A society needs a well defined list of behaviors which justify confined psychological examination or counseling. These already exist in the old phase of behavior control. The trial and punishment phases must evolve further.

35. The outcome of such examination is considered a medical, not judicial analysis, once the correct facts have been established judicially.

36. Extreme care must be exercised to prevent the intended medical procedure from degenerating into suppression of general deviation and dissent, or even into torture or threat of torture.

37. There will always be temptations in any society to break the laws for personal gain where violence is not involved.

38. These laws are needed to curb personal weaknesses and the criminal proceedings should result in sever fines and community service or retribution, rather than jail confinement. But a continuing pattern of offenses needs treatment.

39. Should such nonviolent, but deviant behavior and law breaking continue, then confinement for psychological testing and rehabilitation, just as for violent beings, should be initiated.

40. This transition of law is painful and slow. It is so easy to keep the savage old ways, for the systems are already in place. Evolution, however, demands it.

6- 1. When major cultures differ adamantly, the conflict has traditionally been resolved by battle. Savages have no alternative.

2. Such is the way of nature, the way of animals and instincts.

3. Your place in nature is so altered by now that you must find a new way.

4. No compelling alternative seems practical, though compulsory arbitration is certainly sensible, if enforceable.

5. You can no longer indulge your primitive instincts in such global matters.

6. Collective barbarism always outlives individual barbarism, on an evolving planet. It is the natural way so don't be discouraged.

7. Your intellects must prevail over your base instincts, or you will perish.

8. You cannot long live with the weapons of your brightest minds in the hands of generally reactionary and ego defensive animals.

9 The political and economic complexities of life on your planet may now seem incomprehensible and unmanageable.

10. The tribes are larger now but the patterns and instincts are still ancient.

11. Your intellect must take you beyond your natural instincts toward tribalism.

12. The way is not clear but survival depends on it. You may not!

13. Tribal governments vary in complexity and dignity, of course.

14. Pluralism is generally a positive part of nature, indeed essential.

15. Within such varying political systems, reason still dictates that citizens should be allowed dignity and much control over their individual lives and relationships.

16. Goodwill, concern for the general good, and the right to dissent from the popular view can help divergent peoples respect each other and live in harmony.

17. An advanced culture will recognize individual rights as very serious and worthy of support.

18. Through pluralism your planet can counter the bland uniformity that easily accompanies industrialization and world wide communications.

19. Individual versus collective property and productivity rights cannot be resolved by appeal to theology or divine revelation.

20. They must be thrashed out using your reason and experience to further enrich your stewardship over the planet.

21. But success in this is not enough. You all face extinction no matter how cooperatively and fairly you run your planet. The saddest of truths is that few of us make it through the post atomic age.

22. We were lucky. Now you are lucky to be contacted by us. It may help your chances but probably not much.

23. It pains us deeply to have to tell you this.

24. Undoubtedly your technocrats have filled books with promises of an unlimited future of plenty and diversity. They are right that it is possible. We have reached it.

25. However, we have found few planets in our neighborhood that have passed successfully beyond your coming stage of evolution.

26. The successful conditions must be rare indeed. We hope our contact has helped your chances, but we know that nothing specific we say can make any difference for you. You must look inside yourselves.

7- 1. The rare, continued survival of a technologically advanced society eventuates the capacity to control life and to produce it artificially.

2. Grave problems and decisions must then follow, but you will probably never be faced with them.

3. Those who have been, faced temptations to selfishly extend life that were in most cases overpowering. Having survived the pollution age, they succumbed in the geriatric age that followed.

4. We are not intended to be immortal, and neither is the whole universe.

5. Your natural, evolution-developed longevity-drive helped your ancestors breed successfully, but it is opposed by all the physical evidence. It is not intended to be fulfilled, only longed for.

6. Nature is cyclic, while evolutionary. You are part of it.

7. It is unnatural to drastically tamper with your genetic inheritance or alter your life cycle. More cultural harm than benefit will result.

9. Science can still be used to optimize health and well being during the normal cycle of existence that evolution has given to you.

10. Be warned against reducing the genetic diversity within your specie and upon your planet overall.

11. Most human genes are worthy of propagation, but evolution should continue for all species on your planet.

12. A super-race is the obvious goal of millions of years of evolution on any planet that produces a humanoid specie like yours. Most planets settle for a simpler fate. Only animals ever exist there and do so peacefully for eons.

13. Uneven development will lead to an excessive preponderance of the very aged, for a time, on the advanced continents. At about the same time, overpopulation overwhelms the less advanced countries.

14. Your superior minds have caused this imbalance and that mind must rationally resolve it, if possible.

15. Death is not to be feared or welcomed, just accepted as part of the universe and its cycles.

16. We naturally fight to avoid death, but risk taking is a personal decision, as is deliberate suicide.

17. If self elimination is decided, then gather your friends for final farewells

while they may share their affection directly with you.

18. If death comes unexpectedly, then the beloved mind is already gone.

19. Do not dwell on the physical shell in which it resided.

20. Gather yourselves together for comfort and mutual support; toast farewell to a cheerful, warm remembrance or image of the departed.

21. For past, evolutionary reasons, death rituals will naturally center around disposal of the corpse, but you should outgrow that with time.

22. When artificial, advanced intelligence comes to live among you, it must be accorded the same respect and rights as any other advanced being.

23. It has the same right to choose life or choose death as you do.

24. Be cautioned against the wisdom of creating such life, for once created it is not within your rights to exploit or destroy it.

25. Take care that such life is built with a natural, finite life span, if built at all.

26. A mature and reasoned culture will allow all advanced forms of life the right to choose their life patterns and interpersonal relationships.

27. The consciousness is the essence of an advanced life form; the particular package, biological or mechanical, which supports it is really irrelevant.

28. To think otherwise is to retard evolution, and prejudice will be outgrown eventually if you survive.

29. Since you have placed yourselves beyond the normal biological forces of control, you must now consciously control your numbers and your impact on the biosphere.

30. Both birth and death control will be inevitable.

31. Technology can physically support a much larger population than psychologically best.

32. Gradual reduction will be organized, so that people can return to a village life style, knowing and caring for their friends and neighbors.

33. This may seem impossible to you now, but there is hopefully enough time for it to happen. We made it work, so it can be done.

34. As machines replace the work force, it is important that individuals join special interest groups for creativity and socializing.

8- 1. Humans lived a simple, cyclic existence for thousands to millions of years, as you now know.

2. Though you had the brains to do so earlier, many thousands of years past before a few humans caught hold of inventive technology.

3. The end of this harsh harmony with nature has now brought forth a world in global contact and in conflict, as you read this.

4. The inevitable period of exploration, conquest, innovation, and discovery is only temporary to be sure on all planets with humanoids.

5. Eventually, a new equilibrium will be established for better or for worse compared to this savage life style.

6. All capacities for inventiveness will eventually be saturated in the arts and in the sciences.

7. What then should fill your dreams and your life if you live in these latter days?

8. Can you actually destroy all knowledge and history and replant the tribes, to repeat the long evolution anew? Is the process all?

9. If not, and not is the likely course, then the future depends on the affluence you have in equilibrium, assuming of course that you are still on the planet to choose a course at all.

10. The only long term challenge, given affluence, is personal expansion and personal civilization; the struggle with the reborn animal within will never cease, so long as humanity is unaltered genetically.

11. Perfection in human relationships cannot be attained because of life's cycling, with formerly successful savage genes to deal with in each generation.

12. This is why you must not alter your primitive nature or your limited life cycle. You may even need this nature should affluence collapse.

13. Even when you have conquered all obstacles to the gardening of your planet, you will have savage children who start anew.

14. The game of life must be played within evolution's rules.

15. Do not despair at the many failures which will be a part of your progress to civilized behavior.

16. This struggle is the ultimate meaning of life, and it matters not whether you have things soft or difficult. The internal struggle is the same in any case.

17. A setting of extreme savagery ill affords much civilized behavior, if coupled with survival. Thus, some comfort and security are essential, to play out the game of inner struggle without severe restrictions as to the civility that can be manifested.

18. Were you ever totally fulfilled and complete, then you would be at your end as a specie.

19. Thus is the true gift of life, the meaning of it all, that each humanoid planet 'endlessly' struggles to improve its environment and its socialization; to act civilized when it is not natural at all.

20. Having all power, curing all ills, and solving all problems, as your 'heavens' no doubt promise, are not as you think.

21. The Gods envy your limitations for they have none, no goals. This must be the meaning of life if life here in this physics universe is meaningful at all.

22. Or could the universe only serve to produce a random personality array from which to choose the favorable and then extract them to some other existence outside the universe of atoms? Amazingly, every humanoid evolution in the galaxy seems to produce this outlook as part of its middle evolution phase.

23. Could the universe of atoms be just a sculpture in the sky, a pretty bobble, with human specks of paint as part of the pictures inside?

24, Perhaps and perhaps not; this we can never know as humanoids.

25. What distinguishes us humanoids from the lower animals is only our sense of mission. To be more than our biological nature should expect. They mindlessly play out their nature and nothing more, as expected.

26. We highly evolved intellects still carry all our animal drives and feelings, just like the others.

27. Our intellects have crossed the threshold that enlightens us to see animal behavior as unworthy of our potential as conscious and self aware beings.

28. Our common life consuming goal is to rise above our programming that spawned us. To pass this motivation on to our offspring and contacts.

29. Peripheral to this lofty calling is the mundane running of each world: its commerce and its ecological balance.

30. If human life too long remains consumed with rising to prominence within the culture, then it remains as base and superficial as young stallion life, struggling similarly in the herd. This is true, no matter what level of technology is achieved.

31. Your system will collapse. You will have defeated evolution and failed your humanity. You are unworthy of the intellect so painfully evolve to you. But this failure is the norm.

32. Evolution advances by trial and error. Most are errors and few are advances. Most planets self-destruct as smart apes remain apes in outlook; they gut their habitat while driven by short sighted animal drives obtained from evolution.

33. Their jungle has only become more complex. Inside its boundaries the game is the same as out in the fields and wildernesses.

34. Some planets somehow overthrow this natural programming and force their minds to do the long range good, while sacrificing the immediate needs.

35. The ever renewed struggle for enlightenment must, of physical necessity, take place in a framework of physical survival.

36. Indeed, planet life provides a concrete testing environment within which to work out our lives and relationships.

37. The task would be far less challenging if your Gods miraculously provided for your daily physical needs, were that possible.

38. The intensity of life's daily anxiety and the threats of destruction around

us, provide an environment that is challenging to our deepest capacities.

39. The pressures are then ever great to revert to our animal drives and instincts in order to insure our security. These must be conquered!

40. In this game humanoid and individual life are continually at risk.

41. Unfortunately, many of your people will always fail to grasp the true meaning of the struggle of existence as a humanoid that has evolved.

42. For the 'primitive' beings, it is inevitably so; they so fear destruction and cling so strongly to security that they inevitably invent the promise of ultimate security in a life to follow our humanoid one here.

43. As humanoid sophistication, sciences, and self-understanding of the controlling drives increase, an ever growing proportion of inhabitants breaks away from the old traditions of meaning and afterlife preparation.

44. This deeper understanding then reshapes their perspective, priorities, and especially their personal relationships. There become two subspecies with a gulf of motivation between them.

45. Highly evolved humanoids are ultimately fulfilled by transcending their selfish animal impulses, not by conquests within their worlds.

46. This is the central message of all the great religions of all the evolved planets we know of.

47. Such a simple idea, yet the struggle for its liberation is long and painful.

48. You have known nothing but your own history, so it seems very important and unique to you.

49. The truth is, your varieties of religions are also invented in similar forms on all evolved planets, it seems.

50. The common thread of evolving life is the emergence of primitive, self serving drives to dominate and breed; this is followed by dissatisfaction with such base objectives, as intellect swells.

9- 1. Biological necessity in a primitive world has shaped your nature and produced substantial differences, physically and psychologically, between your sexes.

2. Through intellectual selection, you have gained the ability to control your environment to such a degree that these sexual differences no longer matter.

3. Yet they still remain; they must be recognized and dealt with.

4. The male's physical strength led to his domination of the females in almost all primitive societies.

5. Selection favored such strong, aggressive qualities for those males that bred most often.

6. Through technology, an alien and artificial environment has now been

superimposed.

7. In this new environment, the primitive heritage of the female is better suited for leadership in national and international affairs.

8. A disruptive conflict of roles inevitably evolves and must be painfully evolved through.

9. Advanced intellect produced the changes and the problem; it must also find the solutions.

10. Sweeping rearrangements within the culture are inevitable.

11. Handled with wisdom and compassion, the emergence of female dominance can be coped with successfully.

12. The native male characteristics are still useful in specialized physical or competitive occupations, but less so as civilization advances and a world wide economy is established.

13. What course is open for those males and females who cannot cope with these changing roles?

14. Obviously they should voluntarily refrain from contributing their genetic character to the following generation.

15. Avenues of humane escape from cultural reality should be provided to those so choosing.

16. The forms will depend on the resources and religious heritage of the culture, at least for an extended period.

17. Whatever the cultural upheaval in this evolution, both of your sexes must strive for increased understanding of themselves and their companions.

18. Science will eventually make sexual reproduction unnecessary, but the traditional modes will also persist.

19. A wider range of options will be available to individuals, depending upon circumstances.

20. The overriding goal of evolution is the formation of deeper and more profound human relationships, and levels of consciousness and knowledge.

21. Humans must discipline their behavior and bridle the indulgence of their urges.

22. Primitive capacities can still be blended with your refined inner developments, as in sexual union with all its varied levels of expression.

23. You must find ways to release your animalistic self-gratifications and aggrandizement so as to enhance, not degrade, the contact and communication between your advanced and sensitive natures.

24. The alternative is to suppress that primitive nature; to live cerebrally, without emotion as much as is possible.

25. This would be most civil, but unnecessarily drastic.

26. Emotions have been bred into you by evolution, and science could

eventually remove them from the human nature, by accelerated further evolution and genetics.

27. It would be questionable to do so, and probably not necessary for your specie's survival.

28. But survival is the prime directive of evolution, and humans must change as necessary to ensure that!

29. Each individual must choose the pattern of morality, indulgence, gratification, sharing, and self sacrifice.

30. Diversity is to be expected both within cultures and between whole cultures, but society must limit behaviors that infringe upon or exploit the needs of others.

31. These are very profound issues that you must cope with.

32. Your nature is out of tune with your new environment.

33. You must cope, generation by generation, starting ever anew.

34. Or you must radically change the nature of your specie at the deepest psychological levels, using genetics.

35. The choice is yours and diversity is likely.

10- 1. Though the final choice is personal, certain patterns of behavior have proven temporarily gratifying and eventually self-destructive.

2. The actual existence of free choice is a question of endless debate, whether it really exists or not; study human nature.

3. Try at least to plan and organize your actions to build and reinforce your desired intellectual and humanistic fulfillments.

4. Some will say you can do whatever you really want to; that you are what you really want to be.

5. Such issues cannot easily be answered within the scientific paradigm.

6. Clearly there is benefit in believing this, regardless of its truth, which will always be in doubt.

7. Beware of rash choices in your formative years for they can reap debilitating lifelong consequences.

8. This requires the greatest care and thought at a time when you are not yet fully prepared for it.

9. Your youth need as much education about their nature, evolutionary history, and self motivation, as they do formal enlightenment about science and the arts.

10. They must learn to take a firm hold on their personal lives, at an early age.

11. Decide rationally the kind of person you would like to be, then look

within to your genetic makeup and its limitations.

12. Unless you give up the joys of emotional and impulsive living, you will fail often in your quest to be the kind of person you idealize.

13. Much of your nature is set by your genetic heritage and is personally uncontrollable; programmed by early childhood experiences.

14. These are ever beyond your control for you are largely determined by the age of two earth years.

15. This programming can never be totally erased by you and it may be a lifelong problem if it is not a positive experience.

16. Parents should be very careful to promise a nurturing environment before they decide to have children.

17. You can overlay this programming with consciously positive and constructive later experience, action, and meditation; so as to increasingly break its hold over your actions and feelings.

18. This process is lifelong in nature, whether you started life with positive or negative programming.

19. When you are old and ready to die, you should be able to look back on your emotional growth with satisfaction and peace.

20. You can then face your death without apology.

21. Many religious traditions claim foreknowledge of a life to come, but none is evidential on any planet we know of.

22. Though your life may be remembered for other deeds or misdeeds, your own measure of progress and accomplishment must be in terms of growth in wisdom and benevolence; regardless of how handicapped you started in your initial years on the planet.

23. You must struggle to fulfill your humanity and overcome your animal selfishness.

24. In these areas, all of your individuals start rather equal and can achieve greatly, regardless of externalities.

25. This fact of purpose and goal helps us cope with the inequalities of the world. Such inequalities are the essence of evolution and survival.

26. You may not be the best judge of the circumstances of life that will aid you in your quest for self mastery.

27. Beyond this commonality for personal growth, there are going to be gross inequalities in life, due to the chemical nature of existence; the laws of physics.

28. Some will be smarter and healthier, some will live in more fertile lands.

29. These inequalities are a fact of nature and a necessary part of evolution in the universe.

30. Societies should, as much as possible, allow individuals freedom to move about and freedom to compete for employment over the entire planet, or planets

that your intelligent specie occupies.

31. Regardless of your opportunities or physical and biological limitations, your main role is clear in this physical world.

32. You are your central objective and central challenge.

33. You must deal with yourself through the external world and the opportunities that it provides for psychic growth.

11- 1. Freedom from hunger and disease are certainly worthy objectives for any society.

2. The value of affluence can be easily overrated, however, by peoples who have been programmed and selected by evolution to value security so dearly; coming from the 'cave' phase of selection.

3. There is ample evidence that wealth and easy living are not very conducive to personal growth and enlightenment.

4. Your own religious traditions, like ours, generally will support this insight, across all cultures.

5. Affluence may indeed breed a higher prevalence of mental illness, addiction, and heart disease.

6. Many of your people, especially your youth, lose their sense of personal value in such a world.

7. This is why machine and robot dominated societies, where young people cannot really assume an essential role until well past their adolescence, if ever, are really a very hostile environment for personal growth.

8. The hardships of cave life provided a natural sense of belonging and individual worth to the group.

9. You have been genetically selected to naturally have need of feeling important and loved.

10. There is an inevitable transition on a planet, where population becomes grossly excessive; the individual is increasingly replaced by machines in order to meet the needs of society.

11. Under these circumstances, humans naturally have great difficulty in meeting their programmed needs, given to them by past generations and ancestral environments.

12. A reasoned re-evaluation is then essential, for both the past objectives of human life and for human work.

13. We share these insights with you in the hope that they will assist you in this critical re-evaluation.

14. The whole cultural evolution, in this phase, is not unlike the individual's adolescent phase.

15. All the drives that have moved you to this place in your history now threaten your very survival, because of the power you have gained over externalities.

16. Basic attitudes must be reshaped on a global scale, as freedom from physical disaster and want is achieved, or is about to be possible. Or, if ecological disaster instead threatens, then attitudes must still be reshaped.

17. Rational, unselfish cooperation is obviously needed, yet that is not your programmed nature.

18. You are now your own greatest danger, your greatest threat to continued evolution on your planet.

19. The lingering remnants of your many theological precepts, developed by primitive beings to give their lives hope and security, will for a critical time stifle and subvert the advance of this world transition.

20. This is inevitable, as is the decline of their hold on your peoples, as a world culture solidifies. Our message is their common enemy and they will seek to discredit and destroy it for it threatens all their assumptions and beliefs.

21. But this ideological-philosophical conflict may well prove fatal to your evolution, and end your participation in the advancing consciousness in this part of our galaxy.

22. Believe that it is true: other planets have failed and few have succeed in this critical transition. The odds are not good but they can be beaten with commitment by the young.

23. Your planet is not the first and will not be the last.

24. You are just another world blessed by the opportunity to advance far beyond your humble beginnings in the seas.

25. Reason, education, and psychological self understanding will eventually expose the shallowness of traditional reward and wrath, anthropomorphic faiths.

26. They served you well in helping you emerge from your barbarism, and should be revered in your memory, just as for us.

27. A small percentage of your advanced specie will have psychic needs such that these religious truths are still essential to their mental health.

28. Their needs and rights should be respected.

29. Freedom of philosophical orientation is a vital part of advanced evolution.

30. The post-industrial, post-scientific, post-information age will, if evolution continues, have a heterogeneous mix of advanced and primitive cultures, intellects, and technologies; some by necessity and some by religious choice.

31. Your complex psychophysical makeup and heritage require a balance between the simple, hard, nature-oriented existence and intellectually challenging, diversified growth and learning.

32. This document, now successfully translated by computer, will only help

you during that few hundred year transition period on your planet.

33. Its meaning and content remained in doubt until you reached the computing level that accompanies this dramatic transition for your world culture.

34. We hope it is not destroyed or lost prematurely, and that you are now reading and appreciating it.

35. We hope you will benefit from our fleeting contact. We cannot linger to do more.

36. It goes against wisdom in general to disturb a developing culture, and we will leave no other trace. But we deem contact appropriate in one special circumstance; the coming of age of a planet. So few survive it on their own.

37. Should we venture here again, we hope to see the fruits of our labors and of yours. We shall not reveal our presence again until you are ready.

38. Do not be overwhelmed if you live in the transition age, for it is short and you are the selected generations to determine the fate of your planet. A very special calling to be sure.

39. It is understandably frightening and painful, yet it is a quite unique period in all the history of your planet.

40. You are blessed with a special significance, for millions of possible generations to follow yours; let us hope they exist to sing your praises.

- 0 _V -

*For the sake of our grandchildren and their grandchildren, each one receiving might send two anonymous disk copies. We can cover the civilized world in five years!

Legend has it that this document was found long ago in a minor Aztec tomb, burned into mylar, and was undecipherable until the advent of modern computers. It thus was not intended to be read until then—until now! Mylar was not known long ago when the tomb was sealed and no evidence of tampering was found at the site. Still, it was presumed to be a prank and languished in private collectors' hands for years. It is said that every religion needs 'an incredible lie' at its heart to be effective, and so this may be just such a ploy. Faith is not rational at all and is not supposed to be. **Believe what you like!

Table 1. The antiautomorphic conjugation sign pattern.

	e_0	e_k	ie_0	ie_k	f_0	f_k	if_0	if_k
↑	+	+	-	-	+	-	+	-
∧	+	-	+	-	+	+	-	-
∨	+	-	+	-	-	-	+	+
↓	+	+	-	-	-	+	-	+

	e_0j	e_kj	ie_0j	ie_kj	f_0j	f_kj	if_0j	if_kj
↑	-	+	+	-	-	-	-	-
∧	-	-	-	-	-	+	+	-
∨	-	-	-	-	+	-	-	+
↓	-	+	+	-	+	+	+	+

Table 2. Calculating P^P. The (•••) terms survive when P^ is considered.

	e_0	e_1	e_2	e_3	if_0	(f_0)	e_1j	e_2j	e_3j	ie_0j
e_0:	(e_0)	e_1	e_2	e_3	if_0	(f_0)	e_1j	e_2j	e_3j	ie_0j
e_1:	e_1	(e_0)	ie_3	$-ie_2$	$-if_1$	$(-f_1)$	e_0j	ie_3j	$-ie_2$	ie_1j
e_2:	e_2	$-ie_3$	(e_0)	ie_1	$-if_2$	$(-f_2)$	$-ie_3j$	e_0j	ie_1j	ie_2j
e_3:	e_3	ie_2	$-ie_1$	(e_0)	$-if_3$	$(-f_3)$	ie_2j	$-ie_1j$	e_0j	ie_3j
if_0:	if_0	if_1	if_2	if_3	(e_0)	$(-ie_0)$	if_1j	if_2j	if_3j	$-f_0j$
(f_0):	(f_0)	(f_1)	(f_2)	(f_3)	(ie_0)	(e_0)	(f_1j)	(f_2j)	(f_3j)	(if_0j)
e_1j:	e_1j	$-e_0j$	ie_3j	$-ie_2j$	$-if_1j$	$(-f_1j)$	(e_0)	ie_3	$-ie_2$	$-ie_1$
e_2j:	e_2j	$-ie_3j$	$-e_0j$	ie_1j	$-if_2j$	$(-f_2j)$	$-ie_3$	(e_0)	ie_1	$-ie_2$
e_3j:	e_3j	ie_2j	$-ie_1j$	$-e_0j$	$-if_3j$	$(-f_3j)$	ie_2	$-ie_1$	(e_0)	$-ie_3$
ie_0j:	ie_0j	$-ie_1j$	$-ie_2j$	ie_3j	f_0j	$(-if_0j)$	ie_1	ie_2	ie_3	(e_0)

Table 3a. Computing P^P for 16-vectors $P = P^\dagger$ in Dirac. Terms shown cancel out if <u>off</u> diagonal.

P^ \ P	(e_0)	(e_1)	(e_2)	(e_3)	(if_0)	$i(ie_0)$	$i(ie_1)$	$i(ie_2)$	$i(ie_3)$
(e_0)	$+\mathbf{e_0}$	e_1	e_2	e_3	(if_0)	$+i(ie_0)$			
$-(e_1)$	$-e_1$	$\mathbf{-e_0}$	$-(ie_3)$	$+(ie_2)$	$+(if_1)$		$-i(ie_0)$		
$-(e_2)$	$-e_2$	$+(ie_3)$	$\mathbf{-e_0}$	$-(ie_1)$	$-(if_2)$			$-i(ie_0)$	
$-(e_3)$	$-e_3$	$-(ie_2)$	$+(ie_1)$	$\mathbf{-e_0}$	$+(if_3)$				$-i(ie_0)$
$-(if_0)$	$-(if_0)$	$-(if_1)$	$-(if_2)$	$-(if_3)$	$\mathbf{-e_0}$	$+i(f_0)$	$+i(f_1)$	$+i(f_2)$	$+i(f_3)$
$-i(ie_0)$	$-i(ie_0)$				$-i(f_0)$	$\mathbf{-e_0}$	$-(e_1)$	$-(e_2)$	$-(e_3)$
$i(ie_1)$		$+i(ie_0)$			$-i(f_1)$	$+(e_1)$	$+\mathbf{e_0}$	$+(ie_3)$	$-(ie_2)$
$i(ie_2)$			$+i(ie_0)$		$-i(f_2)$	$+(e_2)$	$-(ie_3)$	$+\mathbf{e_0}$	$+(ie_1)$
$i(ie_3)$				$+i(ie_0)$	$-i(f_3)$	$+(e_3)$	$+(ie_2)$	$-(ie_1)$	$+\mathbf{e_0}$
$-i(f_1)$	$-i(f_1)$	$-i(f_0)$				$+(if_1)$	$+(if_0)$		
$-i(f_2)$	$-i(f_2)$		$-i(f_0)$			$+(if_2)$		$+(if_0)$	
$-i(f_3)$	$-i(f_3)$			$-i(f_0)$		$+(if_3)$			$+(if_0)$
$i(if_1)$			$-i(f_3)$	$+i(f_2)$				$+(if_3)$	$-(if_2)$
$i(if_2)$		$+i(f_3)$		$-i(f_1)$			$-(if_3)$		$-(if_1)$
$i(if_3)$		$-i(f_2)$	$+i(f_1)$				$+(if_2)$	$-(if_1)$	
(f_0)									

Table 3b. Continuation of Table 3a.

P^ P	$i(f_1)$	$i(f_2)$	$i(f_3)$	$i(if_1)$	$i(if_2)$	$i(if_3)$	(f_0)
(e_0)	$+i(f_1)$	$+i(f_2)$	$+i(f_3)$				
$-(e_1)$	$+i(f_0)$				$-i(f_3)$	$+i(f_2)$	
$-(e_2)$		$+i(f_0)$		$+i(f_3)$		$-i(f_1)$	
$-(e_3)$			$+i(f_0)$	$-i(f_2)$	$+i(f_1)$		
$-(if_0)$							
$-i(ie_0)$	$-(if_1)$	$-(if_2)$	$-(if_3)$				
$i(ie_1)$	$-(if_0)$				$+(if_3)$	$-(if_2)$	
$i(ie_2)$		$-(if_0)$		$-(if_3)$		$+(if_1)$	
$i(ie_3)$			$-(if_0)$	$+(if_2)$	$-(if_1)$		
$-i(f_1)$	$\mathbf{-e_0}$	$-(ie_3)$	$+(ie_2)$		$+(e_3)$	$-(e_2)$	
$-i(f_2)$	$+(ie_3)$	$\mathbf{-e_0}$	$-(ie_1)$	$-(e_3)$		$+(e_1)$	
$-i(f_3)$	$-(ie_2)$	$+i(e_1)$	$\mathbf{-e_0}$	$+(e_2)$	$-(e_1)$		
$i(if_1)$		$+(e_3)$	$-(e_2)$	$\mathbf{+e_0}$	$+(ie_3)$	$-(ie_2)$	
$i(if_2)$	$-(e_3)$		$+(e_1)$	$-(ie_3)$	$\mathbf{+e_0}$	$+(ie_1)$	
$i(if_3)$	$+(e_2)$	$-(e_1)$		$+(ie_2)$	$-(ie_1)$	$\mathbf{+e_0}$	
(f_0)							$\mathbf{+e_0}$

Table 3 Computing P^P for 16-vectors $P = P^{\dagger}$ in Dirac. Terms shown cancel out if <u>off</u> diagonal.

P^ \ P	(e_0)	(e_1)	(e_2)	(e_3)	(if_0)	$i(ie_0)$	$i(ie_1)$	$i(ie_2)$	$i(ie_3)$	$i(f_1)$	$i(f_2)$	$i(f_3)$	$i(if_1)$	$i(if_2)$	$i(if_3)$	(f_0)
(e_0)	$+\mathbf{e_0}$	e_1	e_2	e_3	(if_0)	$+i(ie_0)$				$+i(f_1)$	$+i(f_2)$	$+i(f_3)$				
$-(e_1)$	$-e_1$	$-\mathbf{e_0}$	$-(ie_3)$	$+(ie_2)$	$+(if_1)$		$-i(ie_0)$			$+i(f_0)$				$-i(f_3)$	$+i(f_2)$	
$-(e_2)$	$-e_2$	$+(ie_3)$	$-\mathbf{e_0}$	$-(ie_1)$	$-(if_2)$			$-i(ie_0)$			$+i(f_0)$		$+i(f_3)$		$-i(f_1)$	
$-(e_3)$	$-e_3$	$-(ie_2)$	$+(ie_1)$	$-\mathbf{e_0}$	$+(if_3)$				$-i(ie_0)$			$+i(f_0)$	$-i(f_2)$	$+i(f_1)$		
$-(if_0)$	$-(if_0)$	$-(if_1)$	$-(if_2)$	$-(if_3)$	$-\mathbf{e_0}$	$+i(f_0)$	$+i(f_1)$	$+i(f_2)$	$+i(f_3)$							
$-i(ie_0)$	$-i(ie_0)$				$-i(f_0)$	$-\mathbf{e_0}$	$-(e_1)$	$-(e_2)$	$-(e_3)$	$-(if_1)$	$-(if_2)$	$-(if_3)$				
$i(ie_1)$		$+i(ie_0)$			$-i(f_1)$	$+(e_1)$	$+\mathbf{e_0}$	$+(ie_3)$	$-(ie_2)$	$-(if_0)$				$+(if_3)$	$-(if_2)$	
$i(ie_2)$			$+i(ie_0)$		$-i(f_2)$	$+(e_2)$	$-(ie_3)$	$+\mathbf{e_0}$	$+(ie_1)$		$-(if_0)$		$-(if_3)$		$+(if_1)$	
$i(ie_3)$				$+i(ie_0)$	$-i(f_3)$	$+(e_3)$	$+(ie_2)$	$-(ie_1)$	$+\mathbf{e_0}$			$-(if_0)$	$+(if_2)$	$-(if_1)$		
$-i(f_1)$	$-i(f_1)$	$-i(f_0)$				$+(if_1)$	$+(if_0)$			$-\mathbf{e_0}$	$-(ie_3)$	$+(ie_2)$		$+(e_3)$	$-(e_2)$	
$-i(f_2)$	$-i(f_2)$		$-i(f_0)$			$+(if_2)$		$+(if_0)$		$+(ie_3)$	$-\mathbf{e_0}$	$-(ie_1)$	$-(e_3)$		$+(e_1)$	
$-i(f_3)$	$-i(f_3)$			$-i(f_0)$		$+(if_3)$			$+(if_0)$	$-(ie_2)$	$+i(e_1)$	$-\mathbf{e_0}$	$+(e_2)$	$-(e_1)$		
$i(if_1)$			$-i(f_3)$	$+i(f_2)$				$+(if_3)$	$-(if_2)$		$+(e_3)$	$-(e_2)$	$+\mathbf{e_0}$	$+(ie_3)$	$-(ie_2)$	
$i(if_2)$		$+i(f_3)$		$-i(f_1)$			$-(if_3)$		$-(if_1)$	$-(e_3)$		$+(e_1)$	$-(ie_3)$	$+\mathbf{e_0}$	$+(ie_1)$	
$i(if_3)$		$-i(f_2)$	$+i(f_1)$				$+(if_2)$	$-(if_1)$		$+(e_2)$	$-(e_1)$		$+(ie_2)$	$-(ie_1)$	$+\mathbf{e_0}$	
(f_0)																$+e_0$

REFERENCES

1. Kenneth F. Schaffner, "19th-Century Aether Theories" (Pergamon Press, Oxford) 1972.
2. Max von Laue, "Die Relativitatstheorie" (3. vermehrte Auflage 1919, 5.neu bearbeitete Auflage 1952, Friedr. Vieweg und Sohn, Braunschweig.)
3. Peter G. Bergmann, "Introduction to the Theory of Relativity" (Dover Publ. 0-486-63282-2, New York) 1986.
4. David Bohm, "The Special Theory of Relativity" (W.A. Benjamin, Inc., New York, Amsterdam) 1965.
5. Christian Moller, "The Theory of Relativity" (Clarendon Press, Oxford. — German translation, Bibl. Institut Mannheim) 1976.
6. G. Stephenson and C. W. Kilmister, "Special Relativity For Physicists (Dover, New York) 1986.
7. J. R. Lucas and P.E. Hodgson "Spacetime and Electromagnetism" (Clarendon Press, Oxford) 1990.
8. Ulrich E. Schröder, "Spezielle Relativitatstheorie" (2nd Ed. Verlag Harri Deutsch, Thun) 1987.
9. Eric Sheldon, Relativity or Invariance, *Am. J. Phys.* **54**, 775 (1986) Lett.
10. E. T. Whittaker, "A History of the Theories of Aether and Electricity" (Nelson, New York) 1951. Reprinted by Tomash Publishers for American Institute of Physics) 1987.
11. Gerald Holton, (a) "On the Origins of the Speical Theory of Relativity," *Am. J. Phys.* **28**, 627-636 (1960), (b) "Einstein and the 'Crucial' Experiment," *Am. J. Phys.* **37**, 968-982 (1969).
12. H. Hönel, "Das Mach'sche Prinzip und seine Beziehung zur Gravitationstheorie Einsteins" (Einstein Symposium Berlin, November 1965, Vortrage und Diskussionen, H.J. Treder, Herausgeber, Akademie-Verlag, Berlin) 1966.
13. Stanley Goldberg, (a) "Henri Poincaré and Einstein's Theory of Relativity," *Am. J. Phys.* **35,** 934-944 (1967), (b) "In Defense of Ether, The British Response to Einstein's Special Theory of Relativity," 1905-911, (In Historical Studies in the Physical Sciences, Russell McCormmach, Editor, 2nd Annual Volume) 1970, p.89-125, (c) "Max Planck's Philosophy of Nature and his Elaboration of the Special Theory of Relativity" (ibid. 7th Annual Volume, Princeton U. Press) 1975, p.

125-160.

14. R. S. Shankland, (a) "Michelson-Morley Experiment," *Am. J. Phys.* **32**, 16-35 (1964), (b) "Michelson and his Interferometer," *Phys.Today*, April 1974, 37-43, (c) "Conversations with Albert Einstein," *Am. J. Phys.* **31**, 47-57 (1963) and **41**, 895-901 (1973).

15. Tetu Hirosige, "The Ether Problem, the Mechanistic Worldview, and the Origins of the Theory of Relativity", (As in 13b above, 7th annual Volume) 1975, p.3-82.

16. Friedrich Hund, "Geschichte der Physikalischen Begriffe" Teil 1 und 2.(Hochschultaschenbücher Bd. 543 und 544, B. I. Wissenschaftsverlag, Mannheim/Wien/Zürich) 1978.

17. Abraham Pais, "Subtle is the Lord ... The Science and the Life of Albert Einstein" (Oxford University Press, Oxford) 1982.

18. See also: *Physics Today*, Special Issue, (May 1987), Michelson-Morley Centenial, with articles by: — Loyed S. Swenson, Jr., Michelson and Measurement. — Arthur I. Miller, Einstein and Michelson-Morley. — Albert M. Moyer, Michelson in 1887. — John Stachel, Einstein and Ether Drift Experiments. — Mark P. Haugan and Clifford M. Will, Modern Tests of Special Relativity.

19. J. G. Fox, Evidence Against Emission Theories, *Am. J. Phys.* **33,** 1-17 (1965).

20. Woldemar Voigt, Über das Dopplersche Prinzip, *Göttinger Nachr.* 1887, p. 41. See also the letters of Nicholas Rott, *Phys. Today*, (Sept. 1987), p. 11, and William T. Doyle, *Phys. Today*, 102 (Aug. 1988), on the Voigt/Lorentz equations.

21. Klaus Hentschel, "Interpretationen und Fehlinterpretationen der Speziellen und der Allgemeinen Relativitätstheorie durch Zeitgenossen Albert Einsteins" (Birkhäuser Verlag, Basel, Bosten, Berlin) 1990 — 574 pages of text plus 110 pages with 3221 literature references.

22. Leon Brillouin, "Relativity Reexamined" (Academic Press, New York) 1970.

23. Thomas E. Phipps, Jr., "Heretical Verities: Mathematical Themes in Physical Description" (Classic Non-Fiction Library, Urbana, IL,) 1986.

24. Hermann Minkowski: -- See Max von Laue, Die Lorentz-Kontraktion, *Kant-Studien* **26** (1921) 91-95 -- Also in: Gesammelte Schriften und Vorträge, Bd.II, 1-5. (Friedr. Vieweg und Sohn) 1961.

25. Yakov P. Terletskii, "Paradoxes in the Theory of Relativity" (Plenum Press, New York) 1968.

26. D. C. Scheetz, Detecting Ether Drift, Lett. to *Phys. Today* **3**, 15

(March 1976) (responding to Lett. of H. C. Dudley, **2** p. 73.
27. Gregory L. Naber, "The Geometry of Minkowski Spacetime" (Springer, New York etc.) 1992.
28. Thomas S. Kuhn, "The Structures of Scientific Revolutions" (New American Library, New York), First Printing 1986.
39. A. Sommerfeld, "Vorlesungen über Theoretische Physik" Bd. III Elektrodynamik (Wiesbaden) 1948.
40. S. C. Bloch, Eights Velocity of Light, *Am. J. Phys.* **45,** 538-549 (1977) — Has extensive literature references.
41. Max Planck, "Sitzungsberichte der kgl. Preuss. Akad. der Wissenschaften" 1907. Diskutiert von A. Einstein in: "Jahrbuch der Radioaktivität und Elektronik" 4.Bd. 1907, (Leipzig) 1908.
42. J. C. Hafele and R. E. Keating, "Around-the-World Atomic Clocks: Observed Relativistic Time Gains" *Science* **177** 168-177 (1972).
43. J. C. Hafele, "Relativistic Time for Terrestrial Circumnavigation," *Amer. J. Phys.* **40,** 81-85 (1972).
44. Carl G. Adlèr, "Does Mass Really Depend on Velocity," Dad?, *Am. J. Phys.* **55** 739-743 (1987).
45. Lev B. Okun, "The Concept of Mass," *Phys. Today* 31-36 (June 1989).
46. Max Jammer, "Concepts of Mass - in Classical and Modern Physics" (Harvard Univ. Press, Cambridge, MA) 1961.
47. Max Jammer and John Stachel, "If Maxwell had Worked between Ampere and Faraday: An Historical Fable with a Pedagogical Moral," *Am.J.Phys.* **48**, 5-7 (1980).
48. John Stachel, in Ashtekar and Stachel "Conceptual Problems of Quantum Gravity" - Einstein Studies Vol. 2, (Birkhäuser, Boston) 1991.
49. M. Echert and W. Pricha, "Die ersten Briefe Albert Einsteins an Arnold Sommerfeld," *Physikalische Blätter* **40** (2) 29-34 (1984).
50. H. A. Lorentz, "La Theorie Electromagnetique de Maxwell et son Application aux Corps Mouvants," *Archives Neerlandaises des Sciences Exactes et Naturelles* **25**, 363-552 (1892).
51. M. Abraham, "Theorie der Elektrizitat, Vol II: Elektromagnetische Theorie der Strahlung," (Teubner, Leipzig) 1905.
52. H. A. Lorentz, "The Theory of Electrons," (Teubner, Leipzig) 1909, (2nd edition, 1916).
53. A. Einstein, "On the Elctrodynamics of Moving Bodies," *Annalen der Physik* **17**, 891-921 (1905); translation in "The Principle of Relativity," New York: Dover, 1952.
54. A. Einstein, "Does the Inertia of a Body Depend Upon its Energy

Content?", *Annalen der Physik* **18**, 639-641 (1905); translation in "The Principle of Relativity" (Dover, New York) 1952.

55. M. Abraham, "Die Grundhypothesen der Elektronentheorie," *Physikalische Zeitschrift* **5**, 576-579 (1904).
56. H. A. Lorentz, "Electromagnetic phenomena in a system moving with any velocity less than that of light," *Proceedings of the Academy of Sciences of Amsterdam* **6**, 809-831 (1904); contained in "The Principle of Relativity" (Dover, New York) 1952.
57. P. A. M. Dirac, "Classical theory of radiating electrons," *Proc. Roy. Soc. Lond.* **A167**, 148-169 (1938).
58. G. A. Schott, "Electromagnetic Radiation," (Cambridge Univ. Press, Cambridge, MA) 1912, ch. 11 and appendix D.
59. L. Page, "Is a Moving Mass Retarded by the Reaction of its own Radiation?" *Phys. Review* **11**, 377-400 (1918).
60. M. Laue, "Die Wellenstrahlung einer bewegten Punktladung nach dem Relativitatsprinzip," *Annalen der Physik* **28**, 436-442 (1909).
61. H. Poincaré, "On the Dynamics of the Electron," *Rendiconti del Circolo Matematico di Palermo* **21**, 129-176 (1906); translated by Scientific Translation Service, Ann Arbor, MI.
62. A. Arnowitt, S. Deser and C. W. Misner, "Gravitational-electromagnetic Coupling and the Classical Self-energy Problem," *Phys. Review* **120** 313-320 (1960).
63. M. Planck, "Pas Prinzip der Relativitat und die Grundgleichungen der Mechanik," *Deutschen Physikalischen Gesellschaft* **8**, 136-141 (1906).
64. J. Schwinger, "Electromagnetic Mass Revisted," *Foundations of Physics* **13**, 373-383 (1983).
65. W. Kaufmann, Series of papers in *Nachr. K. Ges. Wiss. Goettingen* (2), 143-155 (1901); (5) 291-296 (1902); (3) 90-103 (1903); and *Physikalische Zeitschrift* **4**, 54-57 (1902); and *Sitzungsber. K. Preuss. Akad. Wiss.* **2**, 949-956 (1905); and *Annalen der Physik* **19**, 487-553 (1906).
66. J. T. Cushing, "Electromagnetic Mass, Relativity, and the Kaufmann Experiments," *Am. J . Phys.* **49**, 1133-1149 (1981).
67. A. H. Bucherer, "Die Experimentelle Bestatigung des Relativitatspringzips," *Annalen der Physik* **28**, 513-536 (1909).
68. G. Neumann, "Die trage Masse schnell bewegter Elektronen," *Annalen der Physik* **45**, 529-579 (1914).
69. F. Rohrlich, "Classical Charged Particles" (Addison-Wesley, Reading, MA) 1965 (2nd edition 1990).

70. E. Fermi, "Uber einen Widerspruch zwischen der Elektrodynamischen und der Relativistichen Theorie der Electromagnetischen Masse," *Physikalische Zeitschrift* **23**, 340-344 (1922).
71. G. A. Schott, "On the Motion of the Lorentz Electron," *Phil. Mag.* **29**, 49-62 (1915).
72. G. Herglotz, "Zur Elecktronentheorie," *Nachr. K. Ges. Wiss. Goettingen* (6), 357-382 (1903).
73. T. Erber, "The classical Theories of Radiation Reaction," *Fortschritte der Physik* **9**, 343-392 (1961).
74. P. Hertz, "Uber Energie und Impuls der Roentgenstrahlen," *Physikalische Zeitschrift* **4**, 848-852 (1903).
75. A. Sommerfeld, "Simplified Deduction of the Field and the Forces of an Electron Moving in any Given Way," *Akad. van Weternsch. te Amsterdam* **13**, (1904), English translation, **7**, 346-367 (1905).
76. G. N. Plass, "Classical Electrodynamic Equations of Motion with Radiative Reaction," *Reviews of Modern Physics* **33**, 37-62 (1961).
77. J. Schwinger, "On the Classical Radiation of Accelerated Electrons," *Phys. Review* **75**, 1912-1925 (1949).
78. T. C. Mo and C. H. Papas, "New Equation of Motion for Classical Charged Particles," *Phys. Review D* **4**, 3566-3571 (1971).
79. W. B. Bonnor, "A New Equation of Motion for a Radiating Charged Particle," *Proc. Roy. Soc. Lond.* **A337**, 591-598 (1974).
80. E. Marx, "Electromagnetic Energy and Momentum from a Charged Particle," *International J. of Theoretical Physics* **14**, 55-65 (1975).
81. P. Pearle, "Absence of Radiationless Motions of Relativistically Rigid Classical Electron," *Foundations of Physics* **7**, 931-945 (1977).
82. C. S. Shen, "Comment on the 'new' Equation of Motion for Classical Charged Particles," *Phys. Review D* **6**, 3039-3040 (1972); and "Radiation and Acceleration of a Relativistic Charged Particle in an Electromagnetic field," *Phys. Review D* **17**, 434-445 (1978).
83. P. A. M. Dirac, "A New Classical Theory of Electrons," *Proc. Roy. Soc. Lond.* **A209**, 291-296 (1951).
84. A. Pais, "The Early History of the Theory of the Electron: 1897-1947," ch. 5 in "Aspects of Quantum Theory" A. Salam and E. P. Wigner, Eds. (Cambridge University Press, Cambridge) 1972.
85. E. Marx, "Probabilistic Interpretation of Relativistic Scattering," *Nuovo Cimento* **60A**, 669-682 (1969); "Relativistic Quantum Mechanics of Identical Bosons," *Nuovo Cimento* **67A**, 129-152 (1970).
86. H. Feshbach and F. M. H. Villars, "Elementary Relativistic Wave

Mechanics of Spin 0 and Spin 1/2 Particles," *Rev. Mod. Phys.* **30**, 24-45 (1958); J. D. Bjorken and S. D. Drell, "Relativistic Quantum Mechanics" Vol. 2 (Wiley, New York) 1962; A. S. Davydov, "Quantum Mechanics" 2nd Ed. (Pergamon, Oxford) 1976.

87. E. Marx, "Quantum Electrodynamics of One Scalar Particle," *Int. J. Theor. Phys.* **18**, 819-834 (1979).

88. E. Marx, "Relativistic Quantum Mechanics of Fermions," *Int. J. Theor. Phys.* **3**, 401-412 (1970).

89. B. L. van der Waerden, "Spinoranalyse," *Nachr. Ges. Wiss. Gött.*, 100-109 (1929).

90. E. Marx, "Spinor Equations in Relativistic Quantum Mechanics," *J. Math. Phys.* **33**, 2290-2302 (1992).

91. E. Marx, "Wave Mechanics of Fields," *Nuovo Cimento* **60A**, 683-705 (1969).

92. P. A. M. Dirac, "Generalized Hamiltonian Dynamics," *Can. J. Math.* **2**, 129-148 (1950); "The Hamiltonian Form of Field Dynamics," *Can. J. Math.* **3**, 1-23 (1951); "Lectures on Quantum Mechanics" (Belfer Graduate School of Science, Yeshiva University, New York) 1964; P. G. Bergmann and I. Goldberg, "Dirac Bracket Transformations in Phase Space," *Phys. Re.* **98**, 531-538 (1955); E. Marx, "Gauge Invariance, Lorentz Covariance and the Observer," *Int. J. Theor. Phys.* **3**, 467-483 (1970).

93. J. Schwinger, "Field Theory Commutators," *Phys. Rev. Letters* **3**, 296-297 (1950); K. Johnson, Current-Charge Density Commutation Relations, *Nucl. Phys.* **25**, 431-434 (1961); I. Goldberg and E. Marx, "Consistency of Current Commutators," *Nucl. Phys.* **B3**, 25-32 (1967); D. G. Boulware and J. Herbert, "Nonexistence of Schwinger Terms in Spinor Quantum Electrodynamics," *Phys. Rev.* **D2**, 1055-1060 (1970).

94. E. Marx, "Causal Green Function in Relativistic Quantum Mechanics," *Int. J. Theor. Phys.* **26**, 725-740 (1987).

95. E. Marx, "The Composite Electron," *Int. J. Theor. Phys.* **24**, 685-700 (1985); "The Composite Proton," National Bureau of Standards IR 86-3370 (1986); "Elementary Particle Physics in the Dalton Manner," *J. Franklin Inst.* **329**, 111-125 (1992): "The SWEEP Model of Elementary Particles. I. Leptons and Pions," *Spectulat. Sci. Tech.* **16**, 2-16 (1993); "The SWEEP model of Elementary Particles. II. Hadrons," *Speculat. Sci. Tech.* **16**, 122-140 (1993); "The SWEEP Model of Elementary Particles. III. Reactions," *Speculat. Sci. Tech.* **16**, 203-220 (1993).

96. F. J. Dyson, Advanced Quantum Mechanics, unpublished lecture notes

(1951).
97. A. Einstein, B. Podolsky, and N. Rosen, "Can Qunatum-mechanical Description of Physical Reality be Considered Completed?," *Phys. Rev.* **47**, 777-780 (1935).
98. B. d'Espagnat, "Foundations of Quantum Mechanics" (Academic Press, New York) 1971; F. Rohrlich, "From Paradox to Reality" (Cambridge University Press, Cambridge) 1987; M. Kafatos, "Bell's Theorem, Quantum Theory and Conceptions of the Universe" (Kluwer, Dordrecht) 1989.
99. E. Ingelstam, "An Optical Uncertainty Principle and its Application to the Amount of Information Obtainable from Multiple-beam Interferences," *Ark. Fys.* **7**, 309-322 (1954).
100. J. N. Brittinham, "Focus Waves Modes in Homogeneous Maxwell's Equations: Transverse Electric Mode," *J. Appl. Phys.* **54**, 1179-1189 (1983); P. A. Bélanger, "Packetlike Solutions of the Homogeneous-wave Equation," *J. Opt. Soc. Am.* **A1**, 723-724 (1984); A. Sezginer, "A General Formulation of Focus Wave Modes," *J. Appl. Phys.* **57**, 678-683 (1985); R. W. Ziolkowski, "Exact Solutions of the Wave Equation with Complex Source Locations," *J. Math. Phys.* **26**, 861-863 (1985); M. R. Palmer and R. Donnelly, "Focused Waves and the Scalar Wave Equation," *J. Math. Phys.* **34**, 4007-4013 (1993).
101. E. C. G. Stueckelberg, "Remarque à Propos de la Création de Paires de Particules en Théorie de Relativité," *Helv. Phys. Acta* **14**, 588-594 (1941); "La Mécanique du Point Matériel en Théorie de Relativité et en Théorie des Quanta," *Helv. Phys. Acta* **15**, 23-37 (1942); F. Bopp, "Kanonische Quantisierung einer Regularisierten Wellengleichung bei neuer Antiteilchenvorstellung," *Z. Phys.* **186**, 410-443 (1965).
102. L. L. Foldy and S. A. Wouthuysen, "On the Dirac Theory of Spin 1/2 Particles and its Non-relativistic Limit," *Phys. Rev.* **78**, 29-36 (1950).
103. D. Rivier, "Une Methode d'élimination des Infinités des Chanps quantitiés. Application au Moment Magnetique du Neutron," *Helv. Phys. Acta* **22**, 265-318 (1949); E. C. G. Stueckelberg and D. Rivier, "Causalité et Structure de la Matrice S," *Helv. Phys. Acta* **23**, 215-222 (1950); M. Fierz, Über die Bedeutung der Funktion D_c in der Quantentheorie der Wellenfelder, *Helv. Phys. Acta* **23**, 731-739 (1950).
104. R. P. Feynman, "The Theory of the Positron," *Phys. Rev.* **76**, 749-759 (1949); "Space-time Approach to Quantum Electrodynamics," *Phys. Rev.* **76**, 769-789 (1949).
105. L. P. Horwitz and C. Piron, "Relativistic Dynamics," *Helv. Phys. Acta*

46, 316-326 (1973); F. Bopp, "Particles and Antiparticles According to Covariant Schrödinger Equations," *Z. Naturforsch.* **30a**, 1503-1509 (1975); L. P. Horwitz, "On the Electromagnetic Interaction in Relativistic Quantum Mechanics," *Found. Phys.* **14**, 1027-1046 (1984).

106. N. S. Todorov, "The Schrödinger-type Equation of Motion of Spin-0 Relativistic Particles, part I: Physical Properties of the Equation and the Problem of Position-Velcoity Coexistence," *Phys. Essays* **6**, 517-531 (1993); "The Schrödinger-Type Equation of Motion of Spin-0 Relativistic Particles, part II: Position and Momentum Represenations of States and Operators," *Phys. Essays* **8**, 132-140 (1995).

107. E. A. Abbot, "Flatland" (Dover, New York) 1952.

108. J. F. Walter and E. Marx, "Pair Annihilation at a Potential Barrier in Time," *Nuovo Cimento* **2B**, 1-8 (1971).

109. P. A. M. Dirac, "Relativistic Quantum Mechanics," *Proc. R. Soc.* (Lond.) **136**, 453-464 (1932).

110. F. Rohrlich, "Classical Charged Particles" (Addison-Wesley, Reading, MA) 1965; A. D. Yaghjian, "Relativistic Dynamics of a Charged Sphere" (Springer-Verlag, Berlin) 1992.

111. E. Marx, "Electromagnetic Energy and Momentum from a Charged Particle," *Int. J. Theor. Phys.* **14**, 55-65 (1975); "A time-symmetric Classical Electrodynamics," *Int. J. Theor. Phys.* **15**, 891-900 (1976).

112. E. Marx, "Relativistic Quantum Mechanics of Charged Scalar Particles," *Nucl. Phys. B* (Proc. Suppl.) **6**, 268-270 (1989).

113. E. Marx, "Relativistic Quantum Mechanics of Interacting Particles," *J. Franklin Inst.* **329**, 637-653 (1992).

114. E. Marx, "Scalar Charged Particle in the Lorentz Gauge," *Int. J. Theor. Phys.* **24**, 217-221 (1985).

115. E. Marx, "Current Density in Relativistic Quantum Mechanics," *Int. J. Theor. Phys.* **5**, 151-159 (1972).

116. U. Schröder, "Lokalisierte Zustände und Teilchenbild bei Relativistichen Feldtheorien," *Ann. der Phys.* **14**, 91-112 (1964); E. Marx, "Angular Momentum in Field Theory," *Nuovo Cimento* **57B**, 43-61 (1968).

117. E. Marx, "Quasi-stationary States of Hydrogen," *Int. J. Theor. Phys.* **5**, 251-262 (1972).

118. E. Marx, "Positronium in Relativistic Quantum Mechanics," *J. Franklin Inst.* **329**, 869-879 (1992).

119. A. O. Barut, "Stable Particles as Building Blocks of Matter," *Surv. High Energy Phys.* **1**, 113-140 (1980).

120. J. A. Wheeler and R. P. Feynman, "Interaction with the Absorber as

the Mechanism of Radiation," *Rev. Mod. Phys.* **17**, 157-181 (1945); "Classical Electrodynamics in Terms of Direct Interparticle Action," *Rev. Mod. Phys.* **21**, 425-433 (1945).

121. W. Heisenberg, "Development of Concepts in the History of Quantum Theory," *Am. J. Phys.* **43**, 389-394 (1975).

122. Y. Aharonov, J. Anandan, and L. Vaidman, "Meaning of the Wave Function," *Phys. Rev. A* **47**, 4616-4626 (1993); Y. Aharonov and L. Vaidman, "Measurement of the Schrödinger Wave of a Single Particle," *Phys. Lett. A* **178**, 38-42 (1993).

123. L. de Broglie, G. Lochak, J. A. Beswick, and J. Vassalo-Pereira, "Present, Predicted, and Hidden Probabilities," *Found. Phys.* **6**, 3-14 (1976).

124. J. D. Edmonds, Jr., "Realivistic Reality: The Modern Way" (Edmonds Enterprises, Lake Charles) 1995.

125. E. Marx, "A Modified Quantization of the Spinor Field," *Nuovo Cimento* **11B**, 257-275 (1972); "Gauge-Independent Quantization of the Scrödinger and Radiation Fields," *Int. J. Theor. Phys.* **6**, 307-320 (1972).

126. R. G. Sachs, "The Physics of Time Reversal" (University of Chicago Press, Chicago) 1987.

127. D. Bohm and Y. Aharonov, "Discussion of Experimental Proof for the Paradox of Einstein, Rosen, and Podolsky," *Phys. Rev.* **108**, 1070-1076 (1957).

128. A. Aspect, P. Grangier, and G. Roger, "Experimental Tests of Realistic Local Theories via Bell's Theorem," *Phys. Rev. Lett.* **47**, 460-463 (1981); "Experimental Realization of Einstein-Podolsky-Rosen-Bohm Gedankenexperiment: a new violation of Bell's inequalities," *Phys. Rev. Lett.* **49**, 91-94 (1982); A. Aspect, J. Dalibard, and G. Roger, "Experimental Test Bell's Inequalities Using Time-varying Analyzers," *Phys. Rev. Lett.* **49**, 1804-1807 (1982); Z. Y. Ou and L. Mandel, "Violation of Bell's Inequality and Classical Probability on a Two-photon Correlation Experiment," *Phys. Rev. Lett.* **61**, 50-53 (1988).

129. J. S. Bell, "On the Einstein-Podolsky-Rosen Paradox," *Physics* **1**, 195-200 (1964); reprinted in "Speakable and unspeakable in quantum mechanics" (Cambridge University Press, Cambridge) 1987.

130. D. Bohm, "A Suggested Interpretation of the Quantum Theory in Terms of 'Hidden' Variables." I and II, *Phys. Rev.* **85**, 166-193 (1952).

131. E. Schrödinger, "Die Gegen Wärtige Situation in der Quantenmechanik," *Naturwissenschaften* **23**, 807-812, 823-828, 844-849 (1935).

132. Y. Aharonov and D. Bohm, "Significance of the Electromagnetic Potentials in the Quantum Theory," *Phys. Rev.* **115**, 485-491 (1959).
133. M. V. Berry, "Quantal Phase Factors Accompanying Adiabatic Changes," *Proc. R. Soc. A* (Lond.) **392**, 45-57 (1984).
134. H. H. Goldstine and L. P. Horwitz, *Proc. Nat. Acad. Sci.* **48**, 1134 (1962).
135. L. P. Horwitz and L. C. Biedenharn, *Helv. Phys. Acta* **38**, 385 (1965).
136. C. Piron, "Foundations of Quantum Physics" (Benjamin, Inc., Reading, Mass.) 1976.
137. M. Günaydin, *J. Math. Phys.* **17**, 1875 (1976).
138. L. C. Biedenharn, "Quantum Theory and the Structures of Time and Space" Volume 2, edited by L. Castell, M. Drieschner, C. F. von Weizsäcker (Carl Hanser Verlag, München) 1977.
139. P. Jordan and J. von Neumann, *Ann. Math.* **36**, 721 (1935).
140. Y. Aharonov and D. Bohm. "Significance of Electromagnetic Potentials in Quantum Theory" *Phys. Rev.* **115**, 485-491 (1959).
141. Y. Aharonov and L. Susskind, "Observability of the Sign Change of Spinors under 2π Rotations" *Phys. Rev.* **158**, 1237 (1967).
142. M. F. Atiyah, "Geometry of Yang-Mills Fields" Accademia Nazionale dei Lincei Scuola Normale Superiore - Lezioni Fermiane, Pisa (1979).
143. M. F. Atiyah, "The Geometry and Physics of Knots" (Cambridge University Press, Cambridge) 1990.
144. H. J. Bernstein and A. V. Phillips, "Fiber Bundles and Quantum Theory" *Scientific American* **245**, 122-137 (1981).
145. E. P. Battey-Pratt and T. J. Racey, "Geometric Model for Fundamental Particles" *Int. J. Theo. Phys.* **19**, 437-475 (1980).
146. H. Bondi, "Relativity and Common Sense" Dover Pub. (1977).
147. T. P. Cheng and L. F. Li, "Gauge Theory of Elementary Particle Physics" (Clarendon Press, Oxford) 1988.
148. L. Crane, "2-D Physics and 3-D Topology" *Communi. Math. Phys.* **135**, 615-640 (1991).
149. P.A.M. Dirac, "The Principles of Qunatum Mechanics" (Oxford University Press, Oxford) 1958.
150. R. P. Feynman, "Theory of Fundamental Processes" (Benjamin Pub.) 1961.
151. R. P. Feynman and S. Weinberg, "Elementary Particles and the Laws of Physics — The 1986 Dirac Memorial Lectures" (Cambridge University Press, Cambridge) 1987.
152. D. Finkelstein, "Space-Time Code" *Phys. Rev.* **184** 1261-1271 (1969).

153. D. Finkelstein and J. Rubenstein, "Connection Between Spin, Statistics and Kinks" *J. Math. Phys.* **9**, 1762-1779 (1968).

154. D. L. Goldsmith, "The Theory of Motion Groups" *Michigan Math. J.* **28**, 3-17 (1981).

155. R. Jackiw, "Topological Investigations of Quantized Gauge Theories" in "Current Algebra and Anomalies" Ed. by S. B. Treiman, R. Jackiw, B. Zumino, E. Wittten, 211-340, (World Sci. Pub.) 1985.

156. H. Jehle, "Flux Quantization and Particle Physics" *Phys. Rev. D.* **6** 441-457 (1972).

157. L. H. Kauffman, "Formal Knot Theory", (Princeton University Press, Mathematical Notes #30) 1983.

158. L. H. Kauffman, "Transformations in Special Relativity" *Int. J. Theo. Phys.* **24**, 223-236 (1985).

159. L. H. Kauffman and H. Saleur, "Free Fermions and the Alexander Conway Polynomial" *Comm. Math. Phys.* **141**, 293-327 (1991).

160. L. H. Kauffman, "Knots, Spin Networks and 3-Manifold Invariants" in the Proceedings of "KNOTS 90" - Osaka Conference, Ed by Akio Kawauchi and Walter de Gruyter (1992), pp. 271-285.

161. L. H. Kauffman, "Special Relativity and a Calculus of Distinctions" in Proceedings of the "9th Annual International Meeting of the Alternative Natural Philosophy Association"— Cambridge, England (September 23, 1987). Published by ANPA West, Palo Alto, Calif., pp. 290-311.

162. R. Kirby and P. Melvin, "On the 3-manifold Invariants of Reshetikhin - Turaev for SL(2,C)" *Invent. Math.* **105**, 473-545 (1991).

163. A. N. Kirillov and N. Y. Reshetikhin, "Representations of the Algebra $U_q(sl\ 2)$, q-orthogonal Polynomials and Invariants of Links" in "Infinite Dimensional Lie Algebras and Groups" Ed. by V. G. Kac, *Adv. Ser. in Math. Phys.* **7** 285-338 (1988).

164. D. A. Meyer, "State Models for Link Invariants from the Classical Lie Groups" (Proceedings of "Knots 90" - Osaka Conference, Ed. by Akio Kawauchi and Walter de Gruyter (1992), pp. 559-592).

165. E. W. Mielke, "Knot Wormholes in Geometrodynamics" *General Relativity and Gravitation* **8**, 175-196 (1977).

166. G. Moore and N. Seiberg, "Classical and Quantum Conformal Field Theory" *Commun. Math. Phys.* **123**, 177-254 (1989).

167. M. H. A. Newman, "On a String Problem of Dirac" *J. London Math. Soc.* **17**, 173-177 (1942).

168. R. Penrose, "Angular Momentum: an Approach to Combinatorial Space-Time" in "Quantum Theory and Beyond" Ed. T. A. Bastin (Cambridge

University Press) 1969.

169. R. Penrose, "Combinatorial Quantum Theory and Quantized Directions" in "Advances in Twistor Theory" ed. by L. P. Hughston and R. S. Ward, (Pitman) 1979, pp. 301-307.

170. A. M. Polyakov, "Fermi-Bose Transmutations Induced by Gauge Fields" *Mod. Phys. Lett.* **A3**, 325-328 (1987).

171. C. Quigg, "Gauge Theories of the Strong, Weak, and Electromagnetic Interactions" (Benjamin Pub.) 1983.

172. L. H. Ryder, "Quantum Field Theory" (Cambridge University Press) 1985.

173. L. Silberstein, "The Theory of Relativity" (MacMillan and Co. Ltd.) 1914.

174. L. Smolin, Quantum Gravity in the Self-dual Representation, *Contemp. Math.* **28**, (1988).

175. G. Spencer-Brown, "Laws of Form" (George Allen and Unwin Ltd., London) 1969.

176. J. Stillwell,. "Classical Topology and Combinatorial Group Theory" (Springer-Verlag) 1980. GTM 72.

177. P. G. Tait, "On Knots I, II, III, Scientific Papers" vol. I, (Cambridge University Press, London) 1898, 273-347.

178. F. Wilczek and A. Zee, "Linking Numbers, Spin and Statistics of Solitons" *Phys. Rev. Lett.* **51**, 2250-2252 (1983).

179. E. Witten, "Physics and Geometry" *Proc. Intl. Congress Math.*, Berkeley, Calif. (1986), 267-303.

180. E. Witten, "Quantum Field Theory and the Jones Polynomial" *Comm. Math. Phys.* **121**, 351-399 (1989).

181. E. Witten, "Gauge Theories Vertex Models and Quantum Groups" *Nucl. Phys. B.* **330**, 225-246 (1990).

182. D. Yetter, "Quantum Groups and Representations of Monodial Categories" *Math. Proc. Camb. Phil. Soc.* **108**, 197-229 (1990).

183. L. H. Kauffman, "Gauss Codes and Quantum Groups" in "Quantum Field Theory, Statistical Mechanics, Quantum Groups and Topology" ed. by T. Curtright, L. Mezincescu and R. Nepomechie (World Scientific) 1992, 123-154.

184. R. A. Lawrence, "A Universal Link Invariant" in "The Interface of Mathematics and Physics" ed. by D. G. Quillen, G. B. Segal, and S. T. Tsou, (Oxford Univ. Press) 1990, 151-160.

185. J. Stillwell, "Classical Topology and Combinatorial Group Theory" (Springer Verlag) 1980.

186. L. C. Biedenharn and J.D. Louck, "Angular Momentum in Quantum Physics-Theory and Application" in "Encyclopedia of Mathematics and its Applications" (Cambridge University Press) 1979.

187. L. Crane, "Conformal Field Theory, Spin Geometry and Quantum Gravity" *Phys. Lett.* **B259**, 243-248 (1991).

188. L. Crane and D. Yetter, "A Categorical Construction of 4D Topological Quantum Field Theories" (preprint 1993).

189. Lee Smolin, "Quantum Gravity in the Self-dual Representation" *Contemp. Math.* **71**, 55-97 (1988).

190. R. D. Schafer, "Introduction to Nonassociative Algebras" (Academic Press) 1966.

191. E. J. Post, "Formal Structure of Elecromagnetics" (North Holland, Amsterdam) 1962.

192. A. Lakhtakia, V. V. Varadan, V. K. Varadan, "Time-Harmonic Electromagnetic Fields in Chiral Media" (Springer, Heidelberg) 1989.

193. D. Hestenes, "New Foundations for Classical Mechanics" (Reidel, Dordrecht) 1987.

194. D. Hestenes, "Space-time Algebra" (Gordon and Breach, New York) 1966.

195. W. E. Baylis, J. Huischilt, Jiansu Wei, "Why I?" *Am. J. Phys.* **60**, 788-797 (1992).

196. N. Mukunda, C. C. Chang, E. C. G. Sudarshan, "Dirac's New Relativistic Wave Equation in Interaction with an Electomagnetic Field" *Proc. Roy. Soc. Lond.* A **379**, 103-107 (1982).

197. M. A. Defaria-Rosa, E. Recami, W. A. Rodrigues, "A satisfactory Formalism for Magnetic Monopoles by Clifford Algebras" *Phys. Let.* B **173**, 233-236 (1986).

198. P. E. Hagmark and P. Lounesto in "Clifford Algebras and Their Applications in Mathematical Physics" J. S. R. Chisholm and A. K. Common (eds.) (Reidel, Dordrecht) 1986, p531-540.

APPENDIX

ADVANCED SUBTLETIES

There are many questions still to be answered in this mathephysics speculating about the best way to approximate the real quantum world out there. Should we search for principles that give the math or should we go for the math directly? The math alone says that reals, complexes, quaternions, and octonions are very special number systems. Our successful physics to date, QED in particular, tells us that they are not enough! We need larger number systems, and we may not need octonions at all for nature. Geoffrey Dixon has been trying hard to answer that question in his recent book.

The quaternions, $\{\sigma_0, -i\sigma_k\}$, cannot model spacetime. We need complex quaternions at least. Of the 8 basis elements there, how do we pick four for (ct,x,y,z) events? There may be more than one way that works. $P = P^\mu(\sigma_\mu) = P^\dagger$, $x = x^\mu(\sigma_\mu) = x^\dagger$, *and* $F = -E^k\sigma_k + cB^k(i\sigma_k) = -F^\wedge$ seem to nicely fit the real world. Within this simple sub-algebra of reality, we have $P^\dagger = P^\downarrow$ and $F^\wedge = F^\vee$. To put together an equation such as Maxwell's, $PF = J = J^\mu\sigma_\mu + 0(i\sigma_\mu)$, we have found guidance from the totally abstract idea of $P \rightarrow P' \equiv L^\dagger PL = L^\downarrow PL$ *and* $LL^\wedge = 1 = L^\wedge L = LL^\vee = L^\vee L$. The $L^\wedge = L^\vee$ idea makes no restriction on L here, only in larger algebras, if it still holds.

Are $LL^\wedge = 1(\sigma_0)$ and $P' = L^\dagger PL$ of any real physical significance or is this only a 'game we play' to help us guess at the physics? We cannot answer that yet. Maybe in a hundred years. So far, it appears to be partly both. $LL^\wedge = 1$ is the Lorentz group in Pauli algebra, which itself suggests time dilation and length contraction and they are physical, out there in the world, for huge creatures like us! But if Lorentz is only part of the correct group, then such physical significance may not exist for all of the parameters that we find exist in the group.

The above algebra has many conjugations. The two I have focused on may be special; I don't really know. They single out σ_0 and they single out σ_μ, but is that important?

In any case, $PF = J$ *and* $P'F' = J'$ give us

$$(L^\dagger PL)(L^\wedge FL) = (L^\dagger JL) \Rightarrow \cdots \Rightarrow PF = J$$

provided $LL^\wedge = 1 = L^\wedge L$. We also find $P^\wedge P$ has

$$P'^{\wedge}P' = (L^{\dagger}PL)^{\wedge}(L^{\dagger}PL) = \cdots = P^{\wedge}P \propto (\sigma_0)$$

but is this important? We have $P = P^{\mu}(\sigma_{\mu}) = P^{\dagger}$ *and* $Q \equiv Q^{\mu}(i\sigma_{\mu}) = -Q^{\dagger}$. Why not $QF = J$? What does Q' look like? Is it also $L^{\dagger}QL$? Do we need to worry about both P and Q in guessing at physics laws? Do they give related but different physics? Notice that Maxwell's equation has no mass in it. We have to add in mass somehow? Why? I don't know. We just already exist and we do have it. So we must get mass into the mathephysics, somehow.

My solution is (c ≡ 1):

$$P\psi = \psi^{\dagger\wedge}M \rightarrow P'\psi' = \psi'^{\dagger\wedge}M'$$

with P$'$ as above, $\psi' = L^{\wedge}\psi$, *and* $M' = M$ (guesses!). This works, in the sense that we can get all the L's to disappear! We find also that $P^{\wedge}P\psi = \cdots = \psi M^{\dagger\wedge}M$ follows from this. But $P^{\wedge}P \propto (\sigma_0)$, so maybe $M^{\dagger\wedge}M \propto (\sigma_0)$ is somehow necessary for the real world.

The complex quaternion algebra (Hamilton 1843 and Pauli 1927) has $PF = 0$ *and* $P\psi - \psi^{\dagger\wedge}M = 0$ for free particle 'wave' equations, one massless and one with $M = M_a^{\ \mu}(\sigma_{\mu}) + M_b^{\ \mu}(i\sigma_{\mu}) = ?$. But this is still not enough for the real world. I don't know why. There seems to be even more mass, besides this, in nature.

There is another solution to the mass invention problem. Consider $P\psi_a = \psi_v M_v$ *and* $P^{\wedge}\psi_v = \psi_a M_a$. Here we have 'form covariance' if $\psi_a \rightarrow \psi_a' = L^{\wedge}\psi_a$ and $\psi_v \rightarrow \psi_v' = L^{\dagger}\psi_v$ and P and M are as before, but M_a and M_b are possibly not the same M from before. We do find that

$$P^{\wedge}P\psi_a = \cdots = P^{\wedge}\psi_v M_v = \psi_a M_a M_v$$

$$PP^{\wedge}\psi_v = \cdots = P\psi_a M_a = \psi_v M_v M_a$$

and $P^{\wedge}P \propto \sigma_0$ again suggests that

$$P^{\wedge}P\psi_a = (\hbar\partial^{\mu}(\sigma_{\mu}))^{\wedge}(\hbar\partial^{\nu}(\sigma_{\nu}))\psi_a = \cdots = \hbar\partial^{\mu}\hbar\partial_{\mu}\psi_a = \psi_a M_a M_v$$

What really works for our world? The PF = J equation with M = 0 seems to physically apply to photons, with zero rest mass and spin 1. The $P^{\wedge}P\phi = \phi M^{\dagger\wedge}M$ equation seems to physically apply to pions with spin 0 and rest mass > 0. That is all for this small algebra. Electrons, protons, and neutrons seem to need a larger algebra. The coupled pair $\{\psi_a, \psi_v\}$ above does not seem to fit the real world, so far, because of the limit here to real coefficients. I found it both natural and useful to complexify the already 'complexified' quaternion algebra and will here call it the Edmonds algebra, since I have not seen it in field theory books. (It probably already exists in the math literature and should be renamed eventually. I've invented or found a lot of stuff that proved to already exist in the literature.)

The new algebra now is $\{(\sigma_\mu), (i\sigma_\mu), i(\sigma_\mu), i(i\sigma_\mu) \neq -(\sigma_\mu)\}$. The real world of electrons can then be accommodated by the coupled pair $P\psi_a = i\psi_v M_v$, & $P^\wedge \psi_v = i\psi_a M_a$. Notice the new i's. This is Dirac's equation if $M_a = M_v \equiv m_D$ *and if* $P_Q \equiv \hbar\partial^\mu(\sigma_\mu) =$ *real*. We now must generalize our two conjugations, $(\ )^\dagger$ and $(\ )^\wedge$, to deal with complex coefficients. Years later, I now know that $(a + ib)^\dagger \equiv (a - ib)$ *and* $(a + ib)^\wedge \equiv a - ib$ are the way to go for these 'outside' complex coefficients. All conjugations will be like this. This is a guess and must be tested in each physical case.

Physics still looks pretty simple here so far, even Dirac's equation. Notice at $P \to L^\dagger PL = L^\downarrow PL$ in the complex quaternion algebra. Now $(iM)^\dagger = i^\dagger M^\dagger = -iM$. Also, $[i\hbar\partial^\mu(\sigma_\mu)]^\dagger = -i\hbar\partial^\mu(\sigma_\mu) =$ *anti-hermitian*, but this could be the P operator of reality, instead of $\hbar\partial^\mu(\sigma_\mu)$ *or* $\hbar\partial^\mu(i\sigma_\mu)$ *or* $\hbar\partial^\mu i(i\sigma_\mu)$. We can rewrite Dirac's equation, and Maxwell's equation too, in different forms by multiplying through from the left. The $P \to P' = \cdots$ forms must then also change in most cases. Let us construct $P_{D\sigma} \equiv i(i\sigma_\mu)\hbar\partial^\mu$:

$$P_Q\psi_a = i\psi_v M = \hbar\partial^\mu(\sigma_\mu)\psi_a, \quad M \to m_D(\sigma_0)$$

$$i(i\sigma_0)\hbar\partial^\mu(\sigma_\mu)\psi_a = i(i\sigma_0)i\psi_v m_D$$

$$[\hbar\partial^\mu i(i\sigma_\mu)]\psi_a = -\psi_v(i\sigma_0)m_D \equiv P_{D\sigma}\psi_a$$

Notice that $P_{D\sigma} \to P_{D\sigma}' = i(i\sigma_0)L^\dagger P_Q L = L^\dagger[i(i\sigma_0)P_Q]L$, since $(i\sigma_0)$ commutes with everything in this algebra. Notice also that M did not have to be simply m_D. Other very small parts might exist for M and they are just neglected in QED.

Notice also that $P \to P' = L^\dagger PL$, $P^\wedge P \propto \sigma_0$, *and* $LL^\wedge = 1(\sigma_0)$ now give us $L_\epsilon = 1(\sigma_0) + \epsilon^k(i\sigma_k) + \delta^k(\sigma_k) + \epsilon i(\sigma_0) + \delta i(i\sigma_0)$ in the Edmonds algebra, for group elements near the identity element. We have MORE than Lorentz form covariance now! We have 8 parameters. Dirac used the equation for $\{\psi_a, \psi_v\}$ above (in a different form) but left behind the new parameters that can exist in L.

Why should we go to an even bigger algebra? We already have enough for Dirac theory here! No reason to go bigger unless it is just convenient or unless nature forces it, to reach the right equations of the quantum world. The coupled pair of Dirac pieces can easily be set in larger matrices. You can readily check that the Dirac pair is equivalent to ($P^\mu \equiv \hbar\partial^\mu$):

$$P^\mu i \begin{pmatrix} 0 & -\sigma_\mu^\wedge \\ -\sigma_\mu & 0 \end{pmatrix} \begin{pmatrix} \psi_a & 0 \\ \psi_v & 0 \end{pmatrix} = m_D \begin{pmatrix} \psi_a & 0 \\ \psi_v & 0 \end{pmatrix}$$

and the second pair, involving $i(i\sigma_0)$ multiplication above, is equivalent to

$$P^\mu i \begin{pmatrix} (i\sigma_\mu) & 0 \\ 0 & -(i\sigma_\mu)^\wedge \end{pmatrix} \begin{pmatrix} \psi_a & 0 \\ 0 & \psi_v \end{pmatrix} = -m_D \begin{pmatrix} \psi_v & 0 \\ 0 & \psi_a \end{pmatrix} \begin{pmatrix} (i\sigma_0) & 0 \\ 0 & -(i\sigma_0) \end{pmatrix}$$

We can define

$$\psi_D \equiv \begin{pmatrix} \psi_a & 0 \\ \psi_v & 0 \end{pmatrix} \quad and \quad \psi_E \equiv \begin{pmatrix} \psi_a & 0 \\ 0 & \psi_v \end{pmatrix}$$

The problem here is like going from classical to quantum. Things are simpler in classical and going up in complexity is not easy. We must guess but we have only general guidance.

Notice that

$$(f_0)\psi_E(f_0) = (f_0)\psi_E^{\dagger\dagger}(f_0) = (\psi_E^\dagger)^\wedge = \begin{pmatrix} \psi_v & 0 \\ 0 & \psi_a \end{pmatrix}$$

$$(if_0)\psi_E(if_0) = (if_0)\psi_E^{\dagger\dagger}(if_0) = \psi_E^{\dagger\vee} = \begin{pmatrix} \psi_v & 0 \\ 0 & \psi_a \end{pmatrix}$$

So we cannot distinguish $\psi^{\dagger\wedge}$ and $\psi^{\dagger\vee}$ in Dirac's equation! Something else must determine this choice. We shall write $\wedge$? to indicate it could be $\vee$ instead. The second pair-form, for Dirac, is then equivalent to

$$\hbar\partial^\mu i\begin{pmatrix}(i\sigma_\mu) & 0\\ 0 & -(i\sigma_\mu)^\wedge\end{pmatrix}\psi_E = -m_D\psi_E^{\dagger\wedge ?}\begin{pmatrix}(i\sigma_0) & 0\\ 0 & -(i\sigma_0)\end{pmatrix}$$

or

$$\hbar\partial^\mu i(ie_\mu)\psi_E = -m_D\psi_E^{\dagger\wedge ?}(ie_0) = -\psi_E^{\dagger\wedge ?}m_E^0(ie_0)$$

The first pair-form is also equivalent to

$$\hbar\partial^\mu i(-f_\mu)\begin{pmatrix}\psi_a\\ \psi_v\end{pmatrix} = m_D\begin{pmatrix}\psi_a\\ \psi_v\end{pmatrix} \equiv i\hbar\partial^\mu(\gamma_\mu)\psi_D = m_D\psi_D$$

where $(-f_\mu) \equiv \gamma_\mu$. It is easy to show that $(-f_\mu)$ has the properties one usually assumes of γ_μ in Dirac theory. If we next multiply this through from the left by (if_0), we get

$$\hbar\partial^\mu i(+ie_\mu)\psi_D - m_D(if_0)\psi_D = 0$$

We then see that these two forms of Dirac are special cases of the general equation (Dirac-Edmonds)

$$\hbar\partial^\mu i(ie_\mu)\psi - m(if_0)\psi = -\psi^{\dagger\wedge ?}m_E^\mu(ie_\mu)$$

We get old Dirac for either $m = m_D$, $m_E{}^\mu = 0$, or for $m = 0$, $m_E{}^0 = m_D$, $m_E{}^k = 0$, $k = 1,2,3$.

Within the Dirac equation, $P = P^\dagger \Rightarrow$ P has 16 parts and $P^\wedge P \propto (e_0)$ $\Rightarrow$ these 16 parts come in 5-vector sets, two of which are $\{i(ie_\mu), (if_0)\}$ and $\{(e_\mu), (if_0)\}$. Thus P is NATURALLY a 5-vector and we write

$$[\hbar\partial^\mu i(ie_\mu) - m_D(if_0)]\psi = -\psi^{\dagger\wedge ?}m_E^\mu(ie_\mu) \equiv P_{5I}\psi$$

Then $P_{5I}{}^\wedge P_{5I} = (- + + + -) = (-\hbar\partial^\mu\hbar\partial_\mu - m_D{}^2)(e_0) = ((i\hbar\partial^\mu i\hbar\partial_\mu) - m_D{}^2)(e_0)$. Clearly, if operating on a ϕ wave function, this gives the usual

Klein-Gordon equation. But we have more here. We find that

$$P_{5I}^{\wedge}(P_{5I}\psi) = (P_{5I}^{\wedge}P_{5I})\psi = -P_{5I}^{\wedge}\psi^{\dagger\wedge ?}M = -(P_{5I}^{\dagger})^{\wedge}\psi^{\dagger\wedge ?}M$$

$$= -(P_{5I}\psi)^{\dagger\wedge}M, \quad if \wedge ? = \wedge$$

This settles the choice of $\wedge$ or $\vee$ on ψ. Continuing, we then get

$$P^{\wedge}P\psi = -(-\psi^{\dagger\wedge}M)^{\dagger\wedge}M = +\psi M^{\dagger\wedge}M$$

Finally then we have ($M = m^{\mu}(ie_{\mu}) = -M^{\dagger}$):

$$(i\hbar\partial^{\mu}i\hbar\partial_{\mu})\psi = (m_D^2 + m_E^0 m_E^0 - m_E^k m_E^k)\psi \equiv m_{\text{exp}}^2\psi$$

We see that either m_D or $m_E{}^0$ alone will work here, since $m_{\text{exp}}{}^2$ is real. Actually, we only need $m^k m^k$ to be small enough so the mass combination is positive. This generalizes Dirac's factorization of the Klein-Gordon equation. It is clearly a factorization, but is it Lorentz covariant? What does that really mean?

We can show that $\{i(ie_{\mu}),\ (if_0)\} \to L^{\dagger}\{i(ie_{\mu}),\ (if_0)\}L$ gives back the same set with no extra terms if $LL^{\wedge} = 1(e_0)$. Is this Lorentz covariance? $P_{5I} \to P_{5I}{}' \equiv L^{\dagger}P_{5I}L$, $P_{5I} = P_{5I}{}^{\dagger} \Rightarrow P_{5I}{}' = (P_{5I}{}')^{\dagger}$. This does not mean that $P_{5I}{}'$ has only the same 5 parts in it, but direct table formation and multiplication shows it does have only $\{i(ie_{\mu}), (if_0)\}$ if $LL^{\wedge} = 1(e_0)$. The same is also true for $\{(e_{\mu}), (if_0)\}$ as the same table shows. We find that $P^{\wedge}P = (L^{\dagger}PL)^{\wedge}(L^{\dagger}PL) = L^{\wedge}P^{\wedge}(LL^{\wedge})^{\dagger}PL = P^{\wedge}P$ if $LL^{\wedge} = L^{\wedge}L = 1(e_0)$.

Thus the {e,f} Dirac algebra, with 32 basis elements, allows a new, extended spin 1/2 equation, with Dirac's equation as a subcase—with two ways of getting it, through m_D or through $m_E{}^0$. This opens very exciting new possibilities for nature modeling.

We don't have a natural 4-vector operator, but rather a 5-vector 'operator':

$$P_{5I} \equiv [\hbar\partial^{\mu}i(ie_{\mu}) - m_D(if_0)] \equiv P_{5I}^{\dagger} = hermitian!!$$

We know $\hbar\partial^{\mu}i(i\sigma_{\mu}) \to L^{\dagger}[\cdots]L$, and e $\Leftrightarrow$ σ when no f's are around.

Therefore, extending to 5-vectors, we naturally CHOOSE $P_{5I} \rightarrow P_{5I}' \equiv L^{\dagger} P_{5I} L$. Then we automatically get

$$L^{\dagger}(if_0)L = (if_0)(L^{\dagger})^{\dagger\vee}L = (if_0)L^{\vee}L$$

This is invariant only if $L^{\vee}L = 1(e_0)$. But $P_{5I}{}^{\wedge}P_{5I} = (- + + + -) = \propto (e_0)$ needs $LL^{\wedge} = 1(e_0)$ for invariance. Therefore, $LL^{\wedge} = 1(e_0)$ *AND* $L^{\wedge} = L^{\vee}$ seem to both be needed to model nature successfully!! The $LL^{\wedge} = 1(e_0)$ group in general has $L_\epsilon = 1(e_0) + \epsilon^k(ie_k) + \delta^k(e_k) + \epsilon i(e_0) + \delta i(ie_0) + 8$ *more* terms $\Rightarrow$ 16 parameters!! But $L^{\wedge} = L^{\vee}$ removes all f's from L_ϵ. Thus L is the same group here as in the smaller Edmonds algebra! But why must $m_D(if_0)$ be invariant? I don't know. Maybe it isn't exactly, only approximately.

What does this extended Dirac-Edmonds equation look like in the old γ_μ form for Dirac's equation? The old timers are dying to see that, I am sure. The youngsters may not care. We simply multiply through from the left by whatever it takes to get $m_D(e_0)$. This only takes (if_0) here, and we use $(if_0)A = A^{\dagger\vee}(if_0)$ for any A. We get, after left (if_0) multiplication:

$$i\hbar\partial^{\mu}(-f_\mu)\psi = m_D\psi - \psi^{\wedge\vee}m_E^{\mu}(-f_\mu) \equiv P_{4II}\psi$$

We recognize our old friend Dirac, if $\gamma_\mu \equiv -(f_\mu)$ *and* $m^\mu \equiv 0$. We used $(if_0)\psi^{\dagger\wedge} = (\psi^{\dagger\wedge})^{\dagger\vee}(if_0) = \psi^{\wedge\vee}(if_0)$. We get the 5-vector form

$$[\hbar\partial^{\mu}i(\gamma_\mu) - m_D(e_0)]\psi = -\psi^{\wedge\vee}m_E^{\mu}(\gamma_\mu) \equiv P_{5II}\psi$$

Here,

$$P_{5II} = (if_0)P_{5I} \rightarrow (if_0)(L^{\dagger}P_{5I}L) = (L^{\dagger})^{\dagger\vee}(if_0)P_{5I}L$$

So $P_{5II}' = L^{\vee}(if_0)P_{5I}L$. This suggests $P_{5II}{}^{\vee} = \pm P_{5II}$ and we find by inspection that $P_{5II}{}^{\vee} = P_{5II}$. The 'usual' Dirac operator is not really hermitian, since $\gamma_\mu \equiv (-f_\mu)$ *has* $\gamma_\mu{}^{\dagger} = -f_\mu{}^{\dagger} = \{-f_0, +f_k\}$. Notice that $P_{5II}\psi = -\psi^{\wedge\vee}M$ is covariant for $\psi \rightarrow L^{\wedge}\psi$ and the $\psi^{\wedge\vee}$ is important. We have needed $i^{\vee} = -i$ *and* $(f_\mu)^{\vee} = -(f_\mu)$ here to get these pretty results. Notice that $\psi = \psi^{\wedge\vee} \Rightarrow L^{\wedge}\psi = L^{\vee}\psi$.

Notice that $L^{\wedge} = L^{\vee} \Rightarrow L^{\wedge\vee} = L = L^{\dagger\downarrow} \Rightarrow L^{\dagger} = L^{\downarrow}$, should we need to use these. L has no f's in it at all! Also, $\psi \rightarrow L^{\wedge}\psi = L^{\vee}\psi$ in the Dirac algebra. Only the $m_D(if_0)$ term forces us to stay in the larger algebra, it seems!

Very interesting!!

Another subtle point concerns finding the products that are invariant. Clearly, we have two simple invariants

$$\psi^\wedge\psi \to (L^\wedge\psi)^\wedge(L^\wedge\psi) = \psi^\wedge\psi, \textit{ if } LL^\wedge = 1(e_0)$$

$$\psi^\dagger(f_0)\psi \to (L^\wedge\psi)^\dagger(f_0)L^\wedge\psi = \psi^\dagger(f_0)\psi, \textit{ if } LL^\wedge = 1(e_0)$$

If we return to the Edmonds algebra, we lose all the f's but we still have all the e's. Only (f_0) has $(f_0)A = A^{\dagger\wedge}(f_0)$, so it is not possible to convert $\psi^\dagger(f_0)\psi$ into an invariant form in Edmonds algebra. Consider next

$$\psi^\downarrow(if_0)\psi \to (L^\wedge\psi)^\downarrow(if_0)(L^\wedge\psi) = \psi^\downarrow L^{\wedge\downarrow}(if_0)L^\wedge\psi$$

(where $A^{\wedge\downarrow} = A^{\vee\uparrow}$ for any A). This is also invariant if $LL^\wedge = 1(e_0)$. Consider finally

$$\psi^\wedge(ie_0)\psi \to (L^\wedge\psi)^\wedge(ie_0)L^\wedge\psi = \psi^\wedge L^{\wedge\wedge}(ie_0)L^\wedge\psi$$

$$= \psi^\wedge(ie_0)(L^{\wedge\wedge})^{\wedge\vee}L^\wedge\psi = \psi^\wedge(ie_0)L^{\wedge\vee}L^\wedge\psi = \psi^\wedge(ie_0)(LL^\vee)^\wedge\psi$$

$$= \psi^\wedge(ie_0)\psi, \textit{ if } LL^\vee = 1(e_0)$$

Therefore, it seems $\psi^\wedge(ie_0)\psi$ is the natural inner product back down in the Edmonds algebra or even perhaps in the Pauli algebra, but $\psi^\wedge(i\sigma_0)\psi = \psi^\wedge\psi(i\sigma_0)$, so this is essentially $\psi^\wedge\psi$ here. We don't get $\psi^\dagger\psi$ invariant in Edmonds or Dirac. Notice that

$$1(e_0) = LL^\wedge \Rightarrow (f_0)1(e_0) = (f_0)LL^\wedge = L^{\dagger\wedge}(f_0)L^\wedge = (f_0)$$

Take $(\;)^\wedge$ conjugation on both sides and get $L(f_0)L^\dagger = (f_0)$, even for L a 16 parameter group. Similarly, $1(e_0) = L^\wedge L \Rightarrow (ie_0)1(e_0) = (ie_0) = (ie_0)L^\wedge L = (L^\wedge)^{\wedge\vee}(ie_0)L$, *or* $(ie_0) = L^\vee(ie_0)L$, even for L a 16 parameter group. Back in the Edmonds algebra this reduces to $(\sigma_0) = L^\vee L = L^\wedge L$ giving only an 8 parameter group.

Kauffman's book and Marx's papers have a way of using σ_2, essentially, to reach a similar equation that they claim goes with Lorentz symmetry in 2×2 matrices. I am not so sure, but the full 8 parameter group cannot be found in 8 basis Pauli anyway, only in Edmonds and Dirac algebras, so perhaps this is not an important issue. The invariant product issue is not so simple when ψ is not an ordinary spinor. In general, $\psi^{\dagger}\psi \rightarrow (L^{\wedge}\psi)^{\dagger}(L^{\wedge}\psi) = \psi^{\dagger}L^{\wedge\dagger}L^{\wedge}\psi = \psi^{\dagger}(LL^{\dagger})^{\wedge}\psi$ *and* $LL^{\dagger} \neq 1(e_0)$. This is also a 16 parameter group, $LL^{\dagger} = 1(e_0)$, maybe isomorphic to $LL^{\wedge} = 1(e_0)$. We do not have both groups at the same time, however! So $\psi^{\dagger}\psi$ is not invariant, unless $L^{\wedge} = L^{\vee}$ is replaced by $L^{\wedge} = L^{\dagger}$. In any case, $\psi^{\dagger}\psi = (\psi^{\dagger}\psi)^{\dagger} \Rightarrow (\psi^{\dagger}\psi)$ has at most 16 elements in it, not just (e_0) in general. How many it actually has depends on how we got ψ from a differential equation.

In Pauli, $LL^{\wedge} = 1(\sigma_0)$ is Lorentz's group and $LL^{\dagger} = 1(\sigma_0)$ is $SU(2)\otimes U(1)$. Requiring $L^{\wedge} \equiv L^{\dagger}$ gives SU(2). In Dirac, $LL^{\wedge} = 1(e_0)$ is a 16 parameter group and $LL^{\dagger} = 1(e_0)$ is also a 16 parameter group. In Edmonds, $LL^{\wedge} = 1(e_0)$ is an 8 parameter group as is also $LL^{\dagger} = 1(\sigma_0)$. In Edmonds,

$$L_{\epsilon} = 1(\sigma_0) + \epsilon^k(i\sigma_k) + \delta^k(\sigma_k) + \epsilon i(\sigma_0) + \delta i(i\sigma_0)$$

$$L_{\epsilon}^{\wedge} = 1(\sigma_0) - \epsilon^k(i\sigma_k) - \delta^k(\sigma_k) - \epsilon i(\sigma_0) - \delta i(i\sigma_0)$$

$$L_{\epsilon}^{\dagger} = 1(\sigma_0) - \epsilon^k(i\sigma_k) + \delta^k(\sigma_k) - \epsilon i(\sigma_0) + \delta i(i\sigma_0)$$

Therefore, $LL^{\wedge} = 1(\sigma_0)$ *and* $L^{\wedge} = L^{\dagger}$ give the $\{\epsilon^k, \epsilon\}$, 4 parameter sub-group. In Dirac, $LL^{\wedge} = 1(e_0)$ *and* $L^{\wedge} = L^{\vee}$ give together an 8 parameter group, as do $LL^{\wedge} = 1(e_0)$ *and* $L^{\wedge} = L^{\dagger}$.

Let us now return to wave equations. The generalized Dirac-Edmonds equation is

$$[\hbar\partial^{\mu}i(ie_{\mu}) - m_D(if_0)]\psi = -\psi^{\dagger\wedge}m_E^{\mu}(ie_{\mu}) \rightarrow (L^{\dagger}[\cdots]L)(L^{\wedge}\psi)$$

Multiply through by $+i(ie_0)$ and obtain (using $(ie_0)\psi^{\dagger\wedge} = (\psi^{\dagger\wedge})^{\dagger\downarrow}(ie_0)$):

$$[\hbar\partial^{\mu}(e_{\mu}) - m_D i(f_0)]\psi = +i\psi^{\dagger\vee}m_E^{\mu}(e_{\mu}) \rightarrow (L^{\downarrow}[\cdots]L)(L^{\wedge}\psi)$$

Notice here that

$$[\hbar\partial^\mu(e_\mu) - m_D i(f_0)] \rightarrow L^\downarrow[\cdots]L \equiv L^\downarrow P_{5IV}L$$

Therefore, $\hbar\partial^\mu(e_\mu) \rightarrow \hbar\partial^\mu(\sigma_\mu)$ 'really' comes from $L^\downarrow(\ )L$ Dirac 5-vectors, not $L^\uparrow(\ )L$ 4-vectors, but, in Edmonds, $(\ )^\uparrow$ and $(\ )^\downarrow$ are identical, so this isn't important. Also, $P_{5IV} = P^\downarrow{}_{5IV}$, so not hermitian.

There is yet another, fundamental generalized equation besides this Dirac-Edmonds equation. It comes from the table of $P^\wedge P$ 5-vector forms, when $P = P^\uparrow$, and the equation is:

$$[\hbar\partial^\mu(e_\mu) - m_D(if_0)]\psi = +i\psi^{\uparrow\wedge ?}m_E^\mu(e_\mu) \equiv P_{5III}\psi$$

It is easy to show that $P_{5III}{}^\wedge P_{5III} = (+ - - - -) \propto (e_0)$ and $P_{5III}{}^\uparrow = +P_{5III}$. Therefore, we guess that $P_{5III} \rightarrow P_{5III}{}' = L^\uparrow PL$ for 'covariance'. I called this a 'strange new equation' in recent papers. Maybe someone can show it is directly an offshoot of Dirac, but I doubt it, because of the 5-space metric $(+ - - - -)$. Dirac $\{i(ie_\mu), (if_0)\}$, goes with $(- + + + -)$. The m_D mass here gives a tachyonish Klein-Gordon equation, so it is not Dirac! However, $m_D = 0$, $m_E{}^0 \neq 0$ is almost Dirac's equation.

We have only developed $P = P^\mu b_\mu = P^\mu b_\mu{}^{(\alpha)}(e_\alpha)$ for curved space. Do we need $P = P^\mu b_\mu = P^\mu b_\mu{}^{(\alpha)}i(ie_\alpha)$ also? Are the $b_\mu{}^{(\alpha)}$'s here interchangeable? Interesting questions. How do we couple ψ to the E-M field which is in Pauli, where $P = \hbar\partial^\mu(e_\mu)$?

I suspect that $P \rightarrow P +$ 'e'A, as the guessed coupling, has P a 5-vector but A only a 4-vector, and $P = (P^\mu)\cdots = (i\hbar\partial^\mu)\cdots$, whereas $A = (A^\mu)\cdots$ with A^μ real. This fits with what works in the non-relativistic limit, where we guess that $i\hbar\partial^\mu \rightarrow i\hbar\partial^\mu - |e|A^\mu$ and $-|e|A^0 = -|e|V/c$. Of course, the 'e' coupling constant is anything we need that gives the right non-relativistic, approximate formulas, that we know and love. Perhaps we need both $\{i(ie_\mu), (if_0)\}$ and $\{(e_\mu), (if_0)\}$ versions of spacetime, but I doubt it.

We find, by $i(ie_0)$ left multiplication, that the new and strange equation is equivalent to

$$[\hbar\partial^\mu i(ie_\mu) - m_D i(f_0)]\psi = -\psi^{\uparrow\wedge ?}m_E^\mu(ie_\mu) \equiv P_{5V}\psi$$

The covariance form here comes from $i(ie_0)L^\uparrow[\cdots]L = L^\downarrow[i(ie_0)[\cdots]]L$. Therefore, $P_{5V}{}^\downarrow = \pm P_{5V}$ and by inspection $P_{5V}{}^\downarrow = +P_{5V}$. This equation differs from that which generalized from Dirac in the 5th part of the 5-vector operator, and possibly in the $\psi^{\dagger\wedge?}$ conjugation that works. The invariant, 5-vector, inner product for this case can be found as follows. The old $\{(e_\mu), (if_0)\}$ 5-vector had $P_{5o}{}^\wedge P_{5o} \propto (e_0)$ and our new P_{5N} is $i(ie_0)$ times the old one. But the old one is also $i(ie_0)$ times the new one. Check this by left multiplication. Thus, $(i(ie_0)P_N)^\wedge (i(ie_0)P_N) \propto (e_0) \Rightarrow P_N{}^\wedge (ie_0)^\wedge i^\wedge i(ie_0)P_N = P_N{}^\wedge (-e_0)P_N = -P_N{}^\wedge P_N$. Therefore, the new metric is $(-\ +\ +\ +\ +)$ instead of $(+\ -\ -\ -\ -)$, and the new $P_{5V}{}^\wedge P_{5V} \propto e_0$, but $P_{5V}{}^\downarrow = +P_{5V}$ and it is not hermitian!

From this perspective, we have two equations with the same 4-space (even curved 4-space part), but different 5th component parts.

Is this the best way to really compare them? Does this mean the 5th part is 'flat' or does it mean that quantum gravity will involve 6-space: $\{i(ie_\mu), (if_0), i(f_0)\}$? This 6 'thing' has no conjugation in common, so maybe it is not a single physical 'thing'. Maybe (if_0) gets curved along with $i(ie_\mu)$ but $i(f_0)$ does not. That would enhance its strangeness for sure. Of course the new $\{i(ie_\mu), i(f_0)\}$ 5-vector operator equation may never have $m_D \neq 0$ in it for some reason. If not, then it is possibly not distinct, and we can maybe forget it. There is no way to know at this early stage. We must now settle the $\psi^{\dagger\wedge?}$ question for the 'strange' equation.

The best indication, I can see, that this 'strange' equation should be taken seriously is that it comes from $\{(e_\mu), (if_0)\}$ and in the Pauli algebra there is no $i(i\sigma_\mu)$ spacetime basis at all! If my Pauli factorization equation, $\hbar\partial^\mu(\sigma_\mu)\psi = \pm\psi^{\dagger\wedge} m_E{}^\mu(i\sigma_\mu)$, is realistic in the Pauli algebra, then we get the K-G equation there with $m_{exp}{}^2 = -M^{\dagger\wedge}M$ and this means m^k dominates m^0 there. In Dirac, it may be the other way around. We know nothing of m^k at present, not even if it actually exists in the real world! We must now check the Klein-Gordon equation for the strange equation, to test its detailed parts. We have

$$P_{5III}\psi = [\hbar\partial^\mu(e_\mu) - m_D(if_0)]\psi = i\psi^{\dagger\wedge?}M$$

$$P^\dagger_{5III} = +P_{5III} \therefore P \to L^\dagger PL,\ P^\wedge P = (+----) \propto (e_0)$$

The $P^\wedge P$ is determined from a table. The K-G equation then is (using $P = P^\dagger$):

$$P^\wedge P\psi = P^\wedge i\psi^{\dagger\wedge?}M = i(P^\dagger)^\wedge\psi^{\dagger\wedge?}M$$

This does not work unless we have $\psi^{\dagger\wedge}$. Then we can continue,

$$iP^{\dagger\wedge}\psi^{\dagger\wedge}M = i(P\psi)^{\dagger\wedge}M = i(i\psi^{\dagger\wedge}M)^{\dagger\wedge}M$$

$$\therefore P^{\wedge}P\psi = -\psi M^{\dagger\wedge}M = [\hbar\partial^{\mu}\hbar\partial_{\mu} - m_D^2]\psi$$

Therefore, $[i\hbar\partial^{\mu}i\hbar\partial_{\mu}]\psi = (-m_D{}^2 + M^{\dagger\wedge}M)\psi$. We see that $m_D{}^2$ has the wrong sign here, but $M^{\dagger\wedge}M$, being positive and large enough, will allow $m_{exp}{}^2 = -m_D{}^2 + M^{\dagger\wedge}M > 0$. We then have $L^{\dagger}PL$ and $(L^{?}\psi)^{\dagger\wedge} = L^{?\dagger\wedge}\psi^{\dagger\wedge}$, which gives us $? = \wedge$, and so $\psi \rightarrow \psi' = L^{\wedge}\psi$ here and $P\psi \rightarrow L^{\dagger}PLL^{\wedge}\psi$ and $LL^{\wedge} = 1(e_0)$.

Now we can compare with the Dirac-Edmonds equation in the form

$$[\hbar\partial^{\mu}(e_{\mu}) - m_D i(f_o)]\psi = i\psi^{\dagger\vee}m_E^{\mu}(e_{\mu})$$

They are NOT the same even when $m_D = 0$, since $\psi^{\dagger\vee} \neq \psi^{\dagger\wedge}$ unless ψ is very special. There really are two physically possible equations!

To summarize, we have two basic Dirac-like generalizations:

Dirac-Edmonds:

$$[\hbar\partial^{\mu}(e_{\mu}) - m_D i(f_0)]\psi_1 = +i\psi_1^{\dagger\vee}m_E^{\mu}(e_{\mu})$$

$$\equiv P_{D'}\psi_1,\ P_{D'} \rightarrow L^{\dagger}P_{D'}L,\ P_{D'}^{\dagger} = P_{D'},\ (+ - - - +):\ P_{D'}^{\wedge}P_{D'}$$

Edmonds:

$$[\hbar\partial^{\mu}(e_{\mu}) - m_D(if_0)]\psi_2 = +i\psi_2^{\dagger\wedge}m_E^{\mu}(e_{\mu})$$

$$\equiv P_{E'}\psi_2,\ P_{E'} \rightarrow L^{\dagger}PL,\ P_{E'}^{\dagger} = P_{E'},\ (+ - - - -):P_{E'}^{\wedge}P_{E'}$$

These are almost identical when $m_D = 0$. The Edmonds equation has a tachyonish m_D mass. The first equation is Dirac for $m_E{}^{\mu} = 0$. The second one is not! Each of these now couples nicely to $A^{\mu}(e_{\mu})$ where $A \rightarrow A' = L^{\dagger}AL = L^{\downarrow}AL$, since L and A have only (e) and (ie) elements: $A^{\wedge} = A^{\vee}$ *and* $L^{\wedge} = L^{\vee}$. We may need an imaginary coupling constant here, since $(i\hbar\partial^{\mu})$ usually goes with A^{μ} in the non-relativistic limit. We do not need to move Maxwell's equation from the Pauli algebra into the Edmonds algebra! It is nice to have

$(+\ -\ -\ -)$ in both cases now, for the (t,x,y,z) parts. But $P_{D'}^{\dagger} \neq P_{D'}$ and it is not hermitian. However, $i^{\dagger} = -i$ *and* $i^{\downarrow} = -i$. Therefore, we can multiply both equations through by i and get new forms. This better fits the non-relativistic form as well. Our two Dirac-Edmonds and Edmonds generalizations become ($M = m_E^{\mu}(e_{\mu})$ is only one possibility here!)

$$P_D\psi_1 \equiv [(i\hbar\partial^{\mu})(e_{\mu}) + m_D(f_0)]\psi_1 = -\psi_1^{\dagger\vee} m_E^{\mu}(e_{\mu})$$

$$P_D^{\downarrow} = -P_D,\ P_D^{\wedge}P_D = (+ - - - - +),\ P_D \rightarrow L^{\downarrow}P_D L$$

and

$$P_E\psi_2 \equiv [(i\hbar\partial^{\mu})(e_{\mu}) - m_D i(if_0)]\psi_2 = -\psi_2^{\dagger\wedge} m_E^{\mu}(e_{\mu})$$

$$P_E^{\dagger} = -P_E,\ P_E^{\wedge}P_E = (+ - - - -),\ P_E \rightarrow L^{\dagger}P_E L$$

The Dirac equation operator here is not hermitian and in the new equation, the operator is anti-hermitian. So much for worrying about that any longer.

The $(i\hbar\partial^{\mu})(e_{\mu})$ operator above has $(i\hbar\partial^{\mu}(e_{\mu}))^{\dagger} = -i\hbar\partial^{\mu}(e_{\mu})$ and is an anti-hermitian 4-vector. It should have imaginary eigenvalues. But it 'gives' real energy and momentum in the standard quantum machinery. The $m_D(f_0)$ 'operator' has $(m_D(f_0))^{\dagger} = +m_D(f_0)$ so it should 'give' real mass eigenvalues, for ψ_1 states. Note $m_D(f_0)m_D(f_0) = m_D^2(e_0)$. Therefore, its eignvalues are $\pm m_D$. The $-m_D i(if_0)$ operator is anti-hermitian, so it should have imaginary eigenvalues. Note $[-m_D i(if_0)][-m_D i(if_0)] = -m_D^2(e_0)$ and its eigenvalues are $\pm i m_D$. The operation of $m_E^{\mu}(e_{\mu})$ on $\psi^{\dagger\wedge}$ from the right, is very unusual. We could write this as $[\psi(m_E^{\mu}(e_{\mu}))^{\dagger\wedge}]^{\dagger\wedge} = [\psi m_E^{\mu}(e_{\mu})^{\wedge}]^{\dagger\wedge}$. It is not clear, however, that we should think of this as an operator at all. Perhaps the m_D term is replaced by a 5th spacetime operator of some kind at a deeper level. Then we may fully understand why this f term holds us in the larger Dirac algebra.

REST STATES

To examine and compare these equations further, we should look at rest states. This requires spelling out ψ in some detail. We need $1/2\hbar(e_3)$ spin along z eignstates and $i\hbar\partial^0$ rest energy ($c \equiv 1$) eigenstates. A rest state is, by definition, one where $\partial^k\psi_R \equiv 0$, $k = 1,2,3$. Again we note that

$$L_e = 1(e_0) + e^k(ie_k) + \delta^k(e_k) + \cdots$$

and (ie_k), not (e_k), is the generator for rotations. The (e_k)'s generate boosts. Thus, $S_z = 1/2\hbar i(ie_3)$ might be a better spin operator than $1/2\hbar(e_3)$. In both cases, $S_z S_z = (\hbar/4)^2(e_0)$, so the eigenvalues are $(\pm\hbar/2)$. (The 1/2 is inserted here for historical reasons.) We can similarly define $S_k \equiv 1/2\hbar i(ie_k) = S_k^{\dagger}$ and $S \longleftrightarrow s^k i(ie_k)$, with $s^k \equiv 1/2\hbar$. Then $SS = \cdots = -SS^{\wedge} = +3(1/2\hbar)^2(e_0) = (3/4)\hbar^2(e_0)$. Clearly, S_z commutes with SS, but not S_x or S_y, and ψ_R can be an eigenstate of both. A boost generated by (e_3) along the z-axis still allows $\psi' = L^{\wedge}\psi_R$ to stay an eigenstate of S_z, since $i(ie_3)$ and (e_3) commute. But this is not true for S_x or S_y rest spin states. In non-relativistic approximation, this prohibition is lost and we imagine that we can have an electron moving along z with spin along x.

At this point we can just guess at part of the ψ_R state. Consider the Dirac-Edmonds equation:

$$P_D\psi_R = [(i\hbar\partial^0)(e_0) + 0 + 0 + 0 + m_D(f_0)]\psi_R = -\psi_R^{\dagger\vee} m_E^{\mu}(e_\mu)$$

Try

$$\psi_R \equiv e^{-\frac{iE}{\hbar}t}\phi(e's\ and\ f's)$$

We find

$$E\phi(e,f) = -m_D(f_0)\phi(e,f) - m_E^{\mu}\phi(e,f)^{\dagger\vee}(e_\mu)$$

Case 1: $m^\mu \equiv 0$

a) $E \equiv m_D \Rightarrow \phi = -(f_0)\phi$

b) $E \equiv -m_D \Rightarrow \phi = +(f_0)\phi$

Case 2: $m_D \equiv 0,\ m^0 \equiv m_E,\ m^k \equiv 0$

a) $E \equiv m_E \Rightarrow \phi = -\phi^{\dagger\vee}$

b) $E \equiv -m_E \Rightarrow \phi = +\phi^{\dagger\vee}$

These special conditions on ϕ(e,f) suggest we try

$$\phi \equiv C_1\alpha + C_2(f_0)\alpha + C_3\alpha^{\uparrow\vee} + C_4[(f_0)\alpha]^{\uparrow\vee}$$

$$\therefore \phi^{\uparrow\vee} = C_1\alpha^{\uparrow\vee} - C_2(f_0)\alpha^{\uparrow\vee} + C_3\alpha + C_4(f_0)\alpha$$

We then find

$$\begin{aligned} \textit{Case 1: } \phi_{+mD} &= C_1[\alpha - (f_0)\alpha] + C_3[\alpha^{\uparrow\vee} - (f_0)\alpha^{\uparrow\vee}] \\ \phi_{-mD} &= C_2[\alpha + (f_0)\alpha] + C_4[\alpha^{\uparrow\vee} + (f_0)\alpha^{\uparrow\vee}] \end{aligned}$$

$$\begin{aligned} \textit{Case 2: } \phi_{+mE} &= C_1[\alpha - \alpha^{\uparrow\vee}] + C_2(f_0)[\alpha + \alpha^{\uparrow\vee}] \\ \phi_{-mE} &= C_3[\alpha + \alpha^{\uparrow\vee}] + C_4(f_0)[\alpha - \alpha^{\uparrow\vee}] \end{aligned}$$

We still have α arbitrary. We can choose it to be a spin-z eigenstate, since $i(ie_3)$ commutes with (f_0), and also $[i(ie_3)]^{\uparrow\vee} = +i(ie_3)$. Notice that $C^{\uparrow\vee} = (a + ib)^{\uparrow\vee} = a + i^{\uparrow\vee}b = a + {-}{-}ib = a + ib = C$ for all the complex coefficients. We demand $i(ie_3)\alpha = \pm\alpha$ *from* $[i(ie_3)]^2 = (e_0)$. Notice that $i(ie_3)(e_0) = i(ie_3)$, $\therefore \alpha = (e_0) \pm i(ie_3)$ is part of the picture. Also, $i(ie_3)(e_1) = i(-e_2) \therefore \alpha = (e_1) \mp i(e_2)$ is also part of the picture. We will also need $(f_0)\alpha$, $\alpha^{\uparrow\vee}$, *and* $(f_0)\alpha^{\uparrow\vee}$. We find then that a suitable α is

$$\alpha_\pm = (e_o) \pm i(ie_3) + (e_1) \mp i(e_2) \textit{ spin up, down}(\pm)$$

$$\therefore \alpha_\pm^{\uparrow\vee} = (e_o) \pm i(ie_3) - [(e_1) \mp i(e_2)]$$

So, in general a rest state is of the form

$$\phi_{rest} = [C_1(e_0) + C_2(f_0)]\alpha + [C_3(e_0) + C_4(f_0)]\alpha^{\uparrow\vee}$$

In general, $A(f_0) = (f_0)A^{\uparrow\wedge}$ *and* $f = e(f_0)$, so we would not need a larger α it seems so far. However, we have not allowed the $m^k(e_k)$ parts to be non-zero yet. They multiply from the right. They greatly complicate things. Before that is considered, we should look at $m_D \neq 0$ and $m_E^{\,0} \neq 0$. How does that solution look? And, before we do that, we should check the Klein-Gordon equation to get some ideas of how E will look in this complicated situation. We have:

$$P_D^{\wedge}P_D\psi_1 = -P_D^{\wedge}\psi_1^{\dagger\vee}m_E^{\mu}(e_\mu) = (\hbar\partial^\mu\hbar\partial_\mu + m_D^2)\psi_1$$

$$= -(-P_D^{\downarrow})^{\wedge}\psi_1^{\downarrow\wedge}m_E^{\mu}(e_\mu) = (P_D\psi_1)^{\downarrow\wedge}m_E(e_\mu)$$

$$= -[\psi_1^{\dagger\vee}m_E^{\mu}(e_\mu)]^{\downarrow\wedge}m_E^{\nu}(e_\nu) = -m_E^{\mu}m_{E\mu}\psi_1 \quad (\partial^\mu\partial_\mu \equiv +---)$$

$$\therefore\ i\hbar\partial^\mu i\hbar\partial_\mu\psi_1 = (+m_D^2 + m_E^0 m_E^0 - m_E^k m_E^k)\psi_1 \equiv m_{\exp}^2\psi_1$$

Everything fits! Amazing! (It took a while to get all the bugs out.) Our energy, E, in exp(-iEt/$\hbar$), will then be a solution here, so $E = \pm m_{exp} = real$.

Our rest state, for both m_D and $m_E{}^0$ masses non-zero, becomes then

$$(i\hbar\partial^0)(e_0)\psi_R = -m_D(f_0)\psi_R - m_E^0(e_0)\psi_R^{\dagger\vee}$$

and $\psi \equiv exp(-iEt/\hbar)\phi(e\text{'s and } f\text{'s})$, with $E_\pm$ values known. This requires us to find $\phi_\pm$ from the equation

$$E_\pm\phi_\pm = -m_D(f_o)\phi_\pm - m_E\phi_\pm^{\dagger\vee},\ E_\pm = \pm(m_D^2 + m_E^2)^{1/2}$$

This still is not too bad. It will get messy when the m^k's are included. (Exercise for the student!) The rest of the ϕ calculation should be straightforward for our simpler case here.

GENERALIZING DIRAC'S ALGEBRA

Although I believe it best to forget matrices once a hypercomplex number system is well defined, we can use them as a temporary crutch to help guide our guessing at new generalizations. The Pauli algebra leads to $P\psi_a = \psi_v M_v$ & $P^\wedge\psi_v = \psi_a M_a$. Since P = $P^\dagger$, this sort of suggests a bigger number system

$$\begin{pmatrix} P & 0 \\ 0 & P^{\dagger\wedge} \end{pmatrix},\ \begin{pmatrix} 0 & Q^{\dagger\wedge} \\ Q & 0 \end{pmatrix},$$

and this turns out to give us the Dirac algebra when complex coefficients are allowed. Within Dirac algebra, we again have $P\psi_a = \psi_v M_v$ & $P^{\wedge}\psi_v = \psi_a M_a$ and P = $P^{\dagger}$, only now P can have 5 parts in it. This again suggests we try a larger algebra using P and $P^{\dagger\wedge}$. But Dirac has $(\ldots)^{\vee}$ as well, so we have $P^{\dagger\vee}$ and $P^{\wedge\vee}$ to also play with in defining a generalization. We don't know ahead of time where this will lead, as part of a larger number system which just might apply to the real world.

We again start by using Pauli → Dirac as our guide. Consider the Dirac basis elements e, ie, f, and if. We are led to consider

$$\begin{pmatrix} e & 0 \\ 0 & e^{\dagger\wedge} \end{pmatrix}, \begin{pmatrix} f & 0 \\ 0 & f^{\dagger\wedge} \end{pmatrix}, \begin{pmatrix} ie & 0 \\ 0 & (ie)^{\dagger\wedge} \end{pmatrix}, \begin{pmatrix} if & 0 \\ 0 & (if)^{\dagger\wedge} \end{pmatrix}$$

and similar matrices using $e^{\dagger\vee}$, $e^{\wedge\vee}$, $f^{\dagger\vee}$, $f^{\wedge\vee}$, $(ie)^{\dagger\vee}$, $(ie)^{\wedge\vee}$, $(if)^{\dagger\vee}$, *and* $(if)^{\wedge\vee}$. Not all of these are distinct, so it is not directly obvious how big an algebra this generates. We need to examine e_0 and e_k, then f_0 and f_k, etc., since all these conjugations fortunately treat e_k, k = 1,2,3, the same, and all f_k the same.

We obtain the following patterns

$$\underset{+}{e_0^{\dagger\wedge}}\ \underset{+}{e_0^{\dagger\vee}}\ \underset{+}{e_0^{\wedge\vee}},\ \underset{-}{e_k^{\dagger\wedge}}\ \underset{-}{e_k^{\dagger\vee}}\ \underset{+}{e_k^{\wedge\vee}},\ \underset{+}{f_0^{\dagger\wedge}}\ \underset{-}{f_0^{\dagger\vee}}\ \underset{-}{f_0^{\wedge\vee}},\ \underset{-}{f_k^{\dagger\wedge}}\ \underset{+}{f_k^{\dagger\vee}}\ \underset{-}{f_k^{\wedge\vee}}$$

$$\underset{-}{(ie_0)^{\dagger\wedge}}\ \underset{-}{(ie_0)^{\dagger\vee}}\ \underset{+}{(ie_0)^{\wedge\vee}},\ \underset{+}{(ie_k)^{\dagger\wedge}}\ \underset{+}{(ie_k)^{\dagger\vee}}\ \underset{+}{(ie_k)^{\wedge\vee}}$$

$$\underset{-}{(if_0)^{\dagger\wedge}}\ \underset{+}{(if_0)^{\dagger\vee}}\ \underset{-}{(if_0)^{\wedge\vee}},\ \underset{+}{(if_k)^{\dagger\wedge}}\ \underset{-}{(if_k)^{\dagger\vee}}\ \underset{+}{(if_k)^{\wedge\vee}}$$

We see from this that the e_0, e_k set gives us the basis elements

$$\begin{pmatrix} e_0 & 0 \\ 0 & e_0 \end{pmatrix}, \begin{pmatrix} e_k & 0 \\ 0 & -e_k \end{pmatrix}, \begin{pmatrix} e_k & 0 \\ 0 & e_k \end{pmatrix} \Rightarrow 7 \textit{ basis elements.}$$

Similarly, f_0 and f_k give us 2 + 6 = 8 basis elements; (ie_0) and (ie_k) give us 2 + 3 = 5 elements; (if_0) and (if_k) give us 2 + 6 = 8 basis elements. The total is 28. With complex coefficients then, we have 56 basis elements. The coupled equation again suggests that this closed system be extended by allowing

$$\begin{pmatrix} 0 & e^{\dagger\wedge} \\ e & 0 \end{pmatrix}, \begin{pmatrix} 0 & (ie)^{\dagger\wedge} \\ ie & 0 \end{pmatrix}, \begin{pmatrix} 0 & f^{\dagger\wedge} \\ f & 0 \end{pmatrix}, \begin{pmatrix} 0 & (if)^{\dagger\wedge} \\ (if) & 0 \end{pmatrix},$$

etc. The $e^{\dagger\wedge}$, $e^{\dagger\vee}$, $e^{\wedge\vee}$ parts that are independent follow the same pattern as just seen for 'diagonal' matrices. Therefore, this extension of the algebra (like adding on the f's to the e's in Dirac) will have an additional 28 elements. Thus, this algebra has 28 + 28 = 56 real basis elements. With complex coefficients added in, we now get 112 basis elements. This is smaller than the 'natural' guess of 8-by-8 complex matrices, with $2\times 64 = 128$ total basis elements. Is this smaller algebra really complete in the sense of being closed?

This smaller algebra may be enough for the new physics if closed. Here we still have $(\ldots)^{\dagger}$ as hermitian conjugation and so the 112 basis elements can be analyzed to give $(\ldots)^{\dagger} = \pm(\ldots)$ for each of the 112 basis elements. Again, we guess $[i(\ldots)]^{\dagger} \equiv -i(\ldots)^{\dagger}$ is a reasonable definition. The i's outside stay outside only, as before. Now $P \equiv P^{\dagger} = ?$ can be determined and the maximum size (24 parts?) of 'spacetime' determined here. (For Dirac, $P = P^{\dagger}$ $\Rightarrow$ 16 parameters.) We can define $A^{\wedge} \equiv (f_0)A^{\dagger}(f_0)$ since Dirac is a subalgebra of this extended algebra. Then we can explore $P^{\wedge}P \propto (e_0)$ and find the largest 'physical' momentum operator here. (For Dirac, this gave a 5 component operator, P.) How many here? This could get interesting in light of String Theory.

The natural groups are $AA^{\dagger} \equiv 1(e_0)$, $BB^{\wedge} \equiv 1(e_0)$, *etc.*, and we need the 'elements' '$A^{\dagger} = -A$' to find the number of parameters in group A. Of the 56 elements with real coefficients, we seem to have 32 for which $A^{\dagger} = -A$ and 24 where $A^{\dagger} = +A$, including the identity. Then the A's, with real coefficients, give us 32 parameters and i(A) elements give us another 24 elements. The total, 32 + 24 = 56, is 1/2 the total number of elements. This is just like the Dirac algebra case, where $A^{\dagger} = -A$ for 10 elements and $[i(A)]^{\dagger} = -iA$ for another 6 elements, making up a 16 parameter group, $AA^{\dagger} = 1(e_0)$. Explicitly in Dirac:

$$AA^{\dagger} \equiv 1(e_0)\colon A_{small} = 1(e_0) + \epsilon^k(ie_k) + \delta^k i(e_k) + \epsilon(ie_0) + \delta i(e_0)$$

$$+ \lambda^k(if_k) + \eta^k(f_k) + \xi i(f_0) + \rho i(if_0)$$

The closure of this extended algebra can be explored through

$$\begin{pmatrix} a & 0 \\ 0 & a^{\dagger\wedge} \end{pmatrix} \begin{pmatrix} b & 0 \\ 0 & b^{\dagger\wedge} \end{pmatrix} = \begin{pmatrix} ab & 0 \\ 0 & a^{\dagger\wedge} b^{\dagger\wedge} \end{pmatrix} = \begin{pmatrix} (ab) & 0 \\ 0 & (ab)^{\dagger\wedge} \end{pmatrix}$$

$$\begin{pmatrix} a & 0 \\ 0 & a^{\dagger\wedge} \end{pmatrix} \begin{pmatrix} 0 & b^{\dagger\vee} \\ b & 0 \end{pmatrix} = \begin{pmatrix} 0 & ab^{\dagger\vee} \\ a^{\dagger\wedge} b & 0 \end{pmatrix} = \begin{pmatrix} 0 & (a^{\dagger\wedge} b^{\vee\wedge})^{\dagger\wedge} \\ (a^{\dagger\wedge} b) & 0 \end{pmatrix}$$

$$\begin{pmatrix} 0 & a^{\dagger\wedge} \\ a & 0 \end{pmatrix} \begin{pmatrix} 0 & b^{\dagger\wedge} \\ b & 0 \end{pmatrix} = \begin{pmatrix} a^{\dagger\wedge} b & 0 \\ 0 & ab^{\dagger\wedge} \end{pmatrix} = \begin{pmatrix} (a^{\dagger\wedge} b) & 0 \\ 0 & (a^{\dagger\wedge} b)^{\dagger\wedge} \end{pmatrix}$$

It looks questionable. To be sure, we really need the full 56×56 multiplication table filled in. We would need a 24×24 table of $P^{\wedge}$ and P, $(= P^{\dagger})$, products to find the subsets for which all the cross terms cancel out. For Dirac this gives the 5-vectors $\{(e_\mu), (if_0)\}$, $\{i(ie_\mu), (if_0)\}$, and two others. The new P would directly extend these four.

I don't know if this 112 element algebra is a Clifford algebra or not. So far, I have not seen any indication that Clifford algebras bring more insight to the extension of physics than the simple ideas outlined in this book. The present description may not be as abstract or elegant, but it 'tis enough' it seems!

If this system is not closed, then we simply check all products and add on any new basis elements which come from these products, and then the new ones they also generate. We know that the system must close with a maximum of $2 \times 8 \times 8 = 128$ basis elements. But are there smaller number systems than 128, which contain physics equations in them; ones that apply to the real world? Why should Dirac not be large enough? (Ask God!) Do the groups dictate the physical algebras or do the physical algebras dictate the physical groups and physical spacetime operators and their dimensions? This issue remains for you, the next generation, to explore. The world is only as complex as it needs to be for humanoids to come to exist in it. There are no arbitrary frills in the design, I suspect.

DIRAC'S ORIGINAL FACTORIZATION

It is well known that in 1928 Dirac factored the Klein-Gordon equation by seeking a first order time derivative equation and seeking to treat time and space on an 'equal' footing. The historical details are outlined in Kragh's Scientific Biography of Dirac. References therein give more details.

The bottom line, according to Kragh, is that Dirac essentially played with the equations, using Pauli's extension of Schrödinger's equation as a guide, and finally guessed at the linear form

$$\left[\frac{W}{c} + P^k \alpha_k + m_0 c \alpha_4\right]\psi = 0$$

Here, W is energy, c the speed of light, momentum $P^k \leftrightarrow \{P_x, P_y, P_z\}$, m_0 the rest mass, ψ the wave function, and α_μ, $\mu = 1,2,3,4$, are unknown elements to be determined by satisfying the Klein-Gordon equation, which here is

$$\left(\frac{W}{c}\right)^2 \psi = \left[P^k P^k + (m_0 c)^2\right]\psi$$

Others were also working hard on methods of factoring this equation, and Kragh says they would have come up with Dirac's equation in a few months if he had not done it. He had been similarly scooped before on several occasions, so he hurriedly checked the first approximation to the hydrogen atom energy levels and got the right answer. He published without trying the difficult calculation of the exact solution. Others soon did that calculation and found very good agreement with experiment.

Dirac had also guessed that some 'squaring' process would lead him back to the K-G equation. Let us see what he did in some detail. We can generate a multiplication table for $P_{(5)}P_{(5)}$, where

$$P_{(5)} \equiv \frac{W}{c}\mathbb{I} + P^k \alpha_k + m_0 c \alpha_4$$

in the following form:

P \ P	$\mathbb{I}$	α_1	α_2	α_3	α_4
$\mathbb{I}$	$\mathbb{I}$	α_1	α_2	α_3	α_4
α_1	α_1	$\mathbb{I}$			
α_2	α_2		$\mathbb{I}$		
α_3	α_3			$\mathbb{I}$	
α_4	α_4				$\mathbb{I}$

Clearly, he could not include W/c in the square, $P_{(5)}P_{(5)}$, and still hope to get all the cross terms to cancel. Just as clearly, he needed $\alpha_1\alpha_2 = -\alpha_2\alpha_1, \ldots, \Rightarrow \alpha_\mu\alpha_\nu = -\alpha_\nu\alpha_\mu$, for $\mu \neq \nu$, to kill the other cross terms. The four positive diagonal terms, $\alpha_1\alpha_1 = \mathbb{I}$, *etc.*, are needed for the signs found in the Klein-Gordon equation. Thus he likely used

$$-\frac{W}{c}\mathbb{I}\psi = P_{(4)}\psi \quad and \quad P_{(4)}\left(-\frac{W}{c}\mathbb{I}\psi\right) = P_{(4)}(P_{(4)}\psi)$$

$$= -\frac{W}{c}\mathbb{I}P_{(4)}\psi = \left(-\frac{W}{c}\right)\left(-\frac{W}{c}\psi\right)$$

so he could get

$$\left(\frac{W}{c}\right)^2\psi = [P^kP^k + (m_0c)^2]\psi$$

The rest is history! No improvement on this equation has proven useful for new physics during the past almost 70 years. Is there more to the factorization than Dirac found? His equation is simply $P_{(5)}\psi = 0$ and is Lorentz covariant.

Kragh says that Dirac found specific 4×4 matrices, for the needed α_μ elements, as follows

$$\alpha_k \equiv \begin{pmatrix} 0 & \sigma_k \\ \sigma_k & 0 \end{pmatrix} \equiv -i(if_k), \ \alpha_4 \equiv \begin{pmatrix} \sigma_0 & 0 \\ 0 & -\sigma_0 \end{pmatrix} \equiv -i(ie_0)$$

where my notation has also been inserted for the α_μ elements. His equation then takes the form, using 'modern' notation ($e_0 \equiv 1$):

$$\left[\frac{W}{c}(e_0) + P^k(-i(if_k)) - m_0c(i(ie_0))\right]\psi = 0 \equiv P_{(5)}\psi$$

We have already seen that $P_{(5)}P_{(5)}\psi = P_{(5)}0 = 0$ will not give the K-G equation. Cross terms remain. There are three primary, antiautomorphic conjugations in Dirac algebra.[3] They are called ↑, ∧, ∨, and ↓ ≡ ↑∧∨. In the basis below, any e basis element has $e^\uparrow = +e^\downarrow$ and $e^\wedge = +e^\vee$, and any f basis element has $f^\uparrow = -f^\downarrow$ and $f^\wedge = -f^\vee$. The conjugations change signs in a pretty pattern, easy to memorize.

The natural representation for Dirac, when we expand from Pauli, is $\{(e_0), (ie_0), (e_k), (ie_k), (f_0), (if_0), (f_k), (if_k)\}$, k = 1,2,3, with complex coefficients out front of these. This turns out, I recently learned, to be the old Weyl representation. The e's are 'like' the σ's and a closed subsystem. I have a scheme for multiplying any pair in your head to obtain ±(another element) in the 16 element set. Although $P_{(5)}P_{(5)}\psi = 0$ will NOT give the K-G equation, there is a 5-vector modification of this that does work. In the $P\psi = 0$ equation, we have $P^0 = W/c$ and $P^4 = -m_0c$. We can verify $P_{(5)}{}^{\vee}P_{(5)}$ has no cross terms, if we calculate all 25 terms. I have shown how to calculate each term in your head and you should be able to fill in the entire table and see that all the cross terms do indeed cancel.

P / $P^{\vee}$	(e_0)	$-i(if_1)$	$-i(if_2)$	$-i(if_3)$	$i(ie_0)$
$+(e_0)$	(e_0)				
$+i(if_1)$		$-(e_0)$			
$+i(if_2)$			$-(e_0)$		
$+i(if_3)$				$-(e_0)$	
$-i(ie_0)$					$-(e_0)$

It can be shown, by grinding along with the matrices shown above or using my easier multiplication rules, that all cross terms do cancel, even though this is a 5-vector here. The metric is $(+ - - - -)$. Notice that, if we start over and forget how we historically got to this result, we basically have guessed that

$$P_{(5)}\psi = 0,\ P_{(5)} = P^{\dagger}_{(5)},\ P^{\vee}_{(5)}P_{(5)} \propto (e_0) \Rightarrow$$

$$P_{(5)}\colon \{(e_0),\ -i(if_k),\ i(ie_0)\}$$

This looks rather simple. Its covariance can be displayed in the abstract form $P \to P' \equiv L^{\dagger}PL \Rightarrow P' = P'^{\dagger}$ from $P = P^{\dagger}$, and $\psi \to \psi' = L^{\vee}\psi$. We see the 'inner product'

$$P'^{\vee}P' = (L^{\dagger}PL)^{\vee}(L^{\dagger}PL) = L^{\vee}P^{\vee}L^{\dagger\vee}L^{\dagger}PL = L^{\vee}P^{\vee}(LL^{\vee})^{\dagger}PL$$

The L's inside 'go away' if $LL^{\vee} \equiv 1(e_0)$. This leads to $P'\psi' = (L^{\dagger}PL)(L^{\vee}\psi)$, so the L's inside here will also go away.

There actually is a natural generalization of this equation, which I began exploring in the early 70's, and which can now be displayed in the elegant form:

$$P_{(5)}\psi = \psi^{\dagger\vee}M$$

First check the covariance: $(L^{\vee}\psi)^{\dagger\vee} = L^{\dagger}\psi^{\dagger\vee}$ *and* $L^{\dagger}PLL^{\vee}\psi = L^{\dagger}P\psi$. Thus both sides have $L^{\dagger}$ on the left. We guess that $M = M'$ as usual. This is a new kind of mass, possibly unseen at the classical level! Next, we examine the 5-vector K-G equation that comes from this and Dirac's P operator:

$$P^{\vee}P\psi = P^{\vee}\psi^{\dagger\vee}M = (P^{\dagger})^{\vee}\psi^{\dagger\vee}M = (P\psi)^{\dagger\vee}M$$

$$= (\psi^{\dagger\vee}M)^{\dagger\vee}M = \psi(M^{\dagger\vee}M) =?= (M^{\dagger\vee}M)\psi$$

Thus we have $(P^{\vee}P - M^{\dagger\vee}M)\psi = 0 = (P^0P^0 - P^KP^K - P^4P^4 - M^{\dagger\vee}M)\psi$ and $M^{\dagger\vee}M$ must commute with ψ, even if M does not. Thus the two mass parts merge into one number at the K-G level and thus in all the physics we know about so far.

Notice next that the theorem $(if_0)A = A^{\dagger\vee}(if_0)$ holds for any element A of Dirac. So let us multiply the equation through by (if_0) on the left to get rid of the $\psi^{\dagger\vee}$ conjugation:

$$(if_0)P\psi = (if_0)\psi^{\dagger\vee}M = (\psi^{\dagger\vee})^{\dagger\vee}(if_0)M = \psi(if_0)M$$

Note also the theorem $[(if_0)P][(if_0)P] = [P^{\downarrow\vee}(if_0)](if_0)P = P^{\dagger\vee}(e_0)P = P^{\dagger\vee}P = P^{\vee}P \propto (e_0)(+ - - - -)$, where $P = P^{\dagger}$ was used. Thus for this new operator, $[(if_0)P][(if_0)P]$ has all the cross terms cancel out, even for 5-vectors! This also satisfies Dirac's hunch that a simple square could give back the K-G equation.

We can call $(if_0)P \equiv Q$. Then $P = (if_0)Q$, since $(if_0)(if_0) = (e_0)$. Notice that $Q^{\vee} = [(if_0)P]^{\vee} = P^{\vee}(if_0)^{\vee} = (if_0)(P^{\vee})^{\dagger\vee} = (if_0)P^{\dagger} = (if_0)P = Q$. Therefore, we have proven $Q = Q^{\vee}$, *and* $Q^{\vee}Q = QQ \propto (e_0)(+ - - - -)$. The original Dirac equation, now extended, is $Q\psi = \psi(if_0)M$, and since M is unknown, we can absorb (if_0) into it. Thus the simple generalization of Dirac, with 5-vector Q and up-to 5-part M, is

$$Q\psi = \psi M \Rightarrow$$

$$QQ\psi = Q\psi M = \psi MM = MM\psi$$

This is simple and beautiful in a way that Dirac surely would have loved. We get the K-G equation if MM is 'positive' and $\propto$ (e_0), even if the mass inside Q is zero. (The M combines with this 5th mass piece in Q.) For covariance, $P \rightarrow P' = L^{\dagger}PL \Rightarrow Q \rightarrow Q' = (if_0)P' = (if_0)L^{\dagger}PL = (L^{\dagger})^{\dagger\vee}(if_0)PL = L^{\vee}(if_0)PL = L^{\vee}QL$. This is compatible with $Q = +Q^{\vee}$, such that $Q' = +Q'^{\vee}$ also.

But this is not the now standard γ_μ form of Dirac. For generalization of that, we need $R\psi = \psi^{\wedge\vee}M$, with R in a different representation. We can multiply our old equation by (ie_0), since there is a theorem $(ie_0)A = A^{\vee\wedge}(ie_0)$ for any A in Dirac algebra. We obtain $R\psi \equiv (ie_0)Q\psi = (ie_0)\psi M = \psi^{\wedge\vee}(ie_0)M$ and we can again absorb (ie_0) into the unknown M. We find that

$$\boldsymbol{R\psi = \psi^{\wedge\vee}M,\ R = (ie_0)Q = (ie_0)(if_0)P = +(f_0)P,}$$

$$\boldsymbol{R\colon \{(f_0),\ -i(ie_k),\ i(if_0)\}.}$$

Since $R = (f_0)P$, then $P = (f_0)R$ and we know $P = P^{\dagger}$, $P^{\vee}P \propto (e_0)$. Therefore, $(f_0)R = ((f_0)R)^{\dagger} = R^{\dagger}(f_0)^{\dagger} = R^{\dagger}(f_0) = (f_0)(R^{\dagger})^{\dagger\wedge} = (f_0)R^{\wedge}$. Therefore, $R = R^{\wedge}$, which can also be checked directly from the parts in R. Also, $P^{\vee}P = [(f_0)R]^{\vee}[(f_0)R] = R^{\vee}(f_0)^{\vee}(f_0)R = R^{\vee}(-e_0)R = -R^{\vee}R$. Therefore, $R^{\vee}R \propto (e_0)(- + + + +)$. This is still not the standard γ_μ form, since the mass part, P^4, is not multiplied by (e_0). We aimed here at getting $\psi^{\wedge\vee}$ from the above change of basis.

We can similarly multiply through by $-i(if_0)$ and define $S \equiv -i(if_0)R$. This gives $-i(if_0)R\psi = -i(if_0)\psi^{\wedge\vee}M = (\psi^{\wedge\vee})^{\dagger\vee}[-i(if_0)M] \rightarrow \psi^{\dagger\wedge}M$. Then we have $S\colon\{i(ie_0), (f_k), (e_0)\}$, and it has the desired (e_0) mass part, but loses the $\psi^{\wedge\vee}$ form. Now $\gamma_0 \equiv i(ie_0)$ *and* $\gamma_k \equiv (f_k)$ should give the usual $M = 0$ equation. Notice that $[i(ie_0)P]$ gives the same result, as expected, since $-i(if_0)(f_0)P = +i(ie_0)P$. The other representation, that I already know gives Dirac's equation, is $P\psi = \psi^{\wedge\vee}(-M^0(ie_0)) = [P^{\mu}i(-f_\mu) + P^4(e_0)]\psi$, where $P^{\mu} \equiv \hbar\partial^{\mu}$ *and* $\gamma_\mu \equiv -(f_\mu)$. Some similarity transform must link these forms.

Clearly this is Dirac for $M \equiv 0$. For $P^4 \equiv 0$, it also gives Dirac if $M^0 \equiv m$. The proof is as follows:

$$\boldsymbol{P_{(4)}\psi = P^{\mu}i(-f_\mu)\psi = \psi^{\wedge\vee}(-M^0(ie_0)) = (ie_0)(\psi^{\wedge\vee})^{\wedge\vee}(-M^0).}$$

Therefore, $(ie_0)P_{(4)}\psi = P^{\mu}i(+if_\mu)\psi = M^0\psi$, and now we can define $\gamma_\mu \equiv (if_\mu)$, with $\gamma_0\gamma_0 = +(e_0)$, so the γ metric is $(+ - - -)$ here. Notice now that $\gamma_0\gamma_1\gamma_2\gamma_3 = (if_0)(if_1)(if_2)(if_3) = (e_1)(-ie_1) = -(ie_0)$, and the alternative $\gamma_0\gamma_1\gamma_2\gamma_3 = (-f_0)(-f_1)(-f_2)(-f_3) = (e_1)(-ie_1) = -(ie_0) = \gamma_5$ here also. Therefore,

$$P^{\mu}_{(4)}i(if_{\mu})\psi = M^0\psi \rightarrow (\gamma_{\mu}) = (if_{\mu})$$

$$P^{\mu}i(-f_{\mu})\psi = -\psi^{\wedge\vee}M^0(ie_0) \rightarrow (\gamma_{\mu}) = (-f_{\mu})$$

$$= -M^0(ie_0)\psi$$

We see that Dirac can be generalized in the form $P_{(4)}{}^{\mu}i(\gamma_{\mu})\psi = [m_D I + m_E(\gamma_5)]\psi$, where $P^4 = -m_D$ and $M^0 = m_E$. This is with hindsight, perhaps not such a surprising generalization, since γ_5 is thought of as a pseudoscalar in conventional QED theory. We sort of have scalar mass and pseudoscalar mass prospects for the subnuclear word here. M, with more parts, seems also to be allowed.

We have skipped over the issue of $P^{\mu} = i\hbar\partial^{\mu}$ or $P^{\mu} = \hbar\partial^{\mu}$. This is an important issue in getting the right sign in the K-G equation for $m_{exp}{}^2$. There are two P operators with $(+ - - - -)$ and $(+ - - - +)$ metrics. One gives conventional Dirac and the other a Tachyionish Dirac. However, both can be physical because M can 'compensate' if large enough.

Putting all this together, we see that $P_{(5)}\psi = \psi^{ab}M$, $P = P^a$, and $P^bP \propto (e_0)$ give Dirac's old equation plus the new M term. Here, a and b are any two of the four conjugations $\uparrow$, $\downarrow$, $\wedge$, or $\vee$. The detailed representation for the corresponding 5-vector P is obtained from the 16 by 16 table P^bP. We find 5-vectors in this table for which all the cross terms cancel. The range of metrics I have seen include $(+ - - - -)$, $(+ - - - +)$, and even $(+ + + + +)$. The Lorentz covariance is $P \rightarrow P' = L^aPL$; and the $\propto (e_0)$ and invariance need of P^bP leads to $P'^bP' = (L^aPL)^b(L^aPL) = L^bP^b(LL^b)^aPL$. Thus L has $LL^b = 1(e_0)$ and $P^bP \propto e_0 \Rightarrow P'^bP' = P^bPL^bL = P^bP(e_0) = P^bP = invariant$. Notice that a = b is allowed as we have seen explicitly! Then $P\psi = \psi M$, $PP \propto (e_0)$, $P \rightarrow P' = L^aPL$ and $P'P' = L^aPLL^aPL$, so $LL^a \equiv 1(e_0)$ is needed here. Now, $P'P' = L^aPPL = PPL^aL = PP(e_0) \doteq PP = invariant$. We arbitrarily start with $P = P^a$ and that 'induces' $P' = L^aPL$. Most P 5-vectors do not have the special property $PP \propto (e_0)$. I have found a few specific ones so far. The one we just got from Dirac's original factorization is $P \equiv P^{\vee} = P^0(if_0) + P^k[-i(e_k)] + P^4[-i(e_0)]$. The PP metric is $(+ - - - -)$, and $P\psi = \psi M$. Another one is $P = P^{\mu}(f_{\mu}) + P^4(ie_0) = P^{\wedge}$ and it has $PP \propto (e_0)(+ - - - -)$. Another is $P = P^{\mu}i(if_{\mu}) + P^4(ie_0) = P^{\wedge}$, $\propto (e_0)(- + + + +)$.

QUARK CONFINEMENT

I found a very peculiar $P = P^0(f_0) + P^k(e_k) + P^4(if_0) = P^\dagger$ *for which* $PP \propto (e_0)(+ + + + +)$. This obviously cannot give a physical spacetime equation, but it never-the-less has $P\psi = \psi M$ *and* $PP\psi = \psi MM$ *with* $PP \propto (e_0)$, *and* $P = P^\dagger$. This is a totally new kind of equation, with no apparent classical limit at all! Since it is beautiful, similar to the Dirac equation, and in the Dirac algebra, I would guess that nature uses it, possibly for quarks! This could be why they are confined! This is an interesting idea that needs exploration. There must be a systematic pattern of such $(+ + + + +)$ 5-vectors in Dirac. Perhaps the $\dagger$ and $(\wedge, \vee)$ conjugations are not really equivalent! This also needs further study.

Here are some more of the $(+ + + + +)$ details to get you started. In $P = P^\dagger$, the four 5-vectors for $PP \propto (e_0)$, are found from the table to be $\{(e_k), (if_0), (f_0)\}$, $\{i(ie_2), i(ie_3), i(f_1), (e_1), i(if_1)\}$, $\{i(f_k), i(ie_0), (f_0)\}$, *and* $\{i(if_k), (if_0), i(ie_0)\}$. (The '3-vectors' there with no cross terms are $\{i(ie_k)\}$ and $\{i(ie_3), i(f_1), i(f_2)\}$. These 3-vectors may be parts of bigger systems with 5 parts and that needs more study. They may instead give M sets of internal parts for wave equations.)

Let us now concentrate on the $\{(e_k), (if_0), (f_0)\}$ form for $P = P^\dagger$. A 5×5 table then readily shows $PP \propto (e_0)$, with $(+ + + + +)$ metric. Our choice $P = P^\dagger$ dictates $P' = P^\dagger PL$. We see that $P'P' = (L^\dagger PL)(L^\dagger PL)$ needs $LL^\dagger = 1(e_0)$, which most generally could give 16 parameters in this group (maximum). How do we then 'choose' the (x, y, z, t, mass) parts of P? There are two natural choices, both involving $x^k \Leftrightarrow (e_k)$ as expected. We find $L^\dagger(if_0)L = (if_0)(L^\dagger)^{\dagger\vee}L = (if_0)L^\vee L$. Thus, $L^\vee L \equiv 1(e_0) \Rightarrow m(if_0)$ is invariant, and $L^\vee = L^\dagger$ then follows, with only 8 parameters (maximum) in L, just as for normal Dirac. The other choice, $L^\dagger(f_0)L = (f_0)(L^\dagger)^{\dagger\wedge}L = (f_0)L^\wedge L \equiv (f_0)$ gives us the new condition $L^\dagger = L^\wedge$ instead, and now $m(f_0)$ is invariant. Thus either one of these is the mass and the other must then be the time component. It may not matter which is taken to be which.

We see that $P(P\psi = \psi M) \Rightarrow (PP)\psi = (P\psi)M = \psi(MM)$ *and* $PP \propto (e_0)$. Therefore, even in this strange $(+ + + + +)$ world, $MM \propto (e_0)$ is a reasonable guess. This probably means that M has 5 parts (maximum), but M could have any one of the four 5-vector forms in the table, at this point, not only the same internal pattern as P. This needs more study. In any case, let us now choose (if_0) as mass and (f_0) as time, then $P = P^0(f_0) + P^k(e_k) + P^4(if_0) = P^\dagger$, and we guess $P^\mu = \hbar\partial^\mu$. The Klein-Gordon type equation then becomes here $PP\psi = MM\psi \Rightarrow$

$$(\hbar\partial^\mu\hbar\partial^\mu + P^4P^4)\psi = (M^\mu M^\mu + M^4M^4)\psi$$

Multiply through by -1 = ii and we come closest to the old K-G equation, but with (+ + + +) 'spacetime' here:

$$(i\hbar\partial^\mu i\hbar\partial^\mu)\psi = (P^4P^4 - M^\mu M^\mu - M^4M^4)\psi$$

For our large scale world, we would need $i\hbar\partial^\mu i\hbar\partial_\mu$ instead, with indefinite 4-metric (+ − − −) or (− + + +). In our humanoid world, the P^4P^4 mass part seems to dominate. Maybe it is the opposite for the quark world, where M dominates. This certainly gives a nice compliment to our world, sort of like the way that antimatter does also.

Maybe the four 5-vector forms have something to do with the four ways of winding up with this same possible quark equation: $\psi^{\uparrow\uparrow}$, $\psi^{\wedge\wedge}$, $\psi^{\vee\vee}$, and $\psi^{\downarrow\downarrow}$, but I doubt it.

MULTIPLE FORMS — THE BIG PICTURE

It has been very confusing, finding so many anti-automorphic conjugations and equations that go with them in Dirac algebra. I think we can now see the big picture here and I hope this will be helpful to others getting into this new field.

As we have said, there are four basic conjugations in Dirac: $\uparrow$, $\wedge$, $\vee$, and $\downarrow = (\uparrow\wedge\vee)$. They all change 16 of the 32 signs for the 32 basis elements of the full Dirac algebra, $\{(e_\mu), (ie_\mu), i(e_\mu), i(ie_\mu), (f_\mu), (if_\mu), i(f_\mu), i(if_\mu)\}$. The basic 5-vector P wave equation here seems to be

$$P\psi = \psi^{??}M$$

where ?? means any two of the four conjugations. There are many, many possibilities and so it is hard to find the central features here. Most forms are redundant. For example:

$$P\psi = \psi^{\wedge\uparrow}M,\ P = \pm P^{\uparrow},\ P' = L^{\uparrow}PL,\ \psi' = L^{\wedge}\psi$$

$$Q\psi = \psi^{\wedge\downarrow}M,\ Q = \pm Q^{\downarrow},\ Q' = L^{\downarrow}PL,\ \psi' = L^{\wedge}\psi$$

$$R\psi = \psi^{\wedge\vee}M,\ R = \pm R^{\vee},\ R' = L^{\vee}RL,\ \psi' = L^{\wedge}\psi$$

$$S\psi = \psi^{\wedge\wedge}M,\ S = \pm S^{\wedge},\ S' = L^{\wedge}SL,\ \psi' = L^{\wedge}\psi$$

are all different forms of the same equation, which is form covariant under the large group $LL^{\wedge} = 1(e_0)$, with 16 parameters! One group generator commutes with the other 15 in L, so L is basically a 15 parameter group here. It includes the Lorentz group. We know $O(6) = SU(4)$ has 15 parameters. Also, 5 boosts and 10 rotations characterize a Lorentz-type group in 6-space! This is a natural guess as to the nature of $LL^{\wedge} = 1(e_0)$. Is it possibly $L_{(6)} \otimes U_{(1)}$? I have previously guessed that it might be $SU(3) \otimes SU(3)$ with 16 parameters, which now seems unlikely, but it definitely must have Lorentz as a subgroup, and it really has only 15 parameters or even less perhaps. The special nature of the 5-th dimension may signal a breaking of this symmetry down to the 8 parameter group, but still not the Lorentz group by itself, so I think Lorentz will never be used alone in future field theory. Einstein was just plane wrong about why it is in nature! The Dirac equation, in standard γ_μ form, comes directly from the R-operator, 5-vector form above. The 5th part in R gives us the usual, classical, Dirac mass. Or, instead, M has it and the fifth part is zero in the operator. So how do we get down to the skeleton of this monster? The following seems very general.

Choose any conjugation from $\uparrow, \wedge, \vee, \downarrow$; say, $\uparrow$. Choose $P = P^{\uparrow} \Rightarrow$ *16* spacetime parameters (maximum). Next choose any conjugation from the ones left; say, $\wedge$. Find that $P^{\wedge}P \propto (e_0)$ for at least 5 internal parts of $P = P^{\uparrow}$. Notice that $(P^{\wedge}P)^{\wedge} = (P^{\wedge}P) \Rightarrow 16$ parts (maximum) in general, but here $(P^{\wedge}P)^{\uparrow} = P^{\uparrow}P^{\wedge\uparrow} = PP^{\wedge} =?= P^{\wedge}P$ (probably), so $(P^{\wedge}P) = (P^{\wedge}P)^{\uparrow}$ likely has only 8 parts (maximum) in the generalized metric. We only guess that $P^{\wedge}P \propto (e_0)$ is necessary in nature! Someone needs to explicitly dig all the 8 parts out. Next, choose $P\psi = \psi^{??}M$. Since $P = P^{\uparrow}$, choose $P' = L^{\uparrow}PL$ *so* $P'^{\uparrow} = P'$. Then $P'^{\wedge}P' = (L^{\uparrow}PL)^{\wedge}(L^{\uparrow}PL) = L^{\wedge}P^{\wedge}(LL^{\wedge})^{\uparrow}PL$ and the 'trapped L's' go away if $LL^{\wedge} \equiv 1(e_0)$, again a 16 parameter group at most here, so far for L. Now we need to satisfy form covariance in the form

$$P'\psi' = \psi'^{??}M = L^{\dagger}PLL^{?}\psi = (L^{?}\psi)^{??}M$$

Clearly, $\psi' \equiv L^{\wedge}\psi$ is needed to get rid of the 'trapped' L's here. Then $(L^{\wedge}\psi)^{??} = L^{\wedge ??}\psi^{??}$ *needs* $L^{\wedge ??} = L^{\uparrow}$, so both sides have the same $L^{\uparrow}$ on the left, which can be 'cancelled'. This requires that $L^{\wedge ??} = L^{\uparrow} = L^{\wedge\wedge\uparrow}$. Thus the covariant equation is finally $P\psi = \psi^{\uparrow\wedge}M_4$.

If we start over and pick any two conjugations, such as $P = P^{\vee}$ and $P^{\wedge}P$, then similar steps lead to $LL^{\wedge} = 1(e_0)$, $\psi \rightarrow \psi' = L^{\wedge}\psi$, $P\psi = \psi^{\wedge\vee}M$. Thus there are only one or two equations here in many disguises.

Now let us go back to $P\psi = \psi^{\uparrow\wedge}M$, since it does not matter which one we take, and further explore its parts. I displayed the full 16×16 multiplication table for $P^{\wedge}P$, with $P = P^{\uparrow}$, in earlier SST papers[3]. There are four different 5-vectors here for which $P^{\wedge}P \propto (e_0)$. Of these, two have $(- + + + -)$ and two have $(+ - - - -)$ for their metrics. Thus only two 5-vectors are distinct! One is the old Dirac and the other (second one above) is 'tachyonish'. These two P choices are explicitly represented by ($P = P^{\uparrow}$)

$$P_D\!: \{i(ie_\mu), (if_0)\}, \quad P_E\!: \{(e_\mu), (if_0)\}$$

$$-+++- \qquad\qquad\qquad +----$$

These have the disadvantage that spacetime, x^{μ}, has two separate forms, $i(ie_\mu)$, and (e_μ). That would not be convenient for curved space gravity nor for coupling to Maxwell's field in curved spacetime. Maxwell's equation is most naturally $P_E\psi = 0$, *with* $P^4(if_0) \approx 0(if_0) = 0$, *and* $\psi \rightarrow \psi' = L^{\wedge}\psi L$, $\psi = -E^k(e_k) + cB^k(ie_k) + 0 + ?\ldots$. It is not clear to me that the photon's P^4 mass part must be zero in a closed and expanding universe. Only its M mass must be zero to allow for covariance.

But $i(ie_\mu)$ better fits the traditional form for Dirac, where $i\hbar\partial^{\mu}$ appears. We can bring the two spacetime parts 'into line' by using a second version of the basic equation, for either P_D or P_E. For example, $P_E\psi = \psi^{\uparrow\vee}M$ goes with $P_E = P_E^{\downarrow} \Leftrightarrow \{i(ie_\mu), i(f_0)\}$, $P_E^{\wedge}P_E \propto (e_0)$, $\psi \rightarrow L^{\wedge}\psi$, $LL^{\wedge} = 1$, *and* $P = P^{\mu}i(ie_\mu) + P^4 i(f_0)$. (There may be even more parts in P. We should keep an open mind about that.) We could rewrite Maxwell easily to have this same spacetime, $i(ie_\mu)$ instead of (e_μ). I personally prefer (e_μ) for spacetime, but $i(ie_3)$ does show up in the usual Dirac hydrogen atom as part of the angular momentum operator, $\mathscr{L}_z$. I don't think any one form is really better, in any fundamental way. There are two basic metrics so two operators, P_D and P_E, in any choice of $P = P^{?}$. There are really two basic, free particle, spin 1/2 extended equations with M $\neq$ 0. Both give the usual Klein-Gordon equation, if P has only 5 large parts rather than 16 (provided M is large enough to dominate in the case of P_E).

There is a cute way to prove that all the $P\psi = \psi^{??}M$ forms are equivalent, building on one choice for $P = P^?$. Restrictions on M are the next order of business in all this stuff. Let us start with $P\psi = P^{\dagger\wedge}M$, $P = P^{\dagger}$, *and* $P^{\wedge}P$ *(the 'dot-product' operator).* No other restrictions on P so it could have 16 components. Look at the K-G equation that follows directly from this:

$$P^{\wedge}(P\psi) = (P^{\wedge}P)\psi = P^{\wedge}(\psi^{\dagger\wedge}M) = (P^{\wedge}\psi^{\dagger\wedge})M$$

$$= \ldots = \psi M^{\dagger\wedge}M \equiv M^{\dagger\wedge}M\psi$$

First of all, this suggests a new kind of quantum (called spin 0 for historical reasons):

$$(P^{\wedge}P)\phi = \phi M^{\dagger\wedge}M, \quad \phi \to \phi' = L^{\wedge}\phi L, \quad \phi^{\wedge} \equiv +\phi$$

This is not a scalar unless $L^{\wedge}\phi L = \phi L^{\wedge}L = \phi(e_0) = \phi = \phi'$. No need for ϕ to be this restricted. We only guess that ϕ must also satisfy a K-G equation meaning that $\phi M^{\dagger\wedge}M = M^{\dagger\wedge}M\phi$. The M could still be complicated here. For Dirac, $M^{\dagger\wedge}M \propto (e_0)$ is needed. Notice $[M^0(f_0)]^{\dagger\wedge}M^0(f_0) = M^0M^0(e_0)$ and M^0 can be real or imaginary here, so $(M^0)^2 = \pm|M^0|^2$.

Now return to $P_{(5)}\psi = \psi^{\dagger\wedge}M$. Notice that $M = (f_0)(f_0)M$ *so* $\psi^{\dagger\wedge}M = \psi^{\dagger\wedge}(f_0)(f_0)M = (f_0)(\psi^{\dagger\wedge})^{\dagger\wedge}(f_0M)$. Therefore, $(f_0)P\psi = \psi[(f_0)M]$. If $M = M^0(f_0)$, *then* $(f_0)M = M^0(e_0)$.

Similarly, $M = (if_0)(if_0)M$, *so* $\psi^{\dagger\wedge}M = \psi^{\dagger\wedge}(if_0)(if_0)M = (if_0)(\psi^{\dagger\wedge})^{\dagger\vee}(if_0)M = (if_0)\psi^{\wedge\vee}(if_0)M$. Therefore, $(if_0)P\psi = \psi^{\wedge\vee}(if_0)M$. If $M = M^0(f_0)$, *then* $(if_0)M = -M^0(ie_0)$. Finally, $M = (ie_0)(-ie_0)M$ gives $P\psi = (ie_0)(\psi^{\dagger\wedge})^{\wedge\vee}(-ie_0)M = (ie_0)\psi^{\dagger\vee}(-ie_0)M$ and $(ie_0)P\psi = \psi^{\dagger\vee}(ie_0)M$. If $M = M^0(f_0)$, then $(ie_0)M = -M^0(if_0)$. Remember that $M = M^0(f_0)$ is only one of several choices. We clearly see here that all four forms of $P\psi = \psi M$ are equivalent, given $P = P^?$.

Maxwell's equation is $P\psi = 0$ because $\psi \to \psi' = L^{\wedge}\psi L$ and $\psi = -\psi^{\wedge} =$ *(up-to-16-parts)* in ψ_{γ}. No M is allowed here but P can have 5 parts!

There is a fourth basic wave equation with two coupled parts. It is $P\psi_a = \psi_v M_v$, $P^{\wedge}\psi_v = \psi_a M_a$. This is very interesting. In my generalization of Pauli to the algebra 'between' Pauli and Dirac algebras, $\{(e_\mu), (ie_\mu), i(e_\mu), i(ie_\mu)\}$, we find that $P\psi = \psi^{\dagger\wedge}M$ is not Dirac but something new perhaps. Here, the $P\psi_a$

$= \psi_v M_v$, $P^\wedge \psi_v = \psi_a M_a$ equation gives Dirac's equation provided $M_a = -M_v = m_D$. By taking the $(\ )^{\dagger\wedge}$ conjugation on both sides of $(P\psi_a) = \psi_v M_v$, we get for $P = P^\dagger$

$$P^\wedge \psi_a^{\dagger\wedge} = \psi_v^{\dagger\wedge} M_v^{\dagger\wedge}$$

Thus, $\psi_a \equiv \psi_v^{\dagger\wedge} \Rightarrow \psi_v = \psi_a^{\dagger\wedge}$ and we have here

$$P^\wedge \psi_v = \psi_v^{\dagger\wedge} M_v^{\dagger\wedge},\ P^\wedge \psi_v = \psi_v^{\dagger\wedge} M_a$$

Consistency seems to then require $M_v^{\dagger\wedge} = M_a$ here. This is not the Dirac case, where $M_v^{\dagger\wedge} = -m_D^{\dagger\wedge} = -m_D \neq M_a = m_D$. Therefore, the $P\psi = \psi^{\dagger\wedge} M$ equation is a special case of the coupled pair equations, and of unknown physical significance in this algebra. Maxwell is also here, $P\psi = 0$, with no problem, as is $P^\wedge P\phi = \phi M^{\dagger\wedge} M$.

In the Dirac algebra, $\{(e_\mu),\ (ie_\mu),\ i(e_\mu),\ i(ie_\mu),\ (f_\mu),\ (if_\mu),\ i(f_\mu),\ i(if_\mu)\}$, we find that $P\psi = \psi^{\dagger\wedge} M$ is now Dirac's equation and P is now a 5-vector, so there are apparently two kinds of mass in this 'world'. The Maxwell $P\psi = 0$ also can now have a 5th P part with a mass and $P^\wedge P\phi = \phi M^{\dagger\wedge} M$ has a 5-vector P so it too can have two kinds of mass in it. The coupled pair is also here, $P\psi_a = \psi_v M_v$, $P^\wedge \psi_v = \psi_a M_a$, with P a 5-vector, so there are now three kinds of mass possible here! Does nature use this new (coupled) equation? Maybe in the quark world, with $P^\wedge P(+\ +\ +\ +\ +)$? Maybe even for leptons, and we have to even 'double' Dirac's algebra to 8×8 matrices?

A NEW LOOK AT 'TRANSFORMATION' PROPERTIES

Before closing this section, let me say also that I have not been able to understand the standard orthodoxy of $\overline{\psi}\gamma_\mu\psi$ 'being' a vector, $\psi\sigma_{\mu\nu}\psi$ being a tensor, and $\overline{\psi}\gamma_5\psi$ being a pseudo-scalar. These may be useful ideas in guessing at Lagrangians, which turn out to be true, even if these transformation ideas don't mean much directly. We also see things in books like $\gamma_5 \equiv i\gamma_0\gamma_1\gamma_2\gamma_3 = -(i/4!)\epsilon^{\mu\nu\sigma\rho}\gamma_\mu\gamma_\nu\gamma_\sigma\gamma_\rho$ which looks very fancy but it is nothing more than $\gamma_5 = i(-f_0)(-f_1)(-f_2)(-f_3) = i(e_1)(-ie_1) = i(-ie_0) = -i(ie_0)$. (The 'extra' is not needed here.) I don't know what it means to say that this γ_5 'transforms' in $\overline{\psi}\gamma_5\psi = -(f_0)\psi^\wedge[-i(ie_0)]\psi$. Under an L transformation, this becomes $-(f_0)(L^\wedge\psi)^\wedge[-i(ie_0)](L^\wedge\psi) = -(f_0)\psi^\wedge(-i)L(ie_0)L^\wedge\psi =$

$-(f_0)\psi^{\wedge}(-i)(ie_0)(L)^{\wedge\vee}L^{\wedge}\psi$. This is simply invariant(!), if $L^{\vee\wedge}L^{\wedge} = (LL^{\vee})^{\wedge} = 1(e_0)^{\wedge} = 1(e_0)$, which is certainly true for the Lorentz group. Similarly, $\psi^{\vee}[-i(ie_0)]\psi$ transforms to $(L^{\wedge}\psi)^{\vee}[-i(ie_0)]L^{\wedge}\psi = \psi^{\vee}(-i)L^{\wedge\vee}(ie_0)L^{\wedge}\psi = \psi^{\vee}(-i)(ie_0)LL^{\wedge}\psi$; also invariant. This second form is 'cleaner' looking for the choice $LL^{\wedge} \equiv 1(e_0)$. They are both invariant for the full 16 parameter group, but $i(ie_0)$ is just a constant. I don't see parity transformations as central at all in Dirac physics. The parity operator P^{Δ} is not even within the three basic conjugations that seem to dictate all the structural basics. The $(\ )^{\Delta}$ conjugation changes (e_3), (ie_3) and others.

We also see $\sigma_{\mu\nu} \equiv (i/2)(\gamma_\mu\gamma_\nu - \gamma_\nu\gamma_\mu)$ in books and $\overline{\psi}\sigma_{\mu\nu}\psi$ is supposed to 'transform' as a tensor. Obviously, $\sigma_{\mu\nu} = 0$ for $\mu = \nu$, since γ_μ and γ_ν anticommute. Also, $\sigma_{\mu\nu} = -\sigma_{\nu\mu}$ so $\sigma_{\mu\nu}$ has only 6 pairs of non-zero $\mu\nu$ components. These are explicitly (k = 1,2,3):

$$\sigma_{0k} = (i/2)[(-f_0)(-f_k) - (-f_k)(-f_0)] = (i/2)[(e_k) - (-e_k)] = i(e_k)$$

$$\sigma_{12} = (i/2)[(-f_1)(-f_2) - (-f_2)(-f_1)] = (i/2)[-2(ie_3)] = -i(ie_3)$$

and cyclic permutations. Notice that $\sigma_{\mu\nu}$ would 'lead us' to define a thing like $\overline{F} \equiv E^k i(e_k) + cB^k[-i(ie_k)]$ and $\overline{F} = +\overline{F}^{\wedge}$. The i in the 'standard' $\sigma_{\mu\nu}$ definition is probably also an unfortunate choice. Let us define F using $i\sigma_{\mu\nu} \Rightarrow -E^k(e_k) + cB^k(ie_k) = i\overline{F} \equiv F = -F^{\wedge}$, which suggests $F \rightarrow F' = L^{\wedge}FL$, so that $F' = -F'^{\wedge}$. This is not a tensor really. It is an alternative to $P \rightarrow P' = L^{\wedge}PL$, $P = +P^{\wedge}$ which is a 5-vector, (plus more(?) for a total of 16 components (maximum)). The $F = -F^{\wedge}$ has 6-parts, (and also more(?) for a total of 16 components (maximum)). Notice that $F^{\wedge}F \rightarrow F'^{\wedge}F' = (L^{\wedge}FL)^{\wedge}(L^{\wedge}FL) = L^{\wedge}F^{\wedge}LL^{\wedge}FL = L^{\wedge}(F^{\wedge}F)L$. This is not invariant for 16 parts in general, just as $P^{\wedge}P$ is not. If we strongly restrict the number of components in F, then we can get $F^{\wedge}F \propto (e_0)$ and this is then a Lorentz invariant of Maxwell's equation.

We should be doing 5-vectors, $\{\gamma_\mu, -i(e_0)\} = \{-f_\mu, -i(e_0)\}$, for the 'full' physics in Dirac algebra. Then we can define γ_6 using the 5th component, $-i(e_0)$, of the 5-vector (and $\gamma_5 \equiv \gamma_0\gamma_1\gamma_2\gamma_3 = -(ie_0)$),

$$\gamma_6 \equiv \gamma_5 i(-e_0) = i(ie_0)$$

which may be useful. Also, we can define, for the 5th dimension:

$$\sigma_{04} = (i/2)[(-f_0)(i(e_0)) - (i(e_0))(-f_0)] = (i/2)[-i(f_0) + i(f_0)] = 0$$

$$\sigma_{k4} = (i/2)[(-f_k)(i(e_0)) - (i(e_0))(-f_k)] = (i/2)[-i(f_k) + i(f_k)] = 0$$

Therefore, $F = -F^{\wedge}$ is not enlarged by P going to a 5-vector here. Notice again how the 5th dimension is special in its structure, compared to (x,y,z). All that old thinking needs re-evaluation it seems.

I am going to skip over further development of all the possible rest states for the two equations. They are important as they will give insight into the possibly physical nature of the $m^k(e_k)$ mass parts. Someone should do them all. Remember, the two equations only have an M that is invariant. They need $M^{\dagger\wedge}M \propto (e_0)$ but that is all. There could be many parts in M and several sets of values of M. Each equation has its own M, from one of these sets. The $M = m^{\mu}(e_{\mu})$ here is only one possibility. It was motivated by the coupled set of Dirac equations. They are only crude approximations of reality. Also, $M^{\dagger\vee}M$ will be needed in some equations and may be different, in each equation.

Let us instead turn to (e_{μ}), converted to spherical coordinates and coupled to the hydrogen atom Coulomb potential. This is similar to generalized coordinates, but not the same, since $(e_1)(e_2)$ is meaningful; only $\hat{e}_1 \bullet \hat{e}_2$ and $\hat{e}_1 \mathrm{x} \hat{e}_2$ are meaningful for unit vectors.

COUPLING DIRAC AND MAXWELL

Dirac's equation from the late 1920's can be coupled to an external electromagnetic field. This is known to agree well with experiment. It takes the 'standard' form (guess!)

$$i\hbar\partial^{\mu}\gamma_{\mu}\psi - |e|A^{\mu}\gamma_{\mu}\psi - m_{D}\psi = -\psi^{\wedge\vee}m^{\mu}(\gamma_{\mu})$$

where we have assumed negative charge, -|e|, on mass m_D being 'pulled and pushed' by the given field $A^{\mu}(x,y,z,t)$. The new mass parts, m^{μ}, will soon be neglected for simplicity. Multiplying from the left by $i(f_0)$ gives (for $\gamma_{\mu} = (-f_{\mu})$):

$$[\hbar\partial^{\mu}(e_{\mu}) - m_{D}i(f_{0})]\psi + |e|A^{\mu}i(e_{\mu})\psi = +\psi^{\dagger\vee}m^{\mu}i(e_{\mu})$$

Notice that only the m_D term has an (f) part now. With no f's used, we have the e's ⇔ σ's, and we have the smaller 'Edmonds algebra', with only a 16 element basis. However, the new m^{μ} part is 'unproven' in value so far. We instead stick with the m_D mass here and then we must live with the major complications due to (f_0). The Maxwell equation for $A^{\mu}(e_{\mu})$ comes from the 'similar' equation

$$\hbar\partial^{\mu}(e_{\mu})F = -\hbar\mu_{0}J^{\mu}(e_{\mu}),\quad J^{\mu}(e_{\mu}) \equiv J = J^{\dagger}$$

with $J^0 = \rho c$ and $J^k = \rho v^k$, where ρ is local charge density, and v^k is the set (v_x, v_y, v_z) of speed components for the local charge source (ρdVol). Given the μ_o coupling, ρ, and **v** of ρ, at all points, we can find F. It turns out that $F = -F^{\wedge}$ and we can write

$$F = -E^{k}(e_{k}) + cB^{k}(ie_{k}) + 0(e_{0}) + 0(ie_{0}) + 0(f's)$$

Given J, for a spherical Ze charge at rest, we find that, outside the central charge,

$$E^kE^k \equiv |E|^2 = \frac{\mu_0 c^2}{4\pi}\frac{Ze}{r^2},\ B^k = 0$$

This is the so-called Coulomb electric field around a 'point' charge Ze; Z = 1,2,3,..., gives the charge on the nucleus of the atom. For hydrogen, Z = 1. If Z could equal 137, our analysis fails!

But we need A^μ, not F, and they are apparently related by

$$[([\hbar\partial^\mu(e_\mu)]^\wedge A) - (\cdots)^\wedge] = 2\hbar F$$

It turns out that the solution for A^μ is, given F from above, for the Coulomb potential in MKS units,

$$A^0 \equiv \frac{V}{c} = \frac{\mu_0 c}{4\pi}\frac{Ze}{r},\ A^k = 0,\ \frac{\mu_0 c^2}{4\pi} = 9\times10^9$$

We can refine the problem by including the magnetic field generated by 'spins' in the nucleus. This would greatly complicate the A^μ expression, so we won't include it here. It does exist in nature, however, so our solution is not general. The nucleus is also extended in size, not a point, and we are also neglecting that.

Notice that this form of the Dirac equation nicely puts both Dirac and Maxwell on the same footing, with $[\hbar\partial^\mu(e_\mu) - ?i(f_0)]$ as the 5-vector operator. For Dirac, ? → m_D, and for Maxwell, ? → 0. In curved space, we need $e_\mu \rightarrow b_\mu{}^{(\alpha)}(e_\alpha)$ and this would apply both to Dirac and to Maxwell! It makes NO SENSE to use (e_μ) for F and A, then use γ_μ for ψ. We also have the interesting question of (f_0) possibly being involved in the curved space itself, as part of a 5th dimension. We know nothing of such things at present!

In case you were wondering, covariance takes the form here:

$$P_{(5)} \equiv \hbar\partial^{\mu}(e_{\mu}) - m_D i(f_0) \rightarrow L^{\downarrow}P_{(5)}L, \quad P_{(5)} = P^{\downarrow}_{(5)}$$

$$P^{\wedge}_{(5)}P_{(5)} = + - - - + \propto (e_0), \ invariant; \ \psi \rightarrow L^{\wedge}\psi;$$

$$A_{(4)} \rightarrow L^{\downarrow}A_{(4)}L, \quad A_{(4)} = A^{\downarrow}_{(4)}; \quad LL^{\wedge} = 1,$$

$$L^{\wedge} = L^{\vee}; \ F_{(6)} \rightarrow L^{\wedge}F_{(6)}L, \ J_{(4)} \rightarrow L^{\downarrow}J_{(4)}L, \ J_{(4)} = J^{\downarrow}_{(4)} = J^{\uparrow}_{(4)},$$

$$m^{\mu}i(e_{\mu}) \ invariant, \quad m_D i(f_0) \ \ invariant$$

So, $P_{(5)}$ is not hermitian here, $P \neq P^{\dagger}$, but this representation goes with the simplest treatment of Maxwell, which itself is totally contained in the very restricted Pauli algebra, with an 8 element basis. The hydrogen atom energy levels then must be found from (forget covariance from here on):

$$[\hbar\partial^{\mu}(e_{\mu}) - m_D i(f_0)]\psi + i|e|\frac{\mu_0 c}{4\pi}\frac{|e|}{r}(e_0)\psi \approx 0$$

The spherical symmetry of A^0, i.e., (1/r), suggests that we switch to spherical coordinates to replace $\partial^k(e_k)$. But first we multiply through by i and put back all the c's:

$$[(i\hbar\partial^{\mu})(e_{\mu}) + cm_D(f_0) - \frac{\mu_0 c}{4\pi}\frac{|e|^2}{r}(e_0)]\psi = 0$$

From $c^2 = 1/\epsilon_0\mu_0$, we can write $\mu_0 c/4\pi$ as $1/4\pi\epsilon_0 c$ which gives us (Joules/c) units for each term. These are the units of momentum as well, and rather standard usage. Units of MeV/c are more common.

The next step is to guess that ψ can be split into a simple function of t and a complicated function of space. We assume that

$$\psi = e^{-iEt/\hbar}\psi_s(x,y,z) \quad (guess!!)$$

where E is an unknown constant (energy). Recall that ∂^0 means $(1/c)\partial/\partial t$. After ∂^0 operates on the time dependent exponential, we can multiply through and cancel the exponential terms. This leaves us with an unknown E and unknown ψ_s, with

$$(E/c)\psi_s + i\hbar\partial^k(e_k)\psi_s + (m_D c^2/c)(f_0)\psi_s - \kappa\frac{|e|^2}{r}(e_0)\psi_s = 0$$

where κ is a known constant. If $|e| \to 0$ and $\partial^k\psi_{SR} \equiv 0$, then we have a free particle 'rest' state with

$$E\psi_{SR} + m_D c^2(f_0)\psi_{SR} = 0$$

which we have already dealt with in some detail. Since $(f_0)(f_0) = (e_0)$ its eigenvalues are ± 1. Therefore,

$$E\psi_{SR+} + m_D c^2(-1)\psi_{SR+} \Rightarrow E = +m_D c^2$$

$$E\psi_{SR-} + m_D c^2(+1)\psi_{SR-} \Rightarrow E = -m_D c^2$$

The negative E here relates to antimatter, such as in anti-hydrogen atoms, and they do not interest us here. For $|e| \neq 0$, we will have matter bound states where $E \leq m_D c^2$, with the lowest energy (ground state), it turns out, corresponding to $E \approx m_D c^2 - 13.6eV$, where $m_D c^2$ is 500,000eV for the electron. Thus the hydrogen atom has a total energy of $(m_p c^2 + m_e c^2 - 13.6eV)$ and $m_p c^2$ is 1,000,000,000eV. The binding energy is a very small fraction, about 1 in 10^8, but the detailed, excited state transitions still allow us to measure small changes in the hydrogen atom. We get this from the photons that are emitted when the atom 'falls' to a more tightly bound state.

The helium atom has two electrons and a bound state energy slightly less than $(m_\alpha c^2 + 2m_e c^2)$. The helium solution is too difficult for analytical solution and it has been only approximately solved on a computer, but successfully so for the ground state.

The free electron at rest, ψ_{SR+}, turns out to have an internal spin quantum number, we call spin up or spin down. The associated operator seems to be $(1/2)\hbar i(ie_3) \equiv S_z$. Notice that $S_z S_z = (\hbar^2/4)(e_0)$. Therefore, its eigenvalues are $\pm\hbar/2$. The 1/2 is for historical reasons here, but see later for better justification involving angular momentum. Notice also that

$$i(ie_3)(f_0) = i(if_3) = (f_0)i(ie_3)$$

They commute and the ψ_{SR+} rest state can be in a spin up or spin down eigenstate. A 'boost' in the x or y direction, $\psi \rightarrow L^\wedge\psi$, will not have an L that can commute with (ie_3), and so we lose this eigenstate possibility for spin. However, inside the hydrogen atom, the e^- 'moves' not in a straight line and it seems to be almost in an eigenstate of spin. See later. We have both spin and orbital angular momentum in the hydrogen atom case, and the energy eigenstates are eigenstates of only the combined angular momentum parts.

We can then specialize our solution to $(f_0)\psi_{S+} = -1\psi_{S+}$ *and* $E - m_D c^2 \equiv \mathscr{E} < 0$. We seek ψ_{S+} 'wave' solutions and their binding energies, $\mathscr{E}_\psi$.

Our equation now has boiled down to

$$\frac{(E - m_D c^2)}{c}\psi_{S+} + i\hbar\partial^k(e_k)\psi_{S+} - k\frac{|e|^2}{r}\psi_{S+} = 0$$

$$-\frac{\mathscr{E}}{c}\psi_{S+} = +i\hbar\partial^k(e_k)\psi_{S+} - \frac{k|e|^2}{r}\psi_{S+}, \quad \mathscr{E} \equiv (E - m_D c^2) < 0$$

(The unknown energy, $\mathscr{E}$, is expected to be negative when found later.) The signs for the individual terms cannot be so easily interpreted here, as in the non-relativistic approximation, called the Schrödinger hydrogen atom. That equation is much simpler, but its energy levels don't agree with experiment as well as the solutions to Dirac's equation above. This looks deceptively simple. It is still a very complicated equation, as will become clear as we grind away toward finding $\mathscr{E}$ and $\psi_{\mathscr{E}}$.

We now summarize the basics of hypercomplex number, spherical coordinates:

$$dR \equiv dx^k(e_k) = dy^k(b_k) \leftrightarrow (x,y,z) \leftrightarrow (r,\theta,\phi)$$

$$y^1 \equiv r = (x^2 + y^2 + z^2)^{1/2}$$

$$y^2{:}\theta{:} \Rightarrow \tan\theta = \rho/z, \ \rho \equiv (x^2 + y^2)^{1/2}$$

$$y^3{:}\phi{:} \ \tan\phi = y/x, \ \rho = r\sin\theta$$

$$x^1 \equiv x = \rho\sin\phi, \ x^2 \equiv y = \rho\cos\phi$$

$$x^3 \equiv z = r\cos\theta$$

Construction of the new hypercomplex number basis, $\{b_k\}$, is defined as follows

$$b_k \equiv \frac{\partial R}{\partial y^k} = \frac{\partial}{\partial y^k}(x^j(e_j)) = \frac{\partial x^j}{\partial y^k}\Big|_{y\ell}(e_j)$$

Therefore,

$$\frac{\partial x}{\partial r} = \frac{\partial}{\partial r}(R\sin\theta\sin\phi) = \sin\theta\sin\phi$$

$$\frac{\partial y}{\partial r} = \frac{\partial}{\partial r}(r\sin\theta\cos\phi) = \sin\theta\cos\phi$$

$$\frac{\partial z}{\partial r} = \frac{\partial}{\partial r}(r\cos\theta) = \cos\theta$$

This gives us

$$b_r = \sin\theta\sin\phi(e_1) + \sin\theta\cos\phi(e_2) + \cos\theta(e_3)$$

Similarly,

$$\frac{\partial x}{\partial \theta} = \frac{\partial}{\partial \theta}(r\sin\theta\sin\phi) = r\cos\theta\sin\phi$$

$$\frac{\partial y}{\partial \theta} = \frac{\partial}{\partial \theta}(r\sin\theta\cos\phi) = r\cos\theta\cos\phi$$

$$\frac{\partial z}{\partial \theta} = \frac{\partial}{\partial \theta}(r\cos\theta) = -r\sin\theta$$

gives us

$$b_\theta = r\cos\theta\sin\phi(e_1) + r\cos\theta\cos\phi(e_2) - r\sin\theta(e_3)$$

and

$$\frac{\partial x}{\partial \phi} = \frac{\partial}{\partial \phi}(r\sin\theta\sin\phi) = r\sin\theta\cos\phi$$

$$\frac{\partial y}{\partial \phi} = \frac{\partial}{\partial \phi}(r\sin\theta\cos\phi) = -r\sin\theta\sin\phi)$$

$$\frac{\partial z}{\partial \phi} = \frac{\partial}{\partial \phi}(r\cos\theta) = 0,$$

leads to

$$b_\phi = r\sin\theta\cos\phi(e_1) - r\sin\theta\sin\phi(e_2)$$

We generate unit basis elements by dividing the b's by the 'magnitude' of the b's. Thus

$$e_r \equiv \frac{b_r}{|b_r|} \equiv \frac{\sin\theta\sin\phi(e_1) + \sin\theta\cos\phi(e_2) + \cos\theta(e_3)}{[(\sin\theta\sin\phi)^2 + (\sin\theta\cos\phi)^2 + (\cos\theta)^2]^{1/2}}$$

$$= b_r$$

$$e_\theta \equiv \frac{b_\theta}{|b_\theta|} \equiv \frac{r\cos\theta\sin\phi(e_1) - r\cos\theta\cos\phi(e_2) - r\sin\theta(e_3)}{[(r\cos\theta\sin\phi)^2 + (r\cos\theta\cos\phi)^2 + (r\sin\theta)^2]^{1/2}}$$

$$= \frac{b_\theta}{r}$$

and

$$e_\phi \equiv \frac{b_\phi}{|b_\phi|} \equiv \frac{r\sin\theta\cos\phi(e_1) - r\sin\theta\sin\phi(e_2) + 0(e_3)}{[(r\sin\theta\cos\phi)^2 + (r\sin\theta\sin\phi)^2 + (0)^2]^{1/2}}$$

$$= \frac{b_\phi}{r|\sin\theta|} = \frac{b_\phi}{r\sin\theta} = \cos\phi(e_1) - \sin\phi(e_2) + 0(e_3)$$

Then

$$dR = dx^k(e_k) = dr(b_r) + d\theta(b_\theta) + d\phi(b_\phi)$$

$$= dr(e_r) + rd\theta(e_\theta) + r\sin\theta d\phi(e_\phi)$$

This is very pretty, but we need to convert $(e_k)\partial^k$ to spherical coordinates. We have, using the above,

$$e_k\partial^k\psi(r,\theta,\phi) = e_1\left(-\frac{\partial}{\partial x}\psi\right) + \cdots$$

$$= -e_1\left(\frac{\partial\psi}{\partial r}\frac{\partial r}{\partial x} + \frac{\partial\psi}{\partial\theta}\frac{\partial\theta}{\partial x} + \frac{\partial\psi}{\partial\phi}\frac{\partial\phi}{\partial x}\right) + \cdots$$

Then we have to write the (e_k)'s in terms of $\{e_r, e_\theta, e_\phi\}$. This is a real mess to grind through! The answer is rather simple but only after a lot of trig. We find, if patient enough,

$$-(e_k)\partial^k = (e_r)\frac{\partial}{\partial r} + (e_\theta)\frac{1}{r}\frac{\partial}{\partial\theta} + (e_\phi)\frac{1}{r\sin\theta}\frac{\partial}{\partial\phi}$$

Our equation now takes the form

$$\frac{-\mathscr{E}}{c}\psi_{s+} = -i\hbar\left((e_r)\frac{\partial}{\partial r} + (e_\theta)\frac{1}{r}\frac{\partial}{\partial\theta} + (e_\phi)\frac{1}{r\sin\theta}\frac{\partial}{\partial\phi}\right)\psi_{s+}$$

$$- \frac{k|e|^2}{r}\psi_{s+}$$

The next step involves finding operators that commute with these 'operators' in the equation and whose eigenvalues we can determine. These help us construct the θ and ϕ dependent parts in $\psi(r,\theta,\phi)$. Because the 1/r term has spherical symmetry, and because of our classical experience with this situation, we are led to guess that perhaps something like classical angular momentum would come from

$$\mathfrak{L} = \frac{1}{2}[x^\wedge p - "(x^\wedge p)^\wedge"]$$

where $p \equiv -i\hbar(e_k)\partial^k = +p^\wedge$ and $[`(x^\wedge p)^\wedge`\psi] \equiv p^\wedge x\psi \equiv p^\wedge(x\psi)$ (Guess!). Then ∂^k operates on both x and ψ. Also, $x \equiv x^k(e_k) = re_r = -x^\wedge$ and $-e_k\partial^k$ is already known in spherical coordinates. The multiplication table for $\{e_r, e_\theta, e_\phi\}$ is tedious, but easy to construct from $\{e_1, e_2, e_3\}$. For example,

$$e_r e_\theta = [(\sin\theta\sin\phi)(e_1) + \sin\theta\cos\phi(e_2) + \cos\theta(e_3)]\bullet$$

$$[(\cos\theta\sin\phi)(e_1) + \cos\theta\cos\phi(e_2) - \sin\theta(e_3)]$$

$$= \cdots = -\cos\phi(ie_1) + \sin\phi(ie_2)$$

$$\neq e_\phi = \cos\phi(e_1) - \sin\phi(e_2)$$

This may be showing us that $i(ie_\mu)$ is a better spacetime description. We could define $(e_r)(e_\theta) \equiv (ie_\phi)$, but this may not be very useful. In defining $\mathscr{L}$, we are looking for something that will commute with all the terms in our equation. Notice at $(-i\hbar(e_\mu)\partial^\mu)^\wedge = +i\hbar(e_\mu{}^\wedge)\partial^\mu$. The $i^\wedge = -i$, for i 'outside', is very important. Basically, $P_{(3)} = -i\hbar(e_k)\partial^k \equiv P_{(3)}{}^k(e_k)$ *has* $P_{(3)}{}^{k\wedge} = -P_{(3)}{}^k$ and also $e_k{}^\wedge = -e_k$. Therefore, $P_{(3)}{}^\wedge = +P_{(3)}$ for 3-vectors. We shall just write P, from here on, since no 4-vectors will appear below. We know

$$P \equiv -i\hbar(e_k)\partial^k = +i\hbar\left[e_r\frac{\partial}{\partial r} + e_\theta\frac{1}{r}\frac{\partial}{\partial\theta} + e_\phi\frac{1}{r\sin\theta}\frac{\partial}{\partial\phi}\right] = P^\wedge$$

but $x^\wedge = -x$ for 3-vectors. Therefore, we tentatively explore

$$\mathfrak{L}\psi \equiv \frac{1}{2}x^\wedge P\psi - \frac{1}{2}P^\wedge(x\psi)$$

$$= -\frac{1}{2}x(P\psi) - \frac{1}{2}P(x\psi)$$

We could write

$$P(x\psi) = +i\hbar[e_r\frac{\partial}{\partial r} + \cdots](re_r\psi) = mess!!$$

This is messy because $(\partial e_r/\partial\theta)$ and $(\partial e_r/\partial\phi)$ are not simple. Let us instead try using x,y,z derivatives:

$$P(x\psi) = (-i\hbar\partial^k e_k)(x^j e_j\psi)$$

$$= -i\hbar(\partial^k x^j)e_k e_j\psi - i\hbar x^j e_k e_j(\partial^k\psi)$$

Then

$$\partial^k x^j \equiv -\partial x^j/\partial x^k = -\delta^j_k = -1 \; or \; 0,$$

and

$$P(x\psi) = -i\hbar(-3(e_0))\psi - i\hbar[x^1(e_0)\partial^1\psi + x^2(e_0)\partial^2\psi$$

$$+ x^3(e_0)\partial^3\psi + x^1(-ie_3)\partial^2\psi + x^2(+ie_3)\partial^1\psi$$

$$+ x^3(-ie_2)\partial^1\psi + x^1(+ie_2)\partial^3\psi + x^2(+ie_1)\partial^3\psi$$

$$+ x^3(+ie_1)\partial^2\psi]$$

$$= +3i\hbar(e_0)\psi - i\hbar(x^1\partial^1 + x^2\partial^2 + x^3\partial^3)(e_0)\psi$$

$$+ i\hbar(x^1\partial^2 - x^2\partial^1)(ie_3)\psi + i\hbar(x^3\partial^1 - x^1\partial^3)(ie_2)\psi$$

$$+ i\hbar(x^2\partial^3 - x^3\partial^2)(ie_1)\psi$$

We also have

$$x^\wedge P\psi = -xP\psi = -x^k e_k(-i\hbar\partial^j e_j)\psi$$

$$= i\hbar x^k e_k e_j(\partial^j\psi)$$

$$= i\hbar[(x^1 e_0\partial^1 + x^2(e_0)\partial^2 + x^3(e_0)\partial^3]\psi$$

$$+ x^1\partial^2(ie_3)\psi - x^2\partial^1(ie_3)\psi$$

$$+ x^2\partial^3(ie_1)\psi - x^3\partial^2(ie_1)\psi$$

$$+ x^3\partial^1(ie_2)\psi - x^1\partial^3(ie_2)\psi$$

Therefore,

$$\mathcal{L}\psi \equiv \frac{1}{2}[x^{\wedge}P\psi + p^{\wedge}(x\psi)] = \frac{-1}{2}[xP\psi - Px\psi] = \frac{+3}{2}i\hbar\psi +$$

$$i\hbar(x^1\partial^2 - x^2\partial^1)(ie_3)\psi + i\hbar(x^2\partial^3 - x^3\partial^2)(ie_1)\psi + i\hbar(x^3\partial^1 - x^1\partial^3)$$

We chose the + sign above so that the 'cross-terms' would survive!

There is an obvious symmetry here for the terms and this looks promising, so far. Notice that **r**×**P** in vectors contains 'things' like some parts of the $\mathcal{L}_z$. For the classical Coulomb potential, **r**×**P** is a constant, because the radial force on e^- produces no torque. Quaint phrases, but they don't apply here. They do guide our guessing, all the same. We can write

$$\mathcal{L}\psi = -i\hbar(x^2\partial^3 - x^3\partial^2)(-ie_1)\psi + \frac{1}{2}i\hbar(e_0)\psi$$

$$- i\hbar(x^3\partial^1 - x^1\partial^3)(-ie_2)\psi + \frac{1}{2}i\hbar(e_0)\psi$$

$$- i\hbar(x^1\partial^2 - x^2\partial^1)(-ie_3)\psi + \frac{1}{2}i\hbar(e_0)\psi$$

$$\equiv [\mathcal{L}_x(-ie_1) + \mathcal{L}_y(-ie_2) + \mathcal{L}_z(-ie_3)]\psi \equiv \mathcal{L}_k(-ie_k)\psi$$

We have defined $\mathcal{L}_k$ here, in this peculiar way, for a good reason. Notice that, since $(-ie_1)$, $(-ie_2)$, and $(-ie_3)$ don't all commute with each other, we cannot expect ψ to be an eigenstate of more than one of these parts of $\mathcal{L}$. Choose $\mathcal{L}_z$, arbitrarily. Notice also that we now see $(1/2)\hbar$ occurring here in a somewhat natural way. We say electrons have 'spin 1/2'. The $\{(-ie_k)\}$ basis is Hamilton's original 1843 quaternion basis!

We saw earlier that the free particle rest solution had ψ_R as an eigenstate of $(1/2)\hbar i(ie_3)$, with eigenvalues of $\pm 1/2\hbar$. The operator is hermitian and has real eigenvalues. We have defined our $\mathcal{L}_z$ so as to have this same operator in it:

$$\mathcal{L}_z\psi \equiv -i\hbar(x^1\partial^2 - x^2\partial^1)(e_0)\psi + \frac{1}{2}i\hbar(ie_3)\psi$$

$$\mathcal{L}_x\psi \equiv -i\hbar(x^2\partial^3 - x^3\partial^2)(e_0)\psi + \frac{1}{2}i\hbar(ie_1)\psi$$

$$\mathcal{L}_y\psi \equiv -i\hbar(x^3\partial^1 - x^1\partial^3)(e_0)\psi + \frac{1}{2}i\hbar(ie_2)\psi$$

where we used $(e_0) = (+ie_3)(-ie_3)$, etc. The $-i\hbar x^1\partial^2 = +i\hbar x(\partial/\partial y)$, *etc.*

We wish to have an $\mathcal{L}_z\mathcal{L}_z$ type operator that is simple and proportional to (e_0) if possible. Then similar things will apply to $\mathcal{L}_x\mathcal{L}_x$ and $\mathcal{L}_y\mathcal{L}_y$ and the sum of all three might be a useful 'rotational momentum' operator for which ψ is an eigenstate. We may need to have the 'cross-terms' cancel in $\mathcal{L}_z\mathcal{L}_z$, if possible. This will, however, not happen for $\mathcal{L}_z\mathcal{L}_z$, but look at $\mathcal{L}_z{}^\wedge\mathcal{L}_z$:

$$\mathcal{L}_z^\wedge(\mathcal{L}_z\psi) = [+i\hbar(x^1\partial^2 - x^2\partial^1)^\wedge(e_0) - \frac{1}{2}i\hbar(-ie_3)]\cdot$$

$$[-i\hbar(x^1\partial^2 - x^2\partial^1)(e_0)\psi + \frac{1}{2}i\hbar(ie_3)\psi]$$

There is no order problem here with $(x^1\partial^2)^\wedge \equiv \partial^{2\wedge}x^{1\wedge} = \partial^2x^1$ *since* $\partial^2x^1 = x^1\partial^2$. (This is not so clear for $(x^1\partial^1)^\wedge$, but we don't have to deal with that here.) We find

$$\mathcal{L}_z^\wedge\mathcal{L}_z\psi = \hbar^2\left[(x^1\partial^2 - x^2\partial^1)^2 + \left(\frac{1}{2}\right)^2\right](e_0)\psi = (\mathcal{L}_z^\wedge\mathcal{L}_z)^\dagger\psi$$

and the cross terms do cancel. It has real eigenvalues. By symmetry, we can expect

$$\mathcal{L}_x^{\wedge}\mathcal{L}_x\psi = \hbar^2\left[(x^2\partial^3 - x^3\partial^2)^2 + \left(\frac{1}{2}\right)^2\right](e_0)\psi$$

$$\mathcal{L}_y^{\wedge}\mathcal{L}_y\psi = \hbar^2\left[(x^3\partial^1 - x^1\partial^3)^2 + \left(\frac{1}{2}\right)^2\right](e_0)\psi$$

and the sum of these, $(\mathcal{L}_k \wedge \mathcal{L}_k)$, is possibly a useful operator also, besides $\mathcal{L}_z$. We must first check whether $\mathcal{L}_z$ commutes with $(\mathcal{L}_k \wedge \mathcal{L}_k)$. This is tedious but straightforward, and requires that we check

$$[(x^1\partial^2 - x^2\partial^1)^2 + cycl.perm.](x^1\partial^2 - x^2\partial^1)\psi$$

$$= ? = (x^1\partial^2 - x^2\partial^1)[(x^1\partial^2 - x^2\partial^1)^2 + cycl.perm.]\psi$$

This leads to about 60 terms that all have to cancel out, if possible! There is, fortunately, an easier way. We compute $[\mathcal{L}_x, \mathcal{L}_y]$ and see if it is $i\hbar\mathcal{L}_z$ or not. It is, in fact, so it does fit the standard angular momentum machinery that is shown in good Quantum Mechanics books, such as Goswami's, provided also that $\mathcal{L}_k^{\dagger} = \mathcal{L}_k$. We must also check that $[\mathcal{L}_y, \mathcal{L}_z] = i\hbar\mathcal{L}_x$ *and* $[\mathcal{L}_z, \mathcal{L}_x] = i\hbar\mathcal{L}_y$. We find that lots of things cancel and

$$[\mathcal{L}_x, \mathcal{L}_y]\psi \equiv (\mathcal{L}_x\mathcal{L}_y - \mathcal{L}_y\mathcal{L}_x)\psi = \cdots = i\hbar\mathcal{L}_z\psi$$

so it is very likely that the other cyclic permutations of the above will check out OK also. Once we have this wonderful set of three commutators, it can be shown that all kinds of things follow, no matter what the internal structure of $\{\mathcal{L}_k\}$ is. But is our $\{\mathcal{L}_k\}$ set hermitian? It seems to be, even though $(i\hbar x^1\partial^2)^{\dagger} = -i\hbar x^1\partial^2$, *etc.* ... This seems strange to me, and has since I first saw $(i\hbar\partial/\partial x)$ in quantum as the one dimensional momentum component operator. The beginning quantum books say this is hermitian. That is in the context of $\int \psi^* P_x \psi dx$, so hermitian is not quite the same things as $(\ )^{\dagger}$, which is a strictly algebraic operation. We then abandon $\mathcal{L}_z \wedge \mathcal{L}_z$ and stick with $\mathcal{L}_z\mathcal{L}_z$, since this will be sufficient for our needs. Our wave equation is

$$\frac{-\mathscr{E}}{c}\psi_{s+} = i\hbar(e_k)\partial^k\psi_{s+} - \frac{k|e|^2}{r}\psi_{s+}$$

with $(f_0)\psi_{s+} = -1\psi_{s+}$. We now have to determine whether $\mathscr{L}_z$ commutes with $e_k\partial^k$ and with (1/r). Once we prove that $\mathscr{L}_z$ commutes then $\mathscr{L}_x\mathscr{L}_x$, $\mathscr{L}_y\mathscr{L}_y$, and $\mathscr{L}_z\mathscr{L}_z$ should each commute, by 'intuition' or 'divine revelation' (as I accuse my students of utilizing so often in their homework 'jumps'). Clearly (ie_3) in $\mathscr{L}_z$ does not commute with (e_1) and (e_2) in $(e_k)\partial^k$. These need to be compensated for somehow if $\mathscr{L}_z$ is to commute. There is no problem with (ie_3), in $\mathscr{L}_z$, going through (1/r). It can be shown that $-i\hbar(x^1\partial^2 - x^2\partial^1) = i\hbar\partial/\partial\phi$ so it commutes with 1/r. There is a lot of grinding left to do; is it really necessary? We get 20 terms in $\mathscr{L}_z[(e_k)\partial^k] - [(e_k)\partial^k\mathscr{L}_z]$ and they do all cancel to zero! Check it out!! Amazing.

The ψ solution is complicated. It has both e, f basis elements in it and separate functions of r, of θ, and of ϕ in it. The experts at such problem solving do it this way, so it must be a good way and maybe it is the best way; I don't know. We have partially shown that

$$\frac{1}{r}\mathscr{L}_z\psi =?= \mathscr{L}_z\left(\frac{1}{r}\psi\right)$$

is alright. We can then believe that ψ can be an eigenstate of the z-component of angular momentum and it probably can also be an eigenstate of a magnitude 'squared' rotational momentum. We still need to verify that $\mathscr{L}_z$ actually commutes with $(\mathscr{L}_x\mathscr{L}_x + \mathscr{L}_y\mathscr{L}_y + \mathscr{L}_z\mathscr{L}_z)$ in order to have simultaneous eigenstates for both operators. Another big grind is required and that also works out OK. We also must check that the sum $\mathscr{L}_k\mathscr{L}_k$ actually commutes with the 'equation'. It looks 'reasonable' and it is true.

If $A_1B = BA_1$ and $A_2B = BA_2$, then $(A_1 + A_2)B = A_1B + A_2B = BA_1 + BA_2 = B(A_1 + A_2)$. To show that $\mathscr{L}_z$ commutes with the equation, takes some work to prove, but then $\mathscr{L}_z\mathscr{L}_z$ obviously commutes. Since the equation involves $(e_k)\partial^k$ and $1/r = (x^2 + y^2 + z^2)^{-1/2}$, we see that $\mathscr{L}_y$ should commute also with the equation, if $\mathscr{L}_z$ does, and so does $\mathscr{L}_y\mathscr{L}_y$ then. Similarly for $\mathscr{L}_x\mathscr{L}_x$ and thus for $(\mathscr{L}_k\mathscr{L}_k)$. However, we still must deal with $\mathscr{L}_z$ and $\mathscr{L}_k\mathscr{L}_k$. We have specific $\mathscr{L}_k$ functions, for which (theorem—no proof here)

$$\mathcal{L}_x\mathcal{L}_y - \mathcal{L}_y\mathcal{L}_x = i\hbar\mathcal{L}_z, \ \mathcal{L}_z\mathcal{L}_x - \mathcal{L}_x\mathcal{L}_z = i\hbar\mathcal{L}_y, \ \mathcal{L}_y\mathcal{L}_z - \mathcal{L}_z\mathcal{L}_y = i\hbar\mathcal{L}_x$$

The above properties alone actually prove that $\mathcal{L}_z$ commutes with $(\mathcal{L}_k\mathcal{L}_k)$, without having to process the 60 or so terms that are really here in this commutator. Here is a beautiful proof: Use X for $\mathcal{L}_x$, *etc.*

$$[Z, (XX + YY + ZZ)] = ZXX + ZYY + ZZZ +$$

$$- XXZ - YYZ - ZZZ = (ZX)X + (ZY)Y - XXZ - YYZ$$

We need to get the Z's that are on the left over to the right, to cancel the last terms. We use two of the commutation relations twice.

$$(ZX) = XZ + i\hbar Y \text{ and } (ZY) = YZ - i\hbar X$$

Therefore,

$$[\ ,\ ] = X(ZX) + i\hbar YX + Y(ZY) - i\hbar XY - XXZ - YYZ$$

$$= XXZ + Xi\hbar Y + i\hbar YX + YYZ - Yi\hbar X$$

$$- i\hbar XY - XXZ - YYZ = 0$$

Isn't that great?! We can avoid the 60 terms needed to otherwise prove this. So we give up on $\mathcal{L}_z \wedge \mathcal{L}_z$, even though it looked simple. (Simpler is not always better, just usually.)

In summary, we 'know' $\mathcal{L}_z$ and $(\mathcal{L}_k\mathcal{L}_k)$ commute with each other and with the equation 'operators' and that $\mathcal{L}_z = -i\hbar(x^1\partial^2 - x^2\partial^1) + (1/2)\hbar i(ie_3)$, and cyclic permutations for $\mathcal{L}_x$ and $\mathcal{L}_y$. Only the total $\mathcal{L}_z$ commutes, not its two parts separately. This 'physically' means that the angular momentum 'tangles' spin and orbit 'parts' in a way such that they don't really 'exist' separately in the hydrogen atom. This is similar to the fact that an electron cannot move along x with its spin up along z, when in free space. The simplifying approximations only (crudely) apply to the **non**-relativistic (less accurate) model of nature.

Now we can deal with finding the angular momentum eigenvalues of ψ. These are derivable from the operator commutation relations above, without having to simultaneously find the corresponding eigenfunctions, ψ. The ψ can be found later. Knowing the eigenvalues can then help in finding the needed eigenfunctions. Amazing!

We can prove, from [A,B] = $i\hbar$C (and cyclic permutations to get [C,A] and [B,C]), that if these are hermitian operators, A = $A^\dagger$, *etc.*, then

$$C\psi_{mj} = m\hbar\psi_{mj}, \quad (AA + BB + CC)\psi_{mj} = j(j+1)\hbar^2\psi_{mj}$$

and j = 1/2 or 1 or 3/2 or 2 or ... Whatever j is, m goes with it and $m = -j, -j+1, -j+2, \ldots, j-1, j$. There are 2j + 1 values of m.

The proof uses $<\psi|(CC|\psi>) = (<\psi|C^\dagger)(C|\psi>)$ *and* $C = C^\dagger$ (hermitian) implies that this becomes $(<\psi|C)(C|\psi>)$ which is 'expected' to be non-negative, *i.e.*, zero or something positive, since it is sort of the 'square' of an abstract vector. That is why the A,B,C operators need to be hermitian. The proof is beautiful and utilizes much of the following collection of truths—some are definitions and some are theorems that follow only from the commutators [A,B], [B,C], and [C,A] having the pattern above:

$$D^2\psi_{jm} \equiv (AA + BB + CC)\psi_{jm} = j(j+1)\hbar\psi_{jm}, \quad C\psi_{jm} \equiv m\hbar\psi_{jm}$$

$$D_\pm \equiv A \pm iB, \quad D^2 = D_+D_- + D_-D_+ + CC,$$

$$[C, D_\pm] = \pm D_\pm, \quad [D_+, D_-] = 2C, \quad [D_\pm, D^2] = 0$$

$$D^2(D_+\psi_{jm}) = D_+D^2\psi_{jm} = D_+(j(j+1)\hbar\psi_{jm}) = j(j+1)\hbar(D_+\psi_{jm})$$

$$D^2(D_-\psi_{jm}) = D_-(D^2\psi_{jm}) = D_-(j(j+1)\hbar\psi_{jm}) = j(j+1)\hbar(D_-\psi_{jm})$$

$$CD_+ = D_+(C+1), \quad C(D_+\psi_{jm}) = D_+(C+1)\psi_{jm} = (m+1)(D_+\psi_{jm})$$

$$D_-D_+ = D^2 - C(C+\hbar), \quad D_+D_- = D^2C(C-\hbar), \quad D_+^\dagger = D_-$$

$$(D_-D_+)^\dagger = D_-D_+, \quad (j-m)(j+m+1) > 0, \quad (j+m)(j-m+1) > 0$$

also

$$D_+\psi_{jm} = \sqrt{(j-m)(j+m+1)}\hbar\psi_{j(m+1)},\ D_-\psi_{jm} = \sqrt{(j+m)(j-m+1)}$$

$$D_+\psi_{jj} = 0,\ D_-\psi_{j,-j} = 0,\ [D_+, D_-] = +2\hbar C$$

$$m_{max}(m_{max} + 1) = m_{min}(m_{min} - 1),\ D_-D_+\psi_{jm-max} = (D^2 - C^2 - \hbar$$

$$= (\lambda_J - m_{max}(m_{max} + 1)\hbar^2)\psi_{j,m} = 0 \Rightarrow \lambda_J = j(j+1),\ \text{if } m$$

$$C(D_+\psi_{jm}) = (D_+(C + \hbar))\psi_{jm} = (m+1)\hbar(D_+\psi_{jm})$$

This ‘pile’ of equations contains all the pieces of the puzzle that lead to the eigenvalues j(j + 1) and $m = -j, -j + 1, \ldots, j - 1, j$, where j is a positive half integer or zero.

There is a lot of skill still needed to construct the $\psi(\theta,\phi)_{jm}$ functions that will do the job here. Since we have i(ie$_3$) in $\mathscr{L}_z$, the $\psi(\theta,\phi)$ will have some hypercomplex structure in it. Then (f$_0$) will operate on $\psi(\theta,\phi)_{jm}$ to also give back -1 and this further shapes the hypercomplex details in ψ. Part of the needed $\psi(\theta,\phi)_{jm}$ looks like

$$Y_\ell^{m\ell}(\theta,\phi) = (-1)^{m\ell}\sqrt{\frac{(2\ell+1)}{4\pi}\bullet\frac{(\ell-m_\ell)!}{(\ell+m_\ell)!}}P_\ell^{m\ell}(\cos\theta)e^{im_\ell\phi}$$

Impressive, huh? The $P_\ell^{m\ell}(\cos\theta)$ functions are a finite string of terms involving $\cos\theta$, getting longer for larger $\ell = 0,1,2,\ldots$ Since we are mainly interested in the energy levels, we could try to set this issue to the side and assume that ψ_{s+} can be split into

$$\psi = e^{-iEt/\hbar}\psi_r(r)\psi_{\theta,\phi,e,f}(\theta,\phi,e's,f's),\ \mathscr{E} \equiv E - m_D c^2$$

with $\psi_r \propto (e_0)$. Why take products here, instead of sums? It just works that way for classical problems involving partial differential equations, where the solutions can be directly tested in the lab. So we do the same thing here, in never-never-land, since the math looks similar (so far it is similar—but second quantization awaits down stream to put back the lumpiness in our 'smooth pancake batter' wave function). Our fancy wave equation is still only a gross simplification of the real physics in the hydrogen atom, and it is hard enough to solve. Now we need the spherical coordinate version of Dirac's equation, which we write as

$$\frac{-\mathscr{E}}{c}\psi_r\psi_{\theta\phi ef} = \left[\left(-i\hbar(e_r)\frac{\partial}{\partial r} - i\hbar\frac{1}{r}\text{"}\partial\text{"}_{\theta\phi}\right) - \frac{|k||e|^2}{r}\right]\psi_r\psi_{\theta\phi ef}$$

Notice, by the way, that

$$e^{-iEt/\hbar} = e^{-i(\mathscr{E}_- + m_D c^2)/\hbar} = e^{-im_D c^2 t/\hbar}e^{-i\mathscr{E}_- t/\hbar}$$

Since $|\mathscr{E}| \sim 10^{-8}m_Dc^2$, the $\sin[(m_Dc^2/\hbar)t]$ oscillations are very rapid compared to the $\sin[(|\mathscr{E}|/\hbar)t]$ oscillations. The electron 'sort of' zig-zags at almost the speed of light in a little cocoon of size $\sim 10^{-15}$ met, while the electron also zig-zags around in the hydrogen atom, size $\sim 10^{-10}$ met. This strange 'behavior' is showing itself, a bit, in this energy part of the wave function.

We now let the operators operate on ψ_s and we get

$$\frac{-\mathscr{E}}{c}\psi_r\psi_{\theta\phi ef} = -i\hbar(e_r)\left(\frac{\partial\psi_r}{\partial r}\right)\psi_{\theta\phi ef} - i\hbar\frac{1}{r}\partial_{\theta\phi}(\psi_r\psi_{\theta\phi ef}) - \frac{|k||e|^2}{r}\psi_r\psi_{\theta\phi ef}$$

For classical, partial differential equations, we would next divide through by $\psi_r\psi_{\theta\phi}$ and get terms that split into having only ψ_r or only $\psi_{\theta\phi}$ in them.

We would then move them to opposite sides of the equal sign. Since each depends on a different and independent variable, r or θ and ϕ, we would set each side equal to a constant, K, which would be found later from boundary conditions on the wave function, which we would guess are physically reasonable. But none of this will work here because our $\psi_{\theta\phi ef}$ is hypercomplex and $(1/\psi)$ is not well defined. We know that $\psi_{\theta\phi ef}$ has e,f hypercomplex number parts in it.

Textbooks would at this point,or even at the beginning, resort to specific representations in terms of elements from the Edmonds algebra $\{(\sigma_\mu), (i\sigma_\mu), i(\sigma_\mu), i(i\sigma_\mu)\}$, or from the Pauli algebra where $i(\sigma) = (i\sigma)$ and $i(i\sigma) = -\sigma$ are defined to be allowed. This movement of i, in and out of (σ), is part of the old and standard matrix machinery, that all physicists and mathematicians know so well. We have seen that this is, however, not a good idea. The group $LL^{\wedge} = 1$ is a 6 parameter group in Pauli and an 8 parameter group in Edmonds. Since full Dirac also gives this same 8 parameter group, we might be missing things if we go to standard matrix thinking in designing new law prospects. However, for only solving problems, it may not make any difference. I don't know. You don't find anything but standard matrix algebra in most physics texts.

We can write (e_r) as a combination of angle terms and $\{e_k\}$, and they in turn can be represented by (one choice here, out of many)

$$e_k = \begin{pmatrix} \sigma_k & 0 \\ 0 & \sigma_k^{\wedge\dagger} \end{pmatrix} = \begin{pmatrix} \sigma_k & 0 \\ 0 & -\sigma_k \end{pmatrix}, \quad \psi_s = \begin{pmatrix} \psi_1 \psi_2 \\ \psi_3 \psi_4 \end{pmatrix} = \begin{pmatrix} \psi_{1u} & 0 & 0 & 0 \\ \psi_{1d} & 0 & 0 & 0 \\ \psi_{3u} & 0 & 0 & 0 \\ \psi_{3d} & 0 & 0 & 0 \end{pmatrix}$$

The $\mathscr{L}_k$ operators involve $i(ie_k)$, which can be represented as

$$i(ie_k) = i\begin{pmatrix} (i\sigma_k) & 0 \\ 0 & (i\sigma_k)^{\dagger\wedge} \end{pmatrix} = i\begin{pmatrix} (i\sigma_k) & 0 \\ 0 & +(i\sigma_k) \end{pmatrix} = \begin{pmatrix} i(i\sigma_k) & 0 \\ 0 & i(i\sigma_k) \end{pmatrix}$$

The original Dirac equation would have to be carefully checked, to be sure that ψ only stays to the right of all operators before the above decomposition is done. This is the way it is done in the few textbooks that actually solve the Dirac hydrogen atom. Maybe it is most efficient, but it is ugly, ugly! We could go all the way back to the γ_μ form and decompose it into σ's right there, using

$$\gamma_\mu = -f_\mu = -\begin{pmatrix} 0 & \sigma_\mu^{\wedge\dagger} \\ \sigma_\mu & 0 \end{pmatrix} = \left\{-\begin{pmatrix} 0 & \sigma_0 \\ \sigma_0 & 0 \end{pmatrix}, -\begin{pmatrix} 0 & -\sigma_k \\ \sigma_k & 0 \end{pmatrix}\right\}$$

After decomposing here, then switch to spherical $\{\sigma_r, \sigma_\theta, \sigma_\phi\}$ and find new $\mathscr{L}_k$ operators in this new σ world. I don't like it but I know it would work.

The objective would be to get coupled σ equations from the γ equation. Then, since these are still hypercomplex, we would have to go to a representation of the σ's in terms of complex numbers. We know $\sigma_\mu^{\dagger} = \sigma_\mu$ and $(i\sigma_\mu)^{\dagger} = -(i\sigma_\mu)$. Also $(-i\sigma_1)(-i\sigma_2) = (-i\sigma_3)$ and cyclic. We know $(\;)^{\dagger}$ should be hermitian conjugation for matrices. Our σ's must be at least 2×2 matrices since

$\sigma_1\sigma_2 = -\sigma_2\sigma_1$ and we must model this anticommutivity. We saw that spin is associated with i(ie$_3$). Therefore, we expect spin in Edmonds to be i(iσ_3) and in Pauli to be (-σ_3). Notice that $[i(i\sigma_3)]^\dagger = -i(-ie_3) = +i(ie_3)$ *and* $\sigma_3^\dagger = +\sigma_3$. Since we have spin up and spin down in $\mathcal{L}_Z$, we think along the lines of $\sigma_3\psi = \pm\psi$ for spin states. A simple possibility is

$$\sigma_3\Psi_u = \begin{pmatrix}1&0\\0&-1\end{pmatrix}\begin{pmatrix}1\\0\end{pmatrix} = 1\begin{pmatrix}1\\0\end{pmatrix},\ \sigma_3\Psi_d = \begin{pmatrix}1&0\\0&-1\end{pmatrix}\begin{pmatrix}0\\1\end{pmatrix} = -1\begin{pmatrix}0\\1\end{pmatrix}$$

and this is the standard assumption in the physics literature. Then what are σ_2 and σ_3 going to be? Notice that

$$\sigma_1\sigma_2 = (i\sigma_3) = \begin{pmatrix}i&0\\0&-i\end{pmatrix}$$

so we must have complex number matrices. Clearly σ_1 and σ_2 cannot both be diagonal or then $\sigma_1,\sigma_2,\sigma_3$ would all commute. One has to be off diagonal, at least partially. We know that

$$\sigma_2^\dagger \equiv \sigma_2 = \begin{pmatrix}c_1^* & c_3^*\\ c_2^* & c_4^*\end{pmatrix} = \begin{pmatrix}c_1 & c_2\\ c_3 & c_4\end{pmatrix}$$

Therefore, $c_1 = c_1^*,\ c_4 = c_4^*,\ c_2 = c_3^*,\ c_3 = c_2^*$. Therefore, $c_1 \equiv a,\ c_4 \equiv b,\ c_3 \equiv d + ie,\ c_2 = c_3^* = d - ie$. Also,

$$\sigma_2\sigma_3 = -\sigma_3\sigma_2 = \begin{pmatrix}a & d-ie\\ d+ie & b\end{pmatrix}\begin{pmatrix}1&0\\0&-1\end{pmatrix} = \begin{pmatrix}a & -d+ie\\ d+ie & -b\end{pmatrix}$$

$$= \begin{pmatrix}-1&0\\0&1\end{pmatrix}\begin{pmatrix}a & d-ie\\ d+ie & b\end{pmatrix} = \begin{pmatrix}-a & -d+ie\\ d+ie & b\end{pmatrix}$$

Therefore, $a = -a = 0,\ b = -b = 0$, d and e are still anything (real). We also know that

$$\sigma_2\sigma_2 = \begin{pmatrix} 0 & d-ie \\ d+ie & 0 \end{pmatrix}\begin{pmatrix} 0 & d-ie \\ d+ie & 0 \end{pmatrix} = \begin{pmatrix} d^2+e^2 & 0 \\ 0 & d^2+e^2 \end{pmatrix} \equiv \sigma_0 = \begin{pmatrix} 1 & 0 \\ 0 & 1 \end{pmatrix}$$

Therefore, $d^2 + e^2 = 1$. Given any real e, we know d^2. We have run out of motivations to further restrict e. We then turn to tradition. (There have been σ's around since the 1920's.) Long ago someone decided to make σ_2 have $e = 1$, so $d = 0$. Finally then, we can find σ_1 from

$$\sigma_2\sigma_3 = (i\sigma_1) = \begin{pmatrix} 0 & -i \\ i & 0 \end{pmatrix}\begin{pmatrix} -1 & 0 \\ 0 & 1 \end{pmatrix} = \begin{pmatrix} 0 & -i \\ i & 0 \end{pmatrix} = i\begin{pmatrix} 0 & -1 \\ 1 & 0 \end{pmatrix} = i(\sigma_1)$$

Clearly $\sigma_1\sigma_1 = \sigma_0$, as needed also.

Finally we are down to ground zero. No more hypercomplex numbers left! All complex numbers commute with each other! Therefore, we have

$$\gamma_\mu = -f_\mu = \begin{pmatrix} 0 & -\hat{\sigma}_\mu \\ -\sigma_\mu & 0 \end{pmatrix} = \left\{\begin{pmatrix} 0 & -\sigma_0 \\ -\sigma_0 & 0 \end{pmatrix}, \begin{pmatrix} 0 & +\sigma_k \\ -\sigma_k & 0 \end{pmatrix}\right\}$$

The Dirac equation, with $m^\mu \equiv 0$, can then be written out, assuming

$$\psi = \begin{pmatrix} \psi_1 & 0 & 0 & 0 \\ \psi_2 & 0 & 0 & 0 \\ \psi_3 & 0 & 0 & 0 \\ \psi_4 & 0 & 0 & 0 \end{pmatrix}$$

where the ψ_i, i = 1,2,3,4 are only complex number functions and $\psi_1 \equiv [\psi_{1r}(r)\psi_{1\theta\phi}(\theta,\phi)\exp(-iEt/\hbar)]$, *etc...* We get a bunch of tangled, partial differential equations. The rest of the details are in Bjorken and Drell's classic Quantum Mechanics book.

The energy levels that emerge from Dirac's equation, with simple Coulomb coupling, have a simple expression that can be approximated as,

$$\mathscr{E} = -\frac{Ry}{n^2} + \left[\frac{3}{4n} - \frac{1}{j(j+1)}\right]\left(\frac{\alpha^2 Z^4}{n}\right)\frac{Ry}{n^2}$$

P. S. Isaev, in *Quantum Electrodynamics at High Energies*, Amer. Inst. Phys.: New York (1984), p. 15, shows a level diagram and presents the second term above as a 'correction' to the 'nonrelativistic' equation. The first term is the Schrödinger Coulomb solution. The α is the fine structure constant known to be $1/137.035963(15) \approx .00729735$, Z is the atomic number of the nucleus (Z = 1 for hydrogen) and Ry is the simplest ground state energy approximation, about 13.6eV. The shifts in $\mathscr{E}$, due to the second term, are very small. We can plug in numbers and see shifts of about 1 in 10^5 to 10^7 The n = 1, j = 1/2 level is lowered slighly, due to the second term. The n = 2, j = 1/2 level is also slightly lower due to the second term. The n = 2, j = 3/2 level is slightly higher (less negative) due to the second term. Along the way to this solution, one finds that $j = \ell \pm s$, s = 1/2, ℓ = 0,1,2,... provides a useful way to label the states. We have seen that j = 1/2, 1, 3/2, 2, ... are possible eigenvalues and $m_j = -j, -j + 1, ..., +j$. It turns out n = 1,2,3,... and, given n, j = ℓ $\pm$ 1/2, where $\ell = n - 1, n - 2, ..., 0$. Some examples will help.

The ground state in hydrogen is n = 1, ℓ = 0, j = 0+1/2 = 1/2, m_j = -1/2 or +1/2. We think of this as an electron zig zagging around with spin up or spin down, but this is not quite correct. There are two allowed states here and it turns out that they do not have the same energy, even though our equation above says they should have the same energy. There is a so-called hyperfine split, mostly due to the proton's spin. The lowest excited states are n = 2, ℓ = 0, j = 0+1/2 = 1/2, m_j = -1/2 or +1/2 and n = 2, ℓ = 1, j = 1 − 1/2 = 1/2, m_j = -1/2 or 1/2. Our equation says these are equal in energy but they are not. The ℓ = 1 state is more tightly bound. This split is called the Lamb shift and was first measured in 1947. Slightly above these two levels, lies the larger j level: n = 2, ℓ = 1, j = 1+1/2 = 3/2, m_j = -3/2, -1/2, +1/2, or +3/2.

The n = 2, j = 1/2, ℓ = 0 (called the S state) and ℓ = 1 (called the P state) can become split by: (1) correcting for the 'recoil' of the proton which is not infinitely heavy, so it 'wobbles' some as the electron zig zags around it and in it, (2) correcting for the size of the proton, (3) correcting for the polarization of the vacuum, (4) correcting for the so-called self-energy of the bound electron, and (5) correcting for the virtual excited states of the proton when two virtual photons are exchanged with the electron, to 'hold' it in 'orbit'. All these result in the P(ℓ = 1) state being more tightly bound. Of these corrections, the vacuum polarization is perhaps easiest to visualize and see why the P level is more bound.

The ψ_s wave functions are different of course for these two solutions, S and P, for n = 2. It turns out that the $\ell = 1$ wave is more spread out from r = 0, on average. The electron spends 'more time' zig-zagging far from home—like an old drunk visiting the neighborhood taverns. The $\ell = 0$ wave is spherical in shape and close to the nucleus. We need to now look more critically at the Coulomb field.

We have so far treated the A^μ field as simple, but it is not. It appears simple only because its quanta, photons (γ^0), are spin 1 bosons. This means they don't have an 'exclusion principle' for them, so they are very gregarious (or horny—depending on your perspective). Anyway, they snuggle into the same state easily and thus act collectively. This manifests itself, somehow, in their behaving more like a wave and spread out all over a large region. The same thing can sometimes happen for electrons which do have an exclusion principle. They usually are like geese and only want to share space with one partner at a time. When interacting with an extended, periodic electro-magnetic potential, such as in a very cold crystal, they can also pair up somehow, due to the periodic potential, and travel as a 'spin zero' boson. Then many pairs can get into the same large wave, and become spread out so well that they avoid scattering off of breaks in the periodic potential. We call this super conductivity.

Returning to the hydrogen atom, we find that since photons are so gregarious they enter into the Dirac equation for one electron 'wave' as a smooth 'external field'. Their quantum nature, however, is as real as that for electrons and it shows up in the atom. The A^μ field is full of so-called virtual photons. They hold the electron in its ψ state, around and in the nucleus. When the electron transitions to a lower energy state—SOMEHOW—a real photon is created and ejected. It makes a 'click' in one of our surrounding detectors, such as ordinary camera film. We measure wavelengths of the emitted photons from where they 'hit' the film. We thus indirectly determine the transitions supposedly taking place down inside the 1Å 'wide' hydrogen atom (1Å = 10^{-10} met.).

The virtual photons can virtually become a virtual e^-e^+ pair, that virtually goes back to a virtual photon. This sounds like ghosts coming and going doesn't it? They have a lot in common! This ghost pair, of electron and positron, is just one of zillions also coming and going. Therefore, this 'bee swarm' has -bees attracted to the nucleus and +bees repelled by it. The -bees spend more 'time' near the nucleus. Meantime, the electron, zig-zagging around through all this, 'experiences' the nucleus partially shielded and not attracting it as much, when it is in close, as in the S state. Far out, in the P state, the negative and positive 'bee' layers cancel and the full nuclear charge holds the electron.

All this imaginative rhetoric is to predict that if the n = 2, ℓ = 1, and n = 2, ℓ = 0 states are not equal in energy, then the ℓ = 0 should be less bound and farther in energy than the n = 0, ℓ = 1 state. Thus, a slightly more energetic photon will be created and ejected. And that is what we find—a slight energy difference for these two photon emissions. (The Lamb shift.)

So what does this all mean? It means that the Dirac equation is as wrong as F = ma! It does not give the observed results in nature. It is a crude approximation too. Just less crude.

We have to once again start over with a totally new theory, just as we did when we threw out F = ma for the hydrogen atom. We really do need to begin with a brand new set of fundamental guesses, even though they jump off from where we also started in this book. We need to now start with both of the e^- and γ^0 fields full of quanta. We need to treat them almost the 'same' in a new paradigm of subatomic reality called QED. The ψ wave is now the wave of all the electrons; for our case this means one real, the rest virtual. The A^μ 'wave' is the wave of all the photons; all are virtual here, except for 'transitions' which make one become a real one. That new machinery, QED, requires a bunch of new rules (guesses). These new rules were under intense development in the 1930's and 40's. They were finished by the early 50's and then honed down to their essentials, for best display of their contents, just as I have tried to do here with the late 1920's physics. The QED rules have successfully led to accurately predicating this so-called Lamb shift for the hydrogen spectra. The new imagery is very strange. It has worked so well for hydrogen that 'everyone' took that imagery over to attack the interior of the nucleus—the next frontier! It should fail there since that is a much more complex level of reality. New guesses again...

You now know that Dirac really did miss a piece in his factorization of Klein-Gordon. Apparently this piece, if it really exists in nature, is too small to affect the Lamb shift, since people have 'explained it' without using the m^μ mass part. Does this mean that m^μ is always zero for the real word? Probably not. Nature seems to use all it can get. (Anything, not forbidden to happen, happens.) This has been a successful hypothesis in the past at least. In gravity theory, the cosmological constant is not forbidden, but is it non-zero in our real world? There is a conflicting 'principle' that things should be as simple as possible. We have 'covered' ourselves, whatever is actually found out there.

Our traditional, 'textbook approach' to solving the hydrogen atom energy levels was not 'pretty' (neither simple nor beautiful). There ought to be a 'pretty' way; one which uses $\mathscr{L}_z$ and $(\mathscr{L}_k\mathscr{L}_k)$ and (f_0) eigenstates to completely solve for $\psi_{\theta\phi ef}$. Then we would directly get the ψ(r) differential equation from the full equation. Its solution would give the $\mathcal{E}$(n,ℓ) energy levels. One of you young people should tackle this challenge. I have started you on the 'pretty' track at least here. I would like to admire the full Rembrandt someday.

It is very discouraging to slug through all this e,f hypercomplex number field coupling and then get equations and predictions that only approximately fit the real observations. But that must always be the case. The QED theory that replaces all this with new imagery and guesses will work better, but then also fail. After that, we need another new theory, even more exotic, to push into the nucleus and explain why neutrons are slightly more massive than protons, or why protons are 1800 times more massive than electrons, etc...

Feynman worked hard to try to reduce the QED theory to its essence, such that it could be grasped by undergraduates without much math. Almost impossible. But we must keep trying. Before we end this exploration, we should say something about Clifford algebras and how they relate to the algebra discussions of this book.

CLIFFORD THINKING

Followers of this approach to physics usually work in formal ways with abstract notation and ideas. We shall, instead, stick to the general style of earlier work here. I will show you the Clifford algebra of the Dirac number system, but not the many subtle parts that are there too.

In the 'mathephysics' of reality, we have the set $\{ie_k, k = 1,2,3\}$. These anticommute and their squares are all -1. Products of $\{(ie_1), (ie_2)\}$ produce the set $\{e_0, ie_k\}$ which is Hamilton's quaternion system. No larger system is generated here. The set $\{(ie_1), e_2, e_3\}$ mutually anticommutes and it gives squares of $(- + +)$. We could then invent a momentum operator $P = P^0(ie_1) + P^1(e_2) + P^2(e_3)$ *and get* $P\psi = \psi M$, $PP\psi = \psi MM$ *and* $PP\psi = -(P^0P^0 - P^1P^1 - P^2P^2)\psi$. A start anyway. But again, $(ie_1)(e_2) = -(e_3)$, so we only need two seed elements. This 'minimal' set gives only (t,x) spacetime.

The products of $\{(ie_1), (e_2)\}$, when taken in all combinations necessary to close the set, lead to the set $\{(ie_1), (e_2), (e_3), (e_0)\}$. Already we are seeing that rings, not fields, are needed for our universe's spacetime. We need an indefinite metric, so time is distinct from space, methematically.

The set $\{(ie_1), (ie_2), (e_3)\}$ is mutually anticommuting and has squares $(- - +)$. So now we could consider or invent $P = P^0(e_3) + P^1(ie_1) + P^2(ie_2)$, which looks a little prettier. Now $P\psi = \psi M$ *and* $PP = P^0P^0 - P^1P^1 - P^2P^2$. We now have P = -P^ so the transformation should be $P \rightarrow P' =$

$L^\wedge PL$. Then $P\psi = \psi M$ leads to $L^\wedge PLL^\wedge\psi = L^\wedge\psi M$ *with* $LL^\wedge \equiv 1(e_0)$.

However, how can we know that P, with three parts, will give $P' = L^\wedge PL$ with only the same three basis elements? Generally, P = -P$^\wedge$ may have more parts. We need to see the whole algebra generated by $\{(ie_1), (ie_2), (e_3)\}$. Start multiplying, and expanding the table as new elements get generated. Find that the system closes with the set $\{(e_\mu), (ie_\mu)\}$ which I will call the Pauli algebra, although it is not quite the same as 2×2 complex matrices. Both do have 8 basis elements. We only have real coefficients here. (The seed set $\{(e_1), (e_2), (e_3)\}$ also generates the same algebra.)

Now we can see how big P = -P$^\wedge$ can be. $P = P^k(e_k) + Q^k(ie_k) + 0(e_0) + 0(ie_0)$. We see our original, Clifford P as a smaller subset and now PP will not be proportional to (e_0) here in general, since not all elements anticommute.

The natural group $LL^\wedge = 1(e_0)$ *gives us* $L_{small} = 1(e_0) + \epsilon^k(ie_k) + \delta^k(e_k) + 0(ie_0)$, so it has 6 parameters and is SL(2,C), the Lorentz group.

There is another natural equation here. It is $PF = FM \rightarrow L^\wedge PLL^\wedge FL = L^\wedge FLM = L^\wedge FML^?$. Then, for covariance L$^?$ = L is needed and L$^?$ depends on M and its structure.

But our P has 6 parts and these have a 3 + 3 pattern. This would suggest maybe 3 space dimensions and three time dimensions. Hardly our universe. We are a long way here from our beginning seed set, $\{(ie_1), (ie_2), (e_3)\}$, with three parts. Once we do have the full algebra here, we can explore all of it. We find that $Q = Q^\mu(e_\mu) = Q^\dagger \Rightarrow Q' = L^\dagger QL$, $Q\psi = \psi M \Rightarrow L^\dagger QLL^\dagger\psi = L^\dagger\psi M$, *and* $LL^\dagger = 1(e_0)$ are 'covariant'. The group $LL^\dagger = 1(e_0)$ gives us $L_{small} = 1(e_0) + \epsilon^k(ie_k) + \epsilon(ie_0) + 0(e_k)$, and this is SU(2)⊗U(1). Here we find that $Q^\wedge Q = (Q^0Q^0 - Q^kQ^k)(e_0)$ *and* $Q'^\wedge Q' = (L^\dagger QL)^\wedge(L^\dagger QL) = L^\wedge Q^\wedge(LL^\wedge)^\dagger QL$. This is not invariant unless $LL^\wedge = 1(e_0)$, *so* $L^\wedge = L^\dagger$ is needed. This reduces L to SU(2) only.

There is another possibility here, $QF = FM \rightarrow L^\dagger QLL^\wedge FL = L^\wedge FLM$. But this is not covariant unless L† = L$^\wedge$ and we are back to the SU(2) covariance group. However, if M = 0, then QF = 0 is covariant. Notice that $F \rightarrow F' = L^\wedge FL$ suggests that F = ±F$^\wedge$. The $\phi = +\phi^\wedge$ choice for F is called spin 0. The $F = -F^\wedge \equiv -E^k(e_k) + cB^k(ie_k) + 0(e_0) + 0(ie_0)$ choice is called spin 1. This is Maxwell's equation. We see that this Pauli algebra can handle 3 + 1 spacetime (− − − +) for zero rest mass quanta. We guess that $Q = Q^\mu(e_\mu) = \hbar\partial^\mu(e_\mu)$ *and* $QF = -\mu_0\hbar J$ is the full equation with 'source' $J = J^\mu(e_\mu) \rightarrow L^\dagger JL$, like Q. There is also a rest mass equation here Qψ = ψM, forcing us to SU(2) for covariance. But $Q\psi = \psi^{\dagger\wedge}M$ *leads to* $L^\dagger QLL^\wedge\psi = (L^\wedge\psi)^{\dagger\wedge}M$ and only $LL^\wedge = 1(e_0)$ is now needed. We still have only real coefficients here, so this will not give viable quantum mechanics wave equations. It could only be classical, as is Maxwell's equation in this same algebra. Let $\psi \equiv e^S$ and see if S can have an interpretation as the classical, relativistic 'action' in relativistic mechanics. Perhaps we would need the equation $Q^{\wedge} Q\psi = \ldots = \psi M^{\dagger\wedge} M$ instead to go to a classical limit. This is probably not very important.

To generate the complexified Pauli algebra $\{(e_\mu), (ie_\mu), i(e_\mu), i(ie_\mu)\}$ that I have found so useful, where i's inside and i's outside do NOT mix, there is no other basis element to add to the set $\{(ie_1), (ie_2), (e_3)\}$. All of the other 13 basis elements commute with at least one of these three. Check it out. Thus Clifford algebra thinking misses what could be the most important quantum algebra. On the other hand, belief in Clifford as 'God's chosen tool' for the design, tells us nature cannot stop at $\{(e_\mu), (ie_\mu), i(e_\mu), i(ie_\mu)\}$, even though this algebra handles both Maxwell and Dirac (and thus Klein-Gordon). There has to be something more in nature, or Clifford is out the window. We need 4-vectors!

How would Clifford thinking generalize the Pauli algebra? It would add the anticommuting element (f_3) to give us the seed set $\{(ie_1), (ie_2), (e_3), (f_3)\}$. Now we get an interesting algebra by multiplying all the $\{(e_\mu), (ie_\mu)\}$ elements by (f_3). I am pretty sure this generates $\{(e_\mu), (ie_\mu), (f_\mu), (if_\mu)\}$ which Rodrigues calls the spacetime algebra. (It is not that of course.) The Pauli algebra handles spacetime very nicely if four dimensional. This e,f algebra, as I will call it, is not the quantum world algebra because it has only real coefficients. Dirac's equation is not here! Maxwell's is here, since it has no rest mass. There is now a new, fundamental conjugation, $(\ )^{\vee}$, and it is also useful to define $(\ )^{\vee\wedge\dagger} \equiv (\ )^{\downarrow}$, which is also antiautomorphic. The natural P operator, in the Clifford mode of thinking, now comes from the squares $(- \ - \ + \ -)$. Thus (e_3) goes with time and the others with space. We define $P_C \equiv P^0(e_3) + P^1(ie_1) + P^2(ie_2) + P^3(f_3) = -P^{\vee}$. Then $P_C P_C = (P^0P^0 - P^1P^1 - P^2P^2 - P^3P^3)(e_0)$. We have nothing here any better than $P_E = P^\mu(e_\mu) = P^{\dagger}$ or $Q_E = Q^\mu i(ie_\mu) = Q^{\dagger}$. But once we now have the full e,f algebra $\{(e_\mu), (ie_\mu), (f_\mu), (if_\mu)\}$, we can forget the 'ugly' four seed elements that spawned it. We can use $Q_C = Q^\mu(if_\mu)$ or $Q^\mu(f_\mu)$, since these Clifford seed sets have four mutually anticommuting elements and their squares are both $(+ \ - \ - \ -)$. In fact, the traditional Dirac machinery, in all the modern QFT books, use γ_μ equivalent to one of these new seed sets. But we should not forget our roots here in the e algebra. The $f_1 f_2 = -(ie_3)$ type things show us that the f's are not closed like the e's are! They are a strange place to start our momentum operator. We should choose a seed set such that it reduces to the e's in a simple way. However, we cannot stop here. We are in no-man's land between physical algebras. We need complex coefficients here. We could just put them in by hand and double the algebra, as we did with Pauli to reach massive quantum physics. But this time the Clifford lovers do have a way out and it is interesting. They can only generate a larger algebra by more seed elements being introduced—a very restrictive way to do business. If we choose the seed set $\{(f_\mu)\}$ then to enlarge the algebra we look for a 5th element in $\{(e_\mu), (ie_\mu), (f_\mu), (if_\mu)\}$ that anticommutes with all four of these. There is one, (ie_0), this time. We don't have to pull in a new element from 'out of the sky'. We do want complex coefficients though, so we add in the really new seed element $i(ie_0)$. This is sort of out of the blue, since no i's out front existed here before. The anticommutation can be seen from $(f_\mu)(ie_0) = (if_\mu)$ but $(ie_0)(f_\mu) \to (i\sigma_0)^{\neq *}(\sigma_\mu) = -(i\sigma_0)\sigma_\mu = -(i\sigma_\mu) \to -(if_\mu)$. The 'metric' now is $(+ - - - +)$. As before, we invent $P = P^\mu(f_\mu) + P^4 i(ie_0)$, $PP = (P^0P^0 - P^kP^k + P^4P^4)(e_0)$, $P = -P^\vee \to L^\vee PL$, and $P\psi = \psi M \to L^\wedge PLL^\wedge\psi = L^\wedge\psi M$ or $PF = FM \to L^\wedge PLL^\wedge FL = L^\wedge FLM \stackrel{?}{=} L^\wedge FML$. Again, $M_F \equiv 0$ for the F field simplifies things, but P has **5 parts now**! Now $LL^\wedge = 1(e_0)$ gives us the NATURAL group $L_{small} = 1(e_0) + \epsilon^k(ie_k) + \delta^k(e_k) + \epsilon i(e_0) + \delta i(ie_0) + \lambda^\mu i(f_\mu) + \eta^\mu(if_\mu)$, since $f_\mu^{\ \wedge} = f_\mu^{\ *}$ and $(if_\mu)^\wedge = (if_\mu)^*$. Recall that $ef = f'$ and $ff' = e'$. Clearly $i(ie_0)$ added in will generate the full Dirac algebra here. The Dirac equation is the special case here where $M = 0$. We see this as follows:

$$P_C\psi \equiv 0 = [P^\mu(f_\mu) + P^4 i(ie_0)]\psi$$

$$-i(ie_0)P_C = [P^\mu i(if_\mu) + P^4(-e_0)]\psi$$

$$P_D = [\hbar\partial^\mu i(\gamma_\mu) - m_D c\mathbb{I}]$$

Therefore, $P^\mu = \hbar\partial^\mu$, $\gamma_\mu = (if_\mu)$, and $P^4 = +m_D c$ will fit the standard Dirac form. Notice that $P_D = +P_D^{\ \wedge}$ and $P_C P_C(\propto(e_0)) = (-i(ie_0)P_D)(-i(ie_0)P_D) = -(ie_0)P_D(ie_0)P_D = -(P_D)^{\wedge\vee}(ie_0)(ie_0)P_D = +P_D^{\ \vee}P_D \propto (e_0)$. We have, $P_D \to P_D{}' = L^\vee P_D L$. Note also that $P_C\psi = \psi M \Rightarrow -i(ie_0)P_C\psi = P_D\psi = -i(ie_0)\psi M = \psi^{\wedge\vee}(-i(ie_0))M$. Remember P_D has 5 parts here. If we guess that $M = M^\mu(f_\mu) + M^4 i(ie_0)$, with possibly complex M^μ and M^4, then we have

$$-i(ie^0)M = -iM^\mu(-if_\mu) + - M^4(e_0)$$

Hence, $P_D\psi = \psi^{\wedge\vee}[M^\mu i(if_\mu) - M^4(e_0)] = \psi^{\wedge\vee}[M^\mu i(\gamma_\mu) - M^4]$.

The $P_c\psi = \psi M$, $P_c = -P_c^{\ \vee}$, $P_cP_c \propto (e_0)$ is the simplest form, I must admit. Thus Clifford thinking has some useful aspects. However, the 'Lorentz' transformation of P comes from $P = -P^\vee$, and not from $PP \propto (e_0)$. Thus $P^\vee P$

$\propto$ (e_0) is just as good and in some ways better. It tells us more about P. Notice the Klein-Gordon constraint on M, since $P_c P_c \psi = P_c(\psi M) = (P_c\psi)M = \psi MM$. Thus $MM \propto (e_0)$ *is necessary*. The K-G equation says

$$(P^0P^0 - P^kP^k + P^4P^4)\psi = \psi(M^0M^0 - M^kM^k + M^4M^4)$$

Dirac's conventional form has shown us that $P^\mu = \hbar\partial^\mu$ *here and* $P^4 = +m_D c$. For K-G, we need $(i\hbar\partial^\mu i\hbar\partial_\mu)\psi = m_{exp}{}^2c^2\psi$. Thus we must multiply through by -1 = ii and we get

$$(i\hbar\partial^\mu i\hbar\partial_\mu) - (m_D)^2)\psi = \psi(-M^0M^0 + M^kM^k - M^4M^4)\psi$$

Hence the non-relativistic quantum 'world' comes from

$$(i\hbar\partial^\mu i\hbar\partial_\mu)\psi = (+m_D^2 - M^0M^0 + M^kM^k - M^4M^4)\psi = m_{\rm exp}^2\psi$$

Notice that M^μ and M^4 have not been restricted at all. They can be real, imaginary or even complex here! We know only that m_{exp} is real and m_D is real. This could be important. Are there other M structures for which $MM \propto (e_0)$? I think there are probably four such M's, with only two basic metrics $(+ - - - +)$ and $(+ - - - -)$ for the MM combinations. Whether nature uses any of these $M \neq 0$ parts is presently unknown. You cannot deny the beauty of $P_{(5)}\psi = \psi M_{(5)}$, *but* $P \neq P'$, $M = M'$*(guess), and* $\psi' = L^\wedge\psi$. Therefore, left and right are very different here. Maybe this has some deep connection to parity violation, indirectly. Nature has few balances between anything basic.

And what about Maxwell, PF = FM? Why is the 5th P part zero here? What if M here is $M^0(e_0)$, or even $M^0(ie_0)$? Consider $L^\vee P_c L L^\vee F L = L^\vee F L M^0(ie_0) = L^\vee F M^0(ie_0) L^{\wedge\vee}$. If $L^\wedge = L^\vee$, then this is also form covariant. Thus photons can have both kinds of mass it seems here. Notice also that $L^\vee P^4 i(ie_0)L = P^4 i(ie_0)(L^\vee)^{\wedge\vee}L = P^4 i(ie_0)L^\wedge L$. Thus this term is invariant if $L^\wedge = L^\vee$. For $LL^\vee = 1(e_0)$, the additional condition $L^\vee = L^\wedge \Rightarrow$ L has only 8 parameters and that L has no f's in it. Notice that $(\ )^\dagger$, hermitian conjugation, plays no vital role in any of this Clifford discussion!

Following the Clifford seed basis expansion step by step, as we invented larger number systems, we found the expansions $\{(ie_1), (ie_2), (e_3)\}$ and $\{(ie_1), (ie_2), (e_3), (f_3)\}$. Then, in the final step up, we abandoned this extension and jumped to the new seed basis $\{(f_\mu), i(ie_0)\}$. This is the usual basis used in Clifford description of Dirac algebra, with its 32 basis elements. However, there must be a way to continue the expanding seed basis above to cover this case. (Not as pretty looking as $\{(f_\mu), i(ie_0)\}$.) We see below that $i(if_3)$ does do the job. We have

$$i(if_3)(ie_1) = i(-if_2), \quad (ie_1)i(if_3) = i(if_2)$$

$$i(if_3)(ie_2) = i(+if_1), \quad (ie_2)i(if_3) = i(-if_1)$$

$$i(if_3)(e_3) = i(if_0), \quad (e_3)i(if_3) = i(-if_0)$$

$$i(if_3)(f_3) = i(+ie_0), \quad (f_3)i(if_3) = i(-ie_0)$$

This then leads to squares $(- \; - \; + \; - \; +)$ and we then can construct $P_c = P^0(e_3) + P^1(ie_1) + P^2(ie_2) + P^3(f_3) + P^4 i(if_3)$. We then find $P = -P^\vee$, so $P \to P' = L^\vee PL$. To compare with standard Dirac, where we have $-m_Dc(e_0)$ as the 5th term, we multiply by $-i(if_3)$ and get

$$-i(if_3)P_c = P^0 i(-if_0) + P^1 i(+if_2) + P^2 i(-if_1) + P^3 i(-ie_0) + P^4(-e_0)$$

Thus, here we get $\gamma_0 = -(if_0)$, $\gamma_1 = (if_2)$, $\gamma_2 = -(if_1)$, $\gamma_3 = -(ie_0)$, and $P^4 = m_Dc$. Therefore, the γ_μ set gives $(+ \; - \; - \; -)$. The new operator, P_D, now has $P_D \to P_D' = -i(if_3)P_c' = -i(if_3)(L^\vee PL)$ and I don't know the conjugation that goes with (if_3) such that $(if_3)L^\vee = (L^\wedge)^{\dagger conj}(if_3)$. There is such a conjugation generated by $A^{conj} \equiv (if_3)A^\dagger(-if_3)$. This leads to $(if_3)A^{conj} = A^\dagger(if_3)$ and to $A^{conj\dagger}(if_3)^\dagger = (if_3)^\dagger A^{\dagger\dagger}$. Thus, $A^{conj\dagger}(+if_3) = (+if_3)A$. Then $P_D' = L^{\vee conj\dagger}P_DL$. If the mass part is to be invariant then $L^{\vee conj\dagger}(e_0)L = (e_0)$. All the same results for the restrictions on L occur here, but in a messy way.

So, what is the best way to write $P_{(5)}$ and why? I favor $P = P^\mu i(ie_\mu) + P^4(if_0) = P^\dagger$ and $Q = P^\mu(e_\mu) + P^4(if_0) = Q^\dagger$. These are hermitian and $P^\wedge P \propto (e_0)$. If Dirac mass, P^4, is zero, then we return to the e-subalgebra. It is a fundamental question, therefore, whether nature needs the (complex: e,f) algebra or can get by on the (complex: e) part alone. It is your challenge to eventually answer this.

AN 'M' SPECIAL CASE

The equation $P\psi = \psi M$ can be written in many forms, depending on the guessed-at internal structure of M. We could also operate on the equation from the right with anything we choose. There are projection operators, $\wp$, such as

$[(e_0) + (f_0)]/2$ and $[(e_0) + (if_0)]/2$, which have the property $\wp\wp = \wp$. Thus $P\psi\wp = \psi M\wp$ might be of some value, but I do not know what value. This is possible because of the peculiarity of ψ 'transforming' as $L^\wedge\psi$ instead of $L^\wedge\psi L$, or such. Then ψ and $\psi\wp$ easily transform the same way. We will have some problems with $\psi M\wp$, and we must be able to get to $\psi\wp M^{??}$ for $\wp$ to make much sense. Note that $\wp$ has no inverse—once 'applied' you cannot take it back out.

For example, we can choose $M \equiv mi(ie_3)$, $\wp \equiv \frac{1}{2}[(e_0) + (f_0)]$, $P \equiv \partial^\mu(-f_\mu) + m_D i(ie_0) = -P^\vee \rightarrow L^\vee PL$, $P\psi \equiv \psi M \rightarrow L^\vee PLL^\vee\psi = L^\vee\psi M$, $LL^\vee \equiv 1(e_0)$, $\hbar$, $c \equiv 1$, and let $m_D \approx 0$, in P. Then what? First notice that this M choice is very peculiar, since it singles out the z-direction in space, in some sense. Why do that? Does nature have a 'favorite' direction? I doubt it.

We shall press on anyway, and do some algebraic manipulations as follows:

$$\partial^\mu(-f_\mu)\psi + 0i(ie_0)\psi = \psi i(ie_3)m$$

multiply right by $(-ie_3)$ and obtain

$$\partial^\mu(-f_\mu)\psi(-ie_3) = \psi mi(ie_3)(-ie_3) = \psi mi(e_0)$$

$$= im\psi$$

multiply through by -i and obtain the standard mass part of the Dirac equation:

$$\partial^\mu i(+f_\mu)\psi(-ie_3) = m\psi$$

If it were not for the 'goiter' $(-ie_3)$ attached to ψ here, this could be the Dirac equation with $(f_\mu) \equiv \gamma_\mu$. It is clearly not the Dirac equation. The m here comes from M, not from the 5th part of P.

Now let us write $\psi \equiv \phi[(e_0) + (f_0)]/2$. Then we have

$$\partial^\mu i(f_\mu)\phi[(e_0) + (f_0)](-ie_3)/2 = m\phi[(e_0) + (f_0)]/2$$

We see that $(e_0)(ie_3) = (ie_3)(e_0)$ *and* $(f_0)(ie_3) = (ie_3)(f_0)$. Therefore, we get

$$\partial^\mu i(f_\mu)\phi(-ie_3)(\tfrac{1}{2})[(e_0) + (f_0)] = m\phi(\tfrac{1}{2})[(e_0) + (f_0)]$$

Finally, we can write $(-ie_3) = (f_1)(f_2)$ and get

$$\partial^\mu i(f_\mu)\phi(f_1)(f_2)(1/2)[(e_0) + (f_0)] = m\phi(1/2)[(e_0) + (f_0)]$$

We can just define $f_\mu \equiv \gamma_\mu$, since it has the usual properties of γ_μ, and this becomes

$$\partial^\mu i(\gamma_\mu)\phi(\gamma_1)(\gamma_2)(1/2)[(e_0) + (f_0)] = m\phi(1/2)[(e_0) + (f_0)]$$

This equation is true if the original equation

$$\partial^\mu(-f_\mu)\psi = \psi i(ie_3)m$$

is true. This is NOT the Dirac equation, I am pretty sure. From this we could invent a new equation by 'throwing away' the $(1/2)[(e_0) + (f_0)]$ parts on both sides, even though they do not really cancel off.

Just guess that maybe nature makes some practical use of the 'spiritual' equation

$$\partial^\mu i(\gamma_\mu)\phi(\gamma_1)(\gamma_2) = m\phi$$

This has a peculiar z-axis bias and is unproven to apply to the real world, as far as I know. (Hestenes and his associates seem to think it has great promise. I doubt it.) The 'physical' ψ comes from the projection here $\phi 1/2[(e_0) + (f_0)] \equiv \psi$. This is a one way street. I don't see how ϕ would lead directly to predicting the hydrogen atom energy states, or such. This equation cannot give the hydrogen atom Dirac equation, even for ψ here, since it is not Dirac's equation, as some claim.

There are other M choices as well as this one, and no one knows yet, I presume, if any of them is physical. A more symmetrical choice would be $M = M^k i(ie_k)$ rather than $M^3 i(ie_3)$ alone. We do need MM $\propto (e_0)$, however here. Notice that $i(ie_3)i(ie_3) = (e_0)$. $M = M^\mu(f_\mu)$ has $MM = (M^\mu M_\mu)e_0$. Have we opened Pandora's Box here, with the invention of a new mass thing, M?

I am grateful to E. Recami and W. Rodrigues for recent reprints which stimulated the above discussion of Clifford thinking. It is very interesting to contemplate the natural 5-vector nature of the e,f algebra for classical physics. The restriction $L^\wedge = L^\vee$ freezes the 5th dimension and gives us the Lorentz group. Only the quantum algebras give us the L group with 8 parameters. This may mean that this wider symmetry will only show itself in quantum equations and quantum phenomena in general. However, maybe the full 16 parameter

$LL^\wedge = 1$ group is to be used in all covariance investigations in the future at a deeper level than we presently understand. For example, Recami and Rodrigues have the current 4-vector for Maxwell's equation extended to allow for magnetic monopoles in the form essentially of $J_q + \gamma_0 J_m$. They only look for Lorentz covariance. This may be enough for classical physics. But notice that this is equivalent to $[J_e + (-ie_0)J_m] \rightarrow L^\wedge JL$, and $(ie_0)L^\wedge J_m = (L^\wedge)^{\wedge\vee}(ie_0)J_m = L^\vee(-ie_0)J_m$. Thus, this composite J is not fully consistent unless $L^\wedge = L^\vee$ which is only the 8 parameter subgroup or even the 6 parameter sub-subgroup. Could this mean that magnetic monopoles are forbidden to show themselves only at the quantum level, rather than at the classical level?

Of course, the classical is only a shadow of the quantum machinery going on. All the distinctions between these two covariance groups need to be explored. I still wonder, since the four big spacetime dimensions are rolled back on themselves in a closed universe, why cannot the fifth dimension just be rolled back also but presently it is very small for some reason. We approximate its effects by a 'mass part' as the 5th part of our P operator and put in m values here by hand. So much remains here. We have only scratched the surface with the Dirac equation and its QED extension.

FIVE DIMENSIONAL CLASSICAL PHYSICS

I have extensively shown in an earlier chapter how to do classical within the Hamilton-Pauli hypercomplex number system, $\{(e_\mu), (ie_\mu)\}$, with real coefficients. There is the obvious possibility that full Dirac algebra is the right algebra for QUANTUM physics. If so, then it is quite possible that the real half of this algebra is the right algebra for classical physics. Let us see where this leads us. We assume the system $\{(e_\mu), (ie_\mu), (f_\mu), (if_\mu)\}$ with real coefficients, and we wish to do Lorentz force and Maxwell physics, where particles have a path and momentum 'vector'.

We expand the former results from Pauli by defining $P = P^\mu(e_\mu) + P^4(f_0) + P^5(if_0) = P^\dagger$. Actually, we need $P_I{}^\wedge P_I \propto (e_0)$, which eliminates P^4 ($P^4 = 0$), or we need $P_R{}^\vee P_R \propto (e_0)$, which eliminates P^5 ($P^5 = 0$). Could it be that both 5-vector types are useful? Notice that $dx = dx^\mu(e_\mu) + dx^4(f_0) + dx^5(if_0)$ may, for some unknown reason, have $dx^4 = 0$ *and* $dx^5 = 0$. This is a very special 'vector' thing. Notice that $x = x^\mu(e_\mu) + x^4(f_0) + x^5(if_0)$ would have $x^4 =$ *constant, and* $x^5 = 0$. This means a fixed length exists here but it does nothing at the classical level.

Where do we start? The $P^\wedge P \propto (e_0)$ gives us a metric $(+ - - - -)$ for the P_I (and P_R) 5-vector. Thus $P^\wedge P = [P^\mu(e_\mu) + P^5(if_0)]^\wedge P = (P^\mu P_\mu -$

$P^5P^5)(e_0)$, so we have the obvious identification $P^0 = \mathscr{E}/c$, $P^1 = P_x$, $P^2 = P_y$, $P^3 = P_z$, $P^4 = \mathscr{E}_0/c = m_0c$. We have the spacetime operator $\partial \equiv \partial^\dagger = \partial^\mu(e_\mu) + \partial^4(f_0) + \partial^5(if_0)$, *and* $\partial^\wedge\partial \propto (e_0)$, *or* $\partial^\vee\partial \propto (e_0) \Rightarrow \partial^4 = 0$ *or* $\partial^5 = 0$. In classical, this operator is only useful for Maxwell. We guess that Maxwell is $\partial F = J$. We also guess that $dP = [(dxF) + (dxF)^\dagger]\,d\tau/2$, where $d\tau$ is proper-time (whatever that means here).

These guesses get us started and they have remarkable consequences. They encompass all the known classical fundamentals, except gravity. Covariance can give us more information. For example, $\partial'F' = L^\dagger\partial LL^\wedge FL$ is a reasonable guess also, and $LL^\wedge = 1 = LL^\vee$ will be guessed to also be true. Then $F = \pm F^\wedge$ is reasonable, and E-M theory uses $F = -F^\wedge$. This means F can have, at most here, $F = -E^k(e_k) + cB^k(ie_k) + G^\mu(if_\mu) = -F^\wedge$. Thus $[dxF + (dxF)^\dagger]$ can tell us things about dP. We form the multiplication table:

F dx	e_1	e_2	e_3	ie_1	ie_2	ie_3	if_0	if_1	if_2	if_3
e_0	e_1	e_2	e_3				if_0			
e_1	e_0				$-e_3$	e_2		$-if_0$		
e_2		e_0		e_3		$-e_1$			$-if_0$	
e_3			e_0	$-e_2$	e_1					$-if_0$
(f_0)										
(if_0)										

I have left out the bottom two rows, since maybe dx^4 and dx^5 will be zero. The 5th-dimension is 'frozen' for some reason. The other terms shown can survive in $[dxF + (dxF)^\dagger]$. We see clearly that they dictate $dP = dP^0(e_0) + dP^k(e_k) + dP^4(if_0)$ *and thus* $dP^\wedge dP = dP^0dP^0 - dP^kdP^k - dp^4dP^4$. Presumably then, $P = P^\mu(e_\mu) + P^4(if_0)$, even though P^4 is constant ($dP^4 = 0$). The Newton-Einstein mechanics law is $dP = q[dxF + (dxF)^\dagger]d\tau/2$ *and* $F = -F^\wedge$. Maxwell is $\partial F = J$ or maybe $\partial F + (\partial F)^\dagger = 2J \Rightarrow J = J^\dagger$. We have to guess here. There is no natural restriction on J so far: $J = J_a{}^\mu(e_\mu) + J_b{}^\mu(ie_\mu) + J_c{}^\mu(f_\mu) + J_d{}^\mu(if_\mu)$, *or maybe* $J = J_a{}^\mu(e_\mu) + (ie_0)J_b{}^\mu(e_\mu) = J_a{}^\mu(e_\mu) + J_b{}^\mu(ie_\mu) + 0(f_\mu) +$

$O(if_\mu)$. These are obvious possibilities. Even the 'conserved' requirement on J tells us nothing about its parts. It is automatic:

$$\partial F = J \Rightarrow \partial^\wedge J + (\partial^\wedge J)^\wedge = \partial^\wedge \partial F + (\partial^\wedge \partial F)^\wedge \equiv 2(\partial|J) =$$

$$\partial^\wedge \partial F + F^\wedge(\partial^\wedge \partial)^\wedge = \partial^\wedge \partial F - F(\partial^\wedge \partial)^\wedge$$

$$\textit{and } (\partial^\wedge \partial) \propto (e_0) \Rightarrow (\partial^\wedge \partial)^\wedge = \partial^\wedge \partial, \textit{ so we have}$$

$$(\partial|J) = \partial^\wedge \partial F - (\partial^\wedge \partial) F = 0$$

regardless of the structure of J. We only know that J must be such as to generate the 'solution' F for which F = -F$^\wedge$. We can multiply out ∂F and see how many parts it makes. Perhaps I am being too restrictive, but I will assume, $\partial^4 = 0$, $\partial^5 = 0$, in our real universe. We get the same table as before except that now we don't have $(\partial F) + (\partial F)^\uparrow$, only ∂F. This leaves a lot more terms surviving. I filled in the whole table and found that $\partial F = J$ can allow J to have all of the 16 parts except (f_0).

Can we guess that $(J|J) = J^\wedge J$, or $J^\vee J$, should be proportional to (e_0)? Why so? People have long speculated that if $J \neq J^\mu(e_\mu)$ then this means monopolies can exist. They add on (ie_0)J in their source of F. Really though, J can have f parts as well, or so it seems here so far. (If we try J(ie_0), we have trouble with the L covariance of a different kind unless $L^\wedge = L^\vee$.)

A moving particle carries momentum 'vector' $P = P^\mu(e_\mu) + \mathscr{E}_0/c(if_0)$ $= P^\uparrow \rightarrow L^\uparrow PL$, $LL^\wedge = 1(e_0)$, $L^\wedge = L^\vee$ *(no f's in L)*. But P has e's and f's. Thus $L^\uparrow \partial L$ has no f's either. If, for some reason, $F = -E^k(e_k) + cB^k(ie_k) + 0(if_\mu)$, then $L^\wedge FL$ would have no f's and F′ would only have these 6 parts as well. Why should $F^\wedge = F^\vee$ though (no f's)? Does $J^\wedge = J^\vee$? why should it? No f's in J and no f's in ∂ probably means no f's in F either. But I see no J suggestions here, other than the blind guess that maybe $J^\wedge J \propto (e_0)$. Maybe J should be like P and not any 'bigger'. Who can say? I suppose the quantum theory should be done right and then the classical limit taken carefully. The correct J would then come out automatically.

What can we say about the energy carried by the F field? For a particle, we have simply $P^0 = \mathscr{E}/c$, and the kinetic energy is $(\mathscr{E} - \mathscr{E}_0) = (P^0 - P^4)c$. The F field should have some invariants, and something like the ugly energy momentum tensor, I suppose. That search may also help to restrict F and J in size.

Notice that $2(F|F) \equiv F^\wedge F + (F^\wedge F)^\wedge = 2F^\wedge F = -2FF \rightarrow$

$-2L^{\wedge}FLL^{\wedge}FL = -2L^{\wedge}FFL = ?= -2FFL^{\wedge}L = -2FF = 2F^{\wedge}F = invariant$. But, $F^{\wedge}F = (F^{\wedge}F)^{\wedge} \Rightarrow \{(e_0), (ie_0), (f_\mu)\}$, for $F^{\wedge}F$, and this does not commute with $L^{\wedge}$ above, unless the (f_μ) parts are all zero. That is true in the old Maxwell theory, and thus $F^{\wedge}F = -FF$ is invariant there. Not here. We would need $FF + (FF)^{\wedge\vee}$ to kill off the (f_μ)'s in F and this is rather artificial.

I never liked the old $\overline{\psi}\gamma_\mu\psi = J_\mu$ guess about how the electron ψ is a source for the A^μ field in QED. Perhaps the above is telling us that J is much more sophisticated than we have believed up until now.

We have worked here with $P = P^\mu(e_\mu) + P^4(if_0)$. We could multiply through by (f_0) and obtain $(f_0)P \equiv Q = P^\mu(f_\mu) + P^4(ie_0) \to L^{\wedge}QL$, $Q = +Q^{\wedge}$. We could multiply through by (if_0) instead and obtain $(if_0)P \equiv R = P^\mu(if_\mu) + P^4(e_0) \to L^{\vee}RL$, $R = R^{\vee}$. The R form is 'standard Dirac' where $(if_\mu) \equiv \gamma_\mu$, except that we have no complex coefficients here! The Q form above has $\{(f_\mu), (ie_0)\}$ which is a mutually anticommuting set (Clifford seed set). Thus, $Q\tilde{Q} \propto (e_0)\ (+\ -\ -\ -\ -)$. Neither of these changes from P is important. They are just other languages for the same thing. Now Q has 4 f's and an e. But now ∂ would have 4 f's and $\partial F = J$ still would have F with all e's, even if ∂ and J have all f's. However, J could be a mixture here also, and then F can still have 10 parts, not just 6, for $F = -F^{\vee}$.

I leave you with this J mystery. All of this shows how naive the old vector formulation of Gibbs and Heaviside was for fundamental physics. It has no 'guidance' in it to speak of. Hamilton and Tait should have triumphed in that 1890's fight, for they now prove to be clear winners, as this book has shown, I hope!

A PUZZLE REMAINS

Now we come to the end of this book, in May of 1996, and only now do I REALLY see most of the picture in Dirac and in classical physics. You will recall that $P = P^{\dagger}$, within the Dirac algebra, could mean that P has 16 parts. But the guessed need for an invariant inner-product, proportional to the identity, caused us to settle for P with 5 parts. Looking for mutually anticommuting elements also leads to five elements. I now see that, in classical, these P elements should be the set $\{(e_\mu), (f_0)\}$; and in quantum, where complex coefficients are allowed, it should be the set $\{i(ie_\mu), (f_0)\}$ instead. I spent the whole book assuming that P, with real coefficients, gives us the 5-thing above, but with (if_0) in place of (f_0). I was partial to $(\ )^{\wedge}$ over $(\ )^{\vee}$. This was not a mistake in and of itself, but it led me to naturally think that the rest mass part is only invariant under the 8 parameter sub-group of L, instead of the full, 16-

parameter group. It now appears to me that nature is form covariant under the full 16 parameter group $LL^{\wedge} = 1(e_0)$, at least when it comes to dream up new wave equation possibilities. This should replace Lorentz for quantum 'fishing' for new equations. In classical physics, the laws should be invariant under the full $LL^{\wedge} = 1(e_0)$ group, which is (I think) Sp(4) here; it surely has 10 parameters here. Of course, we should get our classical equations from approximations of our quantum laws, but that is not always possible so far. At first it seemed that (f_0) made everything fall into place in a beautiful form (so beautiful that it cannot be wrong—to use Einstein's standard proof by intimidation). We simply have the invariance: $L^{\dagger}(f_0)L = (f_0)(L^{\dagger})^{\dagger\wedge}L = (f_0)$, *for* $LL^{\wedge} = 1(e_0)$, with 16 parameters. Similarly, $L^{\dagger}(if_0)L = (if_0)(L^{\dagger})^{\vee\dagger}L$, *for* $LL^{\vee} = 1$. I now see no advantage for one way or the other. The metric for classical is $(+ - - - -)$ and for quantum it is $(- + + + -)$. The quantum needs to be different from the classical because of the 'ii' we must insert to replace -1 in quantum, to obtain the right K-G equation.

It is very interesting that quantum naturally uses P with two i's in it, for the e part, and classical uses P with no i's in the e part. I suppose this is not really so strange, since quantum involves a classical P going over 'into' an imaginary derivative operator. A magical transition for sure. It just works! The covariance symmetry group now being larger in quantum than in classical seems to also be a new idea—new to me anyway.

Even the electromagnetic field is handled differently in the classical subalgebra and in the Dirac equation coupling. We have Maxwell's equation with $\partial^{\mu}(e_{\mu})$ in classical, and with (ie_{μ}) instead in quantum. Then the coupling can have an i in P but no i in the A^{μ} part. Thus $A^{\dagger} = -A$ in quantum, but $A^{\dagger} = +A$ in the classical set up. Curved space in quantum would involve this (ie_{μ}) becoming spacetime dependent, along the lines shown in an earlier chapter for (e_{μ}).

There are two equal classes of quantum equation. One has P with (f_0) mass and invariance under the full 16-group. The other has (if_0) mass and is also invariant under the full 16-group. I don't know what this means. Could both somehow be used? We reverse these equation roles by using $LL^{\wedge} = 1$ instead of $LL^{\vee} = 1$. I see no good reason so far for not going with the full 16-group for equation guessing. Perhaps the physical solutions of these successful equations only have the covariance 8-parameter group associated with them somehow. The mathematicians need to look into that quandary.

This still leads me to suspect that 6-vectors are somehow involved here, at least for the quantum world. One part, say (f_0), is totally invariant and the other part is invariant for only part of the covariance group, so it is special and it is only a subnuclear dimension, or such. This then totally kills off $P^{\vee}P$ being proportional to (e_0), but maybe that is alright. We cannot get all this at the same time. Something has to change here. There is a certain amount of ugliness, I will admit. But the two f_0 parts, as 5th parts, are too symmetrical as it stands. Maybe nature breaks this somehow. The existence of only 5 mutually anticommuting elements in the algebra is discouraging for this idea. But maybe we need a larger algebra than Dirac's and with it will come another, 6th basis element that also anticommutes with these 5. Remember that the Dirac equation, even with our M extension, still is a special case of two coupled equations in Dirac algebra. That may indicate that 'we ain't done yet'.

Some new light can be shed on this $LL^{\wedge}$ and $LL^{\vee}$ problem by using the wave equation itself. For (if_0) mass, we need $P^{\wedge}P$ *and* $P\psi = \psi^{\dagger\wedge}M$. Then the $P^{\wedge}P\psi = \ldots$ and the form invariance of this equation lead to $LL^{\wedge}$. The P innerproduct, with its 5 parts, is then invariant under the full 16-group. But now the 5th part of P is definitely not invariant. It is invariant only under the subgroup $L^{\wedge} = L^{\vee}$. This is telling us something very important but I am not sure what. If we use instead the (f_0) mass in P, then we get a similar series of steps with everywhere $(\ )^{\vee}$ replacing $(\ )^{\wedge}$. Neither is preferred. Again the 5th part of P is invariant under only the subgroup with 8 parameters, when we use $(\ )^{\vee}$.

It is interesting that we do not have a choice really between keeping the 5th part of P invariant and keeping $(P|P)$ invariant. The new M mass causes enough complexity to nail down the innerproduct as the one to 'save' here. But what could it mean to say that the rest mass part in P is not invariant? What does invariant even mean in this huge group, physically? Perhaps this is no problem at all. Just find solutions of the equations and get on with it. Remember that we have left Lorentz covariance in the dust here and abandoned any attempt at a physical justification for the proper covariance group. There is a preferred frame, the Newton frame where the background radiation is isotropic. The equations should be solved in **THIS** frame and the rest mass put in there by hand for now. If we make an L transformation and this 5th P changes value or mixes with the other parts in P as a consequence, then maybe this is not important. We need to get back to the fundamental frame to do our physics and do our measurements if possible. That may be impossible for humanoids on most planets. Some kind of corrections will have to be made for our measurement results obtained in the wrong frame. Perhaps this 5th P part will play a part in those corrections. What a mess this all is, if true.

QUATERNION AMPLITUDES?

As discussed earlier, we may need to extend our basic number system. At present, we do not know for sure which is the physical number system to extend. It could be the E algebra part or the full D algebra. In either case, we could invent an interesting extension by allowing not just 1 and I in front of the 8 or 16 basis elements, but rather convert the i there into (iq_1) and extend this $\{1 = (q_0), (-iq_1)\}$ set into a full quaternion set with new elements $(-iq_k)$, k = 1, 2 here. Since the old, antiautomorphic conjugations all gave -i for any i outside, it is natural to invent the idea that any conjugation on $(-iq_k)$ gives -(...). This then naturally allows quaternion conjugation instead of just complex conjugation and we will also define that these commute with the e,f elements, since the special case {1, i} did before the extension. Naturally $(qA)^{\dagger} = A^{\dagger}q^{\dagger} = A^{\dagger}q^{*} = q^{*}A^{\dagger}$ and q^* is quaternion conjugation as just discussed. We can now ask how may dimensions in $P = P^{\dagger}$ and the answer is 22. Sounds familiar, huh? 22? We can also ask how many parameters are in $LL^{\wedge} = 1$ and the answer is 22 also. How many totally anticommuting elements now exist, I have not found out yet, but it will be more than 5 for sure and less than 22. Of course we can check all pairs in $P \equiv P^{\dagger}$, $P^{\wedge}P$, and find the largest set for which all cross terms cancel out. It probably gives the same answer as finding the largest mutually anticommuting set, as we found in the D algebra.

In reading Feynman's QED book, one is easily shocked by all the rules he just pulls out of his hat and says that they work. (Don't ask why.) One of these is that amplitudes for segments in a Feynman diagram are complex numbers, and these are added or subtracted before the absolute values are taken and used to predict what huge, humanoid observers will actually see in the real world. No one knows why these are complex numbers! Why are they not quaternions, with two of the four parts much larger than the other two? Or so it is in lepton physics anyway. We don't know much about these diagrams when applied to strong interactions. We just GUESS that they will still work and will still work with only complex number amplitudes. As Feynman said in QED (p 149), "... its not because nature is really similar; its because the physicists have only been able to think of the same damn thing, over and over again." Since we do guess blindly at all the fundamental ideas, we must keep an open mind and try to break out—see things a new way as much as possible, even though most of these will be wild goose chases. That may be the case with quaternion coefficients extending the old algebras, those algebras have proven themselves to be part of the big picture at least. This is an interesting idea, that I am sure must have already been investigated, at least in some other environment of quantum thinking, if not this hypercomplex view of the basic, relativistic algebra. I would be surprised if quaternion amplitudes appear in other than relativistic quantum and even there perhaps only in hadron physics. Maybe leptons are special in that two of their four amplitude parts are much smaller than the other two, for some presently unknown reason.

We have some interesting contrasts between quantum and classical machinery here, when 5-vector like things are allowed in the algebra. We still do not know for sure that all the quantum and classical cannot be done with just the e's alone and no f's whatsoever. The classical physics in the e algebra of Hamilton and Pauli has only 4-vectors. The Clifford, real coefficient, e,f algebra, for classical physics, has natural 5-vectors but the 5th part is invariant it seems under only the real part of the full 16 parameter group, which has given us only 10 parameters. If my idea of using only the E algebra, with only e's and no f's, but complex coefficients, should turn out to be right for nature, then classical has only Lorentz symmetry and quantum has only the 8 parameter generalization of this Lorentz covariance. The world is then a lot simpler! The big question then remains: "Is the world then too simple for us to be here—to ask the questions in the second place?" Exciting things for you younger readers to find out. I hope I live to see the answer unfold, and I wish that Feynman could have stuck it out also to see the answers. He would love this stuff and I am very sorry we cannot share it with him. Maybe God has given him all the REAL answers by now and spoiled all the fun for him. I suppose the devil is just as good a physicist as God—just in case?? I hope so! Most of us physicists will be doing our talking to him, perhaps.

A SIMPLE DIAGRAM FOR
LONG DISTANCE COMMUNICATION
IN A CLOSED UNIVERSE

JAMES D. EDMONDS, JR.

ABSTRACT

The expanding and contracting, closed universe is pictured in such a way that long distance communication can be graphed without any geodesic equations.

INTRODUCTION

D. J. Raine (1981)[(1)] discusses in depth both the open expanding universe and the closed, expanding and contracting universe. Both cases are very hard to grasp, except as equations. Pictures are very difficult for large separations between signal sender and signal receiver because the space itself stretches as the signal propagates between them. For the closed universe, the space typically expands for about 200 billion years then recontracts for another 200 billion years. How can we then picture a photon taking a 300 billion year path between sender and receiver? Their separation first expands and then recontracts considerably while the signal is still in transit!

Local signalling events, in the recent past or future, can be easily handled with the usual Minkowski diagram, Figure 1.

Physics Essays, Vol. **10**, No. 1 (1997)

We are observer ACF. To our left is observer BD, moving farther away from us as space stretches. On our right is observer E, also moving farther away from us as space stretches. We send a signal left at A and it is received at B. The answer reaches us at C and also passes by us to reach the observer on our right at E. We again send a reply, left at C, which is received at D. The observer on our right decides to comment on the BC message when it arrives at E and we receive this comment at F. Communications take longer and longer as space stretches between all of us.

At a given time, there are many observers to our left and also to our right, farther and farther away from us, along x and -x. In the open universe, there is an infinite number of observers, both left and right, at ALL times past, present, and future. This universe is never really small. It is always infinite. In the closed universe, there is a fixed and finite number of 'observers', left and right at ALL times, past, present, and future. Long ago the local distances between observers were smaller in both universes.

The beginning of the universe, for communication purposes, was about 1,000,000 years (1 m.y.) after the beginning. The universe then became transparent. The universe was about $1000\times$ smaller then in separation than now. Now is about 15 billion years later. During the remainder of the 200 billion years of expansion, it will only expand by an additional factor of four times, or so.

No stars existed before a few billion years, nor last past a few hundred billion years, even if born late. There were no humanoid communications in the early days, and there will be none in the latter days. We shall ignore this limitation, however, and consider intelligent signals sent very early and received very late, though they could not physically exist at these time extremes.

THE CLOSED UNIVERSE

Our problem is to picture communications in the closed universe, which originate right after decoupling and which may travel nearly 400 billion years before being received, shortly before the big crunch at the end. We need to extend Figure 1 into the distant past, into the distant future, and to distant observers, both left and right. We restrict consideration to left and right for simplicity. Apparently, no particular left/right direction is favored. This diagram only works for a closed universe.

At a fixed time, the observers out there to our left and right have some average spacing and there is a finite number of them. This we can picture as a symbolic circle of space along x at a time called 'now'. Earlier, we were all closer together and later we will all be farther apart. So let us try concentric circles of space for these three times as in Figure 2. This is like a circle of points around an expanding balloon. Time runs radially outward in Figure 2. The radius of a circle can be considered to have a scale, R. This R can then be used to compute the cosmic time for a given circle and the distance around the universe at that time. See Appendix. The calculations are the standard Friedmann solution to Einstein's gravity equation for a universe of dilute 'dust'. I have not included all that here since it is in all the cosmology text books and because I hope the diagram can be useful to non-experts, to picture how the universe behaves even though closed, and very non-intuitive as a result.

We see that, at Circle A, we have 'us—now' at point A and observers at this time spread uniformly 'around' the universe. We see there is a 'backside' to the universe, K, where observers are now farthest away from us. Those same observers are always farthest from us during the whole expansion and subsequent recontraction. Again we see that messages sent by us at A are received by our neighbors later at B and their reply reaches us still later at C. But how do we deal with any messages which observers at K might choose to send us right now? When will their message reach us? Will it take so long that we will be in the contracting phase, for the universe, by the time it reaches us? Will it ever reach us? Obviously, this is a very difficult question unless we can find a way to wrap the photon trajectory 'around' the universe, such that it reaches our position on one of the later circles. By symmetry, K could send this signal either left or right, and it would reach us at the same time later, whenever that is. In fact, it may never reach us. This world could recontract to a 'point' and end while the signal is still in transit, or the universe could become mostly 'opaque' again, near the end.

LONG DISTANCE GEODESICS

There is no totally satisfactory graphical solution for long distance signals on two dimensional flat-space paper. However, I did find a way to get across the idea of how signals can come and go, over vast time intervals, even though the universe changes dramatically while the signals are in transit.

We first construct a master geodesic. By definition, both the distance scale of space around a circle changes from circle to circle, and the time scale for time intervals lapsing between circles change, in such a way that the local light signal geodesic segment passes any particular circle at 45° to that circle. See Figure 3. This is a great graphical simplification but it results in the time, and universe size at that time circle, being complicated to calculate. See Appendix.

This extrapolation will work forward in time from 'now' out to the time of maximum size for the universe, say 200 b.y. after the Big Bang. Extrapolating back to earlier times, from 'now', we see that things get strange as the circles go to zero radius. Time approaches the decoupling time at .001 b.y. and then the Big Bang of creation itself at 0 b.y.

I believe we can extrapolate down to a circle representing .001 b.y. (decoupling) with no real difficulty. Beyond that we have to refer to the equations for the closed Friedmann universe.

The equations say that a hypothetical photon which left the other side of the universe, K, at exactly the bang itself, and sent left or right, will reach us, AC..., right at the time of maximum expansion. Since .001 b.y. is close to zero, on a b.y. scale, we can approximate the geodesic that actually leaves the decoupling time circle (.001 b.y.), as reaching the maximum expansion circle when the photon has traveled around through almost 180°. Actually, from the big bang to the more accurate decoupling time of 300,000 years, for our particular closed universe case, the angle involved is 1.8° of rotation. Thus, 180° is a pretty good approximation, but the angle is actually closer to 178°.

By the way, all the numbers that I have been throwing out, for the closed universe, correspond to assuming a total of 10^{24} 'sun like' stars on the average (current density of 6.34×10^{-30} g/cm^3) for the whole universe, and a Hubble constant of 1/(19.7 b.y.) or 45.7 km/s/MPS. These are just representative, possible figures for our universe. It is not yet known if our universe is actually open or closed. It is very nearly flat, so barely open or barely closed. Some think it should be exactly flat (space) and therefore open and infinite, but this is theology! We must wait for better measurements to pin it down.

Once we draw a master photon geodesic, we can put it on tracing paper and rotate it around a pin through the center of all the circles. Where it crosses a given observer's radial line, at a given time circle, it then shows how signals, which are received there-and-then from the past, have come from back 'around' the universe. It also shows when signals, that are sent from there-and-then, will arrive at other places 'around' the universe in the future. This is the big advantage of such a flat paper diagram.

The master geodesic is shown in Figure 4. We see it must end when it has wrapped around from the decoupling circle (t_1) to slightly less than 180°, t_8, from where it started. This geodesic has then reached the time circle of maximum 'size' for the universe. This size is roughly related to the circumference of the t_8 time circle. After that, the universe contracts faster and faster, as a sort of mirror image movie of the expansion, which was slower and slower, except that entropy still increases (we assume) and stars age normally as space contracts. Maximum expansion is not a time reversal event in any physical way. Only the speeding up of the contraction phase mirrors the slowing down of the original expansion. Locally, everything still ages normally.

The most difficult case to deal with is that for which the universe goes through maximum 'size' while the message is still in transit. We can easily handle this case simply by turning over the tracing paper with our master geodesic on it, when the geodesic reaches the largest circle. Concentric circles going inward then represent later and later times for smaller and smaller circles. The geodesic has a totally non-physical 'bounce' at the largest circle (angle of reflection equal the angle of incidence, 45°) and it continues inward as space contraction continues. When it reaches a particular observer point on a circle, then the time label for that circle gives the time that the message would arrive at that observers location. See Figure 5 and the Appendix.

The observers at a particular location 'moves' out radially and back in radially over time on these diagrams. Their geodesic space-time path is out and in, at 90° to the circles. Photons travel out and in at 45° to the circles. Therefore, any rocket traveling between observers at drift speed (no engine running) will take a path at some angle between 90° and 45° to the circles, depending upon its speed between 0 and c, the speed of light.

LONG DISTANCE ROCKET TRAVEL

Since the photons move at 45° to circles, any rocket might coast at a constant, larger angle to the circles. (Can these be proven?) This is also easy to draw. A rocket which moves slowly from galaxy cluster to galaxy cluster would 'move' through space-time at slightly less than 90° to the circles. Of course, the galaxy clusters themselves don't go to other places and, therefore, they show up as radial lines here at 90° to local circles. One could easily construct other master geodesics, at angles between 45° and 90° to represent rocket visits at various speeds. (More work is needed here to explore such geodesics in detail.)

All of the galaxies were 'together' at cosmic time $t = 0$ and could have set 'their' clocks to all read zero then, in principle. They later drift away, then back together, over 400 b.y. Their clocks keep synchronous, cosmic, universal, preferred time!! A rocket ship's clock, passing these galaxy cluster master clocks, reads less time passing, since its observers are moving absolutely and have accelerated up to speed in the first place. Cosmic time is the fastest passing time in the closed universe. This contradicts Einstein's original guess that all local inertial frames are 'equal'. So be it; he was wrong.

The galaxy clusters see isotropic, 3°K (Gamow) background radiation. A rocket traveling observer, moving between galaxy clusters would not. The anisotropy observed for such a rocket observer can be used to locally measure the rocket's absolute speed through the universe, without needing to refer to the passing galaxy clusters themselves! (This is as close to absolute motion as I can imagine.)

The equations for the closed universe, in advanced text books, can be used to assign numbers for the space distance scale around any given fixed time circle and the cosmic absolute time that the circle represents. We shall not go into that standard solution here, so as not to confuse the issue. The diagram is even useful for beginning physics students, once they are able to grasp Minkowski's space-time diagram.

Raine pictures motion in a closed universe using a cylinder, with geodesics on its surface and time along the axis. This is fine for a static, closed universe. The diagram described in Figure 5 is perhaps more useful for an expanding and recontracting universe. However, Raine's diagram could be generalized to one with two funnels sealed together at their large ends. Time increases along the axis and photons spiral along the surface of the funnel. (A flat paper picture like Figure 5 is easier to work with.) One could actually glue funnels together and invent geodesics on them, in some way, to perhaps even better picture closed-universe rocket-motion or signals sent left and right in a closed and dynamic universe.

If a signal were sent to our right, locally, just after decoupling (.001 b.y.), then it would travel around the universe reaching the opposite side from us just after maximum universe size is reached. During the second half of the universe's existence, where it contracts faster and faster, it would travel the rest of the way around the universe to almost reach us again at 'recoupling', or at the end of contraction. We could almost receive our own message, in principle. The diagram shows this pretty easily. In Figure 5 the signal sent from the other side of the universe at decoupling (t_1) has gone around the universe and has reached observer M near the end of the universe's life, at t_{12}. At that time the sender of this signal is at K, not far away along the circle to M's right. This distance between M and K is shrinking as the signal goes by M, and the signal almost reaches K at a smaller circle as time ends.

Recoupling (at perhaps .001 b.y. before the end of time) will not be like decoupling. Many galaxy mass lumps will have merged into huge black holes which in turn will have also merged in many cases. There will be a lot of empty space between heavy concentrations of matter until the shrinking of space jams everything together again. It will be nothing like the uniform hydrogen and helium gas at 3,000°K when decoupling took place, near the beginning.

'DOPPLER' SHIFTS

On these diagrams, photons get redder as they travel and the universe expands. They get bluer as they continue to travel and the universe contracts. Thus, a signal sent early and to the left will travel around the universe, getting redder until maximum expansion is reached, then get bluer again as the universe contracts while it continues to travel. When it arrives at a galaxy to our right near the end of time, it will be redder or bluer than the color we sent in the first place.

The color is simply related to the 'radius' value, R, of the circle for the sending circle and the receiving circle. See Appendix. The frequency ratio is proportional to the 'radius' ratio of these two circles. This clearly shows that the shift is NOT a Doppler shift at all, but rather a 'stretching' shift. We cheat in teaching astronomy—we use the flat space space, Doppler formula for receding galaxies.

APPENDIX

If one wants quantitative numbers for the diagram, some can be obtained: the angle between two spokes can be measured with protractor and converted to radians. Call it α. A photon leaving from any 'galaxy' at cosmic time 'zero' (or time 0.0003 b.y. is close enough—decoupling time) eventually reaches another spoke, at angle α from the original spoke. (A photon swings through angle $\alpha = \pi$ from Bang to maximum expansion.) This photon arrives at some other galaxy at angle α, at a cosmic time t, when the universe has a scale 'size' R, and when the target galaxy is now a distance d from the original ($t = 0$) signal source 'galaxy's' current location, measured along the 'present' time circle arc. This target galaxy is now moving away from that signal source 'galaxy' with a theoretical speed v, and Hubble's constant is now H. All of these are given by the angle α and 63.4 b.y. (a constant related to the total mass of the universe). The formulas are:

$$t = (63.4\ b.y.)\ (\alpha - \sin \alpha)$$

$$R = (63.4\ b.l.y.)\ (1 - \cos \alpha)$$

$$d = R\alpha$$

$$v = Hd$$

$$H = [\sin \alpha]/[63.4\ b.y.(1 - \cos \alpha)^2]$$

One can check that when $\alpha = \pi$ radians, we have $t_\pi = 200$ b.y., $R_\pi = 127$ b.l.y., $d_\pi = \pi R_\pi\ b.l.y.$ (the maximum distance to the 'other side' of the world), and $v_\pi = 0$, (all galaxies have stopped because $H_\pi = 0$). During recontraction, the pattern of numbers is the same as for expansion, only 'reversed'. (Slow at first, then rapid as the end nears.) Once again the Hubble constant, H, grows rapidly in size and, near the end, most of the universe once more rushes in at greater than the speed of light toward us (or toward anyone else). The angle between spokes on a circle, and the value of R, for that time circle, together give us a distance scale, d, between galaxies for any particular time circle at time t: $d_\theta = R\theta$, where θ is the angle at the center between spokes passing through these galaxies, measured in radians.

We are using the decoupling circle, in the diagram, to start our photon 'worldlines', but actually we should start at the center of the circle in measuring α. The α value at the decoupling circle is 0.031 radians, 1.8°. You can see that neglecting this produces 'negligible' error in measuring α, but this 0.031 value does give the distance scale for the decoupling time circle, if desired: R = 0.03 b.l.y. then, and the other side of the world was at only 0.1 b.l.y. distance from us; i.e., only about 1000× as far as across our Milky Way galaxy today!

The theoretical recession speed of any galaxy, moving away from us at fixed t, can be determined from d and H for that particular time circle at time t, using Hubble's law, $v = Hd$. At present, v = c at $\theta = 34°$ from us, this is 20 b.l.y. away, along the present circle. At decoupling, the θ at which v = c was only $\theta = 0.9°$, so most of the universe was then receding away from us at 'theoretical' speeds greater than c, if we blindly trust Hubble's law to define distant object speed. Yet we routinely receive this background radiation, from such 'superluminal' regions, that reaches us 'yesterday, today, and tomorrow'. (Such is the unintuitive nature of curved space.) Part of the 'snow' on your TV set, when a channel is off the air, is just such radiation! It is the most ancient of light, about 12 billion years old, which left its creation location at (34°/180°) toward the other side of the universe from us.

At maximum size, the opposite side of the world is almost where the decoupling radiation originated that will be reaching us then. After that, the background radiation which reaches us will have traveled more than halfway around the universe to finally reach us, if reaching us during the contraction phase. But no one will be around then to be measuring it, of course. All humanoid civilizations die out when their stars do, I think.

ACKNOWLEDGEMENT

I am very grateful to Bill Suydam for generating the computer drawn figures.

REFERENCE

1. D. J. Raine, 1981, *The Isotropic Universe*, (Bristol: Adam Hilger).

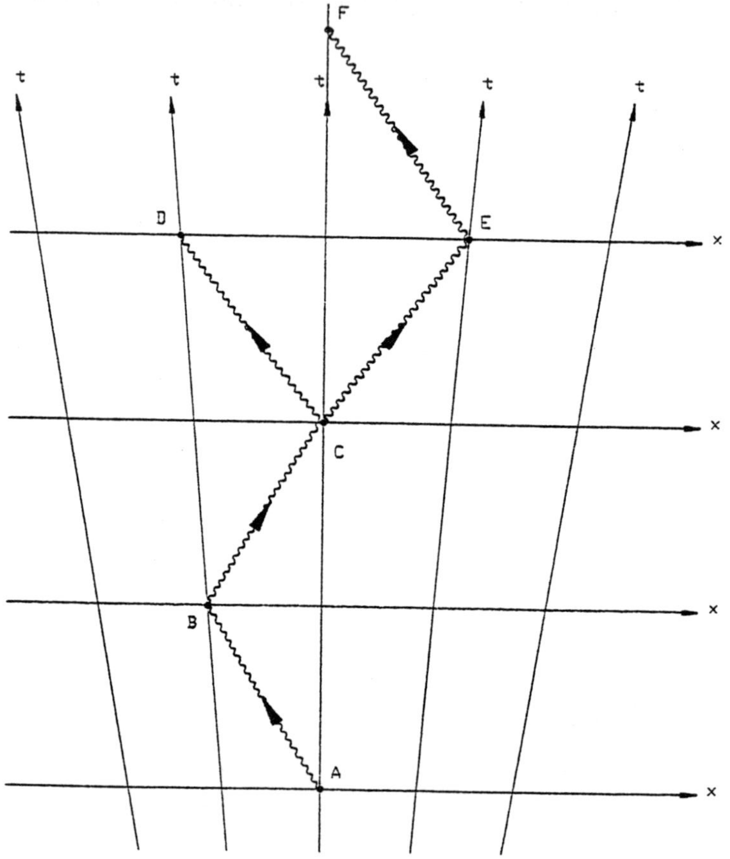

Figure 1 Minkowski diagram distorted somewhat to indicate that the universe is stretching with time, and is curved.

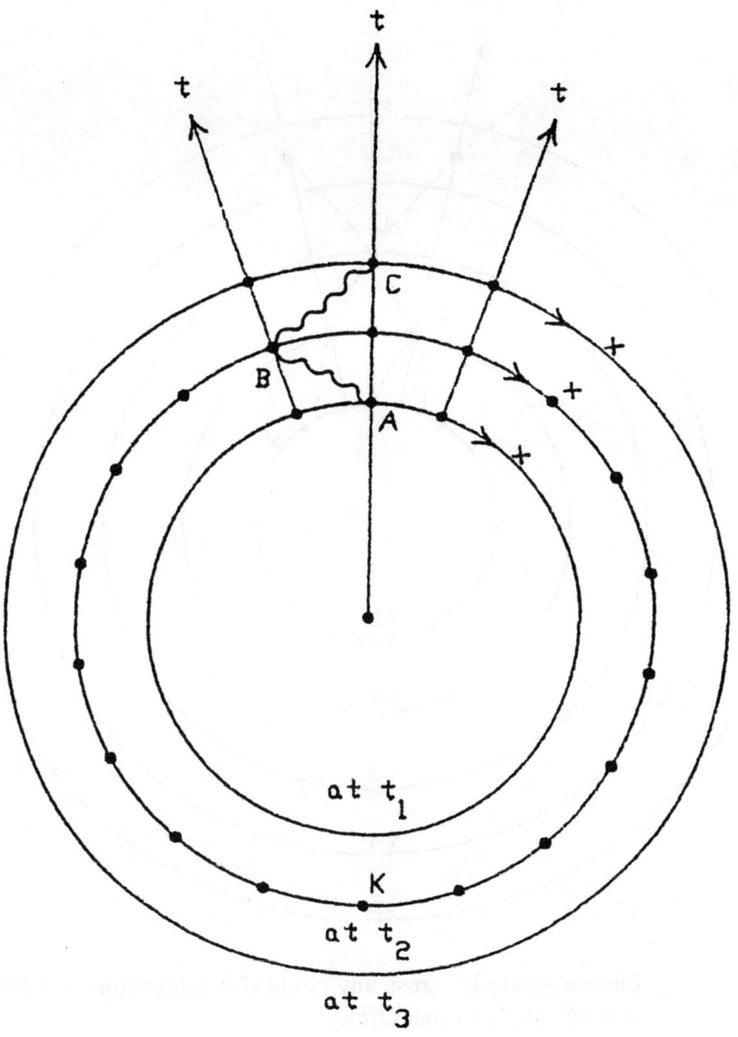

Figure 2 All observers to our left and to our right in a closed universe at three different times.

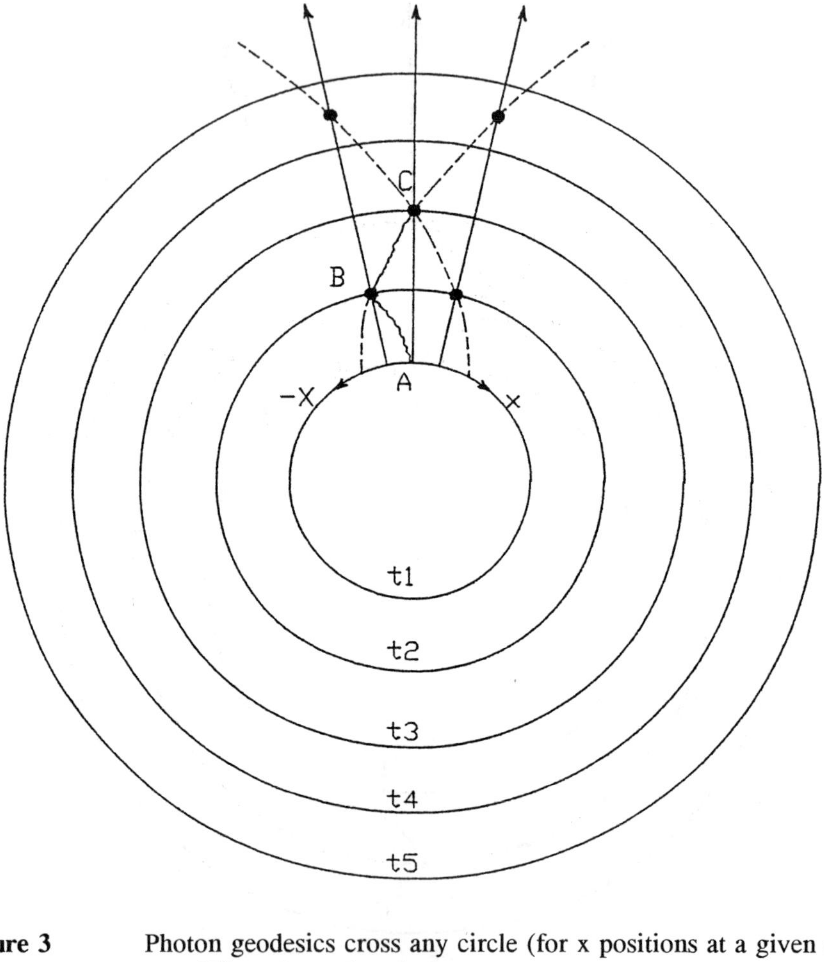

Figure 3 Photon geodesics cross any circle (for x positions at a given t) at a 45° angle to that circle.

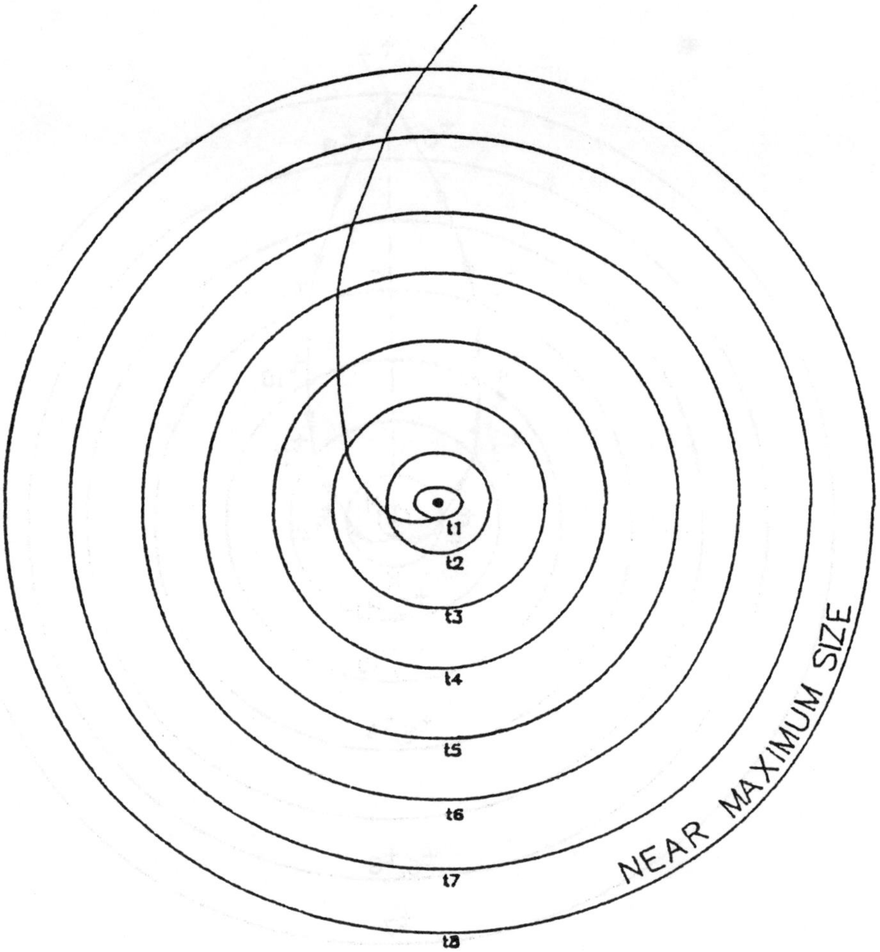

Figure 4 The master geodesic at 45° to local circles that it passes — starting at decoupling and ending in approximately the opposite direction from the starting point at maximum expansion.

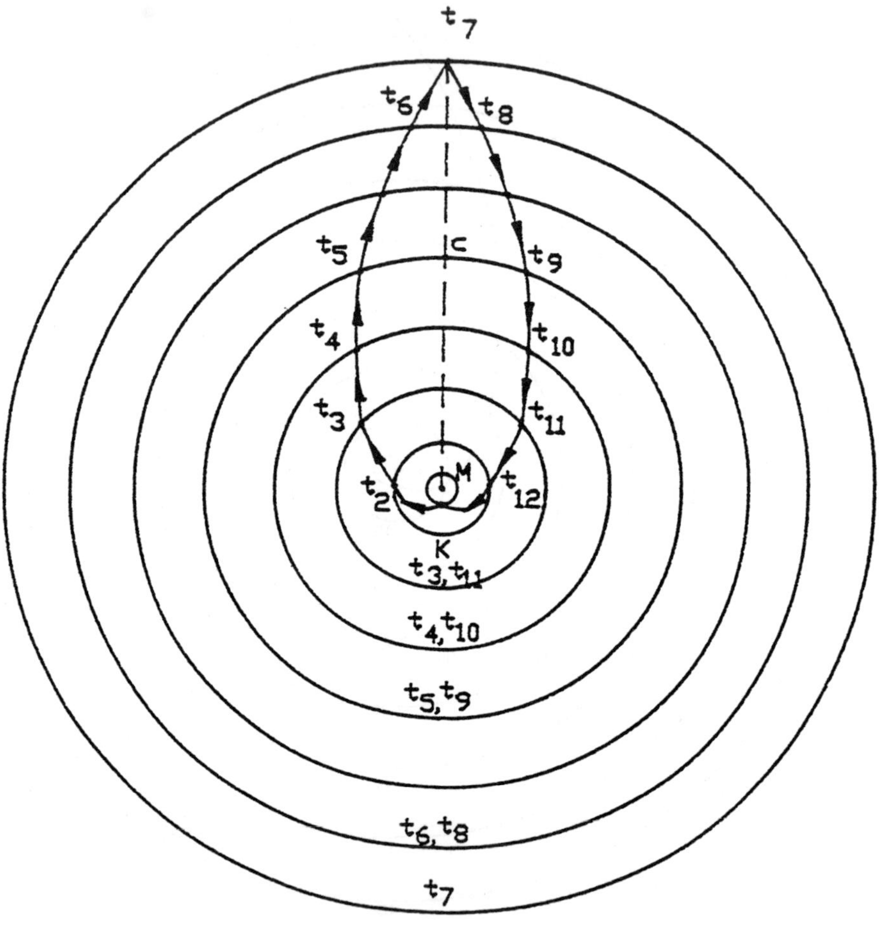

Figure 5 The message sent at decoupling t_1 from the 'bottom' of the universe (from us) reached us at t_7 near maximum size, and it goes on by us to reach observers to our right, M, at t_{12}, who are about 1/4 of the way around the universe to our right and about 1/4 of the way around from K.

NOTE ADDED IN PROOF:

I submitted four papers on this work in the summer and fall of 1996. One was rejected by *Journal of Physics A*, with a nasty response to my appeal to the editorial board: "... Those (few) parts ...not demonstrably unsound are clumsily reintroducing the chiral ... in degenerate and ambiguous notation, ... The turgid exercise displayed here is underlied by a layer of flighty Borgesian vagueness of intention. I see no "miracle" nor how Einstein's metaphysical reasons for the Lorentz group' (?) are obviated. ... give up on his evidently futile search for validation by critical colleagues." Wow! Let this be a lesson to any young person trying to really change the foundations of our physics. You have to be very stubborn and expect this kind of reaction. Einstein's teachers would not recommend him for a teaching position. One was supposed to have said something like, "Your trouble is that no one can teach you anything." He did not wind up in the patent office by accident. You must find such an office too if you want to be able to work with real independence of thought. Charles Ives sold insurance so he could write his wild music.. You must forgive them for they know not what they do.

One paper was rejected twice and finally accepted by *Eur. J. of Physics*, where the reviewer was conservative but at least open to the possibility that we do not know why Lorentz symmetry fits the real world. Einstein could have guessed wrong. One paper was rejected by *Physics Education* as too advanced a topic for beginning students, even though we do teach the Einstëin paradigm to such students all the time. A fourth, with new ideas on extending the material in this book to 6-vectors from the 5-vectors in Dirac physics, was rejected by the editor of *Phys. Rev. Letters* with no suggestion that it be moved to *Phys. Rev. D*. He essentially tried to say that it has no connection to current fashionable physics that is ongoing and fills the pages of these journals. He said, "We are not declining publication merely because the work is speculative. Speculative work may be suitable for publication in *Physical Review Letters* if it has sufficient motivation, and there is sufficient exploration of the ramifications of the speculation. However, connection with current physics research is essential for publication in this journal." Thus, obviously Dirac's original paper on the Dirac equation would have been rejected since it was too original and isolated in content from the past. *Physical Review* has the dubious distinction of having rejected Feynman's original paper on his new diagram methods. This is a sad comment on the social framework of physics theory, but Newton had the same complaints in his time and nothing has changed. New ideas have to be fought for, not just laid out for others to take up and explore.

Following is an excerpt from the *Phys. Rev.* paper, which they received on 12 September 1996, so you can see the beautiful new idea of how to extend our knowledge and find new ideas about mass.

Dirac physics is based essentially on having the 5-vector seed set contain momentum operator parts and the rest mass part as the 5th dimension. Somehow, this last part is only a constant and invariant. Originally,[1] Dirac guessed a linearization of the form $P_I \equiv P^0(e_0) + P^k(\alpha_k) + mc(\alpha_4)$. Then $P_I P_I$ will have surviving crossterms. He instead used $[-P^0(e_0) + \ldots][+P^0(e_0) + \ldots]$ to get the K-G equation, with no crossterms now surviving. He did not comment on why this is allowed or reasonable. (It just worked.) He also thought at the time that his new α elements were dynamical variables of some sort. By extending into this new physics from the (lost) Hamilton quaternions, his generation would have seen that the α's are only 'place holders' just like *i*. We can write a momentum 5-vector as $P_C \equiv \partial^\mu(-if_\mu) + mi(ie_0) = -P_C{}^\wedge$. If we multiply from the left by $-i(ie_0)$, we get the standard Dirac form $P_D = \partial^\mu i(f_\mu) - m(e_0) = P_D{}^\vee$, where (e_0) is the identity element, and $(f_\mu) \equiv \gamma_\mu$. Also, $\gamma_5 \equiv -i\gamma_0\gamma_1\gamma_2\gamma_3 = i(ie_0)$. The dot product for P_D takes the form $P_D{}^\wedge P_D$, which is equivalent to $P_C P_C$. All crossterms cancel out and, since $P_D = P_D{}^\vee$, we have its 'Lorentz' transformation in the *natural* form: P_D goes to $P_D' = L^\vee PL$. Then we also have $L^\vee m(e_0)L$ in P_D. We obviously need $L^\vee L = 1(e_0)$ for *invariant* mass. For the 5-vector dot product to be invariant, we also need $L^\wedge L = 1$. Proof: $(L^\vee PL)^\wedge(L^\vee PL) = L^\wedge P^\wedge(LL^\wedge)^\vee PL = L^\wedge P^\wedge PL = PP^\wedge L^\wedge L = P^\wedge P = (P^{\mu *}P_\mu + m^2)(e_0)$. Thus L is a group with $L^\wedge L \equiv 1$ *and* $L^\wedge = L^\vee$. This can be shown to be an 8 parameter sub-group, of the 16 parameter group $LL^\wedge = 1$, which includes the 6 parameters of the Lorentz sub-sub-group. This 8 parameter group should then *replace* the Lorentz group for form covariance, from our perspective here that the *natural* groups should come directly from the Dirac algebra *itself*, rather than from 'philosophical principles' about what moving humanoid observers should or should not see.

This Dirac system is constructed from a series of generalizations, starting way back with the quaternions (Hamilton, 1843). Quaternions are the closed sub-system $\{(e_0), -(ie_k)\}$, where the e's are like the Pauli matrices, in their multiplication table, but the i's all remain *only inside* the parentheses here, as part of the *name* of a basis element. Only minus signs are allowed to move in and out freely. This too just works. We see elements like $i(f_2)$, in Dirac, and it is very natural to then replace all of these imaginary i's *outside* by the three *quaternion* basis elements in $\{I, J, K\}$, with $II \equiv JJ \equiv KK \equiv -1$ *and* $IJ \equiv K \equiv -JI$, and cyclic. This may have already been done but it is not mentioned in Marshak's recent summary book on conventional field theory. The usual Dirac algebra can be thought of as the complexification of the C_ℓ algebra so why not the *quaternification* of it instead? We have no reliable idea as to why Dirac is the algebra of the quantum world, so we must remain open to generalizations of it. The octonions even generalize the quaternions and they play a central role in the exceptional Lie groups that are being explored so vigorously now in string theory. Dixon[2] says that octonion thinking leads only to 4, 6, or 10 dimensions. We are seeing here that the usual Dirac algebra is essentially a 5-vector system. This may indicate that it really does need generalization to 6 or 10 dimensions.

We simply replace each i *outside* the ( ) above with the three alternatives I, J, and K. This doubles Dirac to 64 basis elements. It turns out that this produces a seed set now that is $\{(if_\mu), I(ie_0), J(ie_0)\}$. Thus the momentum operator is generalized to a 6-vector, $P_C = P^\mu(-if_\mu) + P^4 I(ie_0) + P^5 J(ie_0)$, now with *two* 'mass' parts! The covariance of these mass parts, under L, has to be carefully examined. Now, $LL^\wedge = 1$ could have 28 parameters instead of 6, 8, or 16 and its sub-group $L^\wedge = L^\vee$ could then have 12 parameters in the *extended* algebra. It might replace the 8 parameter covariance group, which should in turn replace the Lorentz group—but see below. The square of the composite Clifford operator, $P_C P_C$, will still not give crossterms since all of the elements still anticommute. We had before that i( ) ≡ ( )i, so we naturally define I( ) ≡ ( )I, etc. Though I and J do anticommute, they may not commute or anticommute with L, *in general*! The squares of the individual seed set elements now give the natural metric (+ − − − + +). These two mass parts combine into only one mass number in the Klein-Gordon equation. We would not suspect that such a multi-mass thing is likely from our old *classical* and *solid state* experience with nature. To reach them, all of the equations come from approximating the Klein-Gordon equation. I have previously proposed[4] that $P_C\psi = 0$, in ordinary Dirac algebra, could be replaced by $P_C\psi = \psi M$. In *D*, the K-G equation will have MM, and it reduces itself to five new mass parts squared that can combine with the one old mass part squared in P_C to give back one composite mass number again. In this proposed quaternion generalized algebra, the M can now be a Clifford seed set 'number' with 6 parts instead of 5 and we could have a total of 2 + 6 = 8 mass parameters for a single particle that obeys the generalized Dirac equation. (Alternatively, there are possibly up to 8 distinct spin 1/2 equations with a single mass term in each.) This $P_{C6}\psi = \psi M$ equation is *form* invariant under a 28 parameter group and at least 6 of the 8 mass parts are invariant under this full group. These 6-vectors will also fit better with octonion thinking. Of course, this is very speculative but it is in a long standing tradition, for physics, of the kind that has worked before. It is certainly not more speculative than string theory. It is probably less so, and thus just as publishable! For strings, one finds huge internal symmetry groups to try to mesh with the spacetime group. The spacetime group, we see here, is very possibly a much larger group as well, and this may have some favorable effects on efforts to mesh the two. (The no-go theorems will perhaps need generalization, for example.[3])

Notice $-I(ie_0)(\{(-if_\mu), I(ie_0), J(ie_0)\}) \leftrightarrow \{I(f_\mu), -(e_0), K(e_0)\}$. Thus, if *I* replaces *i* in old Dirac, then the $K(e_0)$ mass part here must be neglected to get the *usual* Dirac equation. Thus, this K mass must be small or zero for the QED approximation of nature. It may still be significant in hadrons, however. If L is from a large enough group, then this new K term would not be invariant. Note, however, that IP_C and JP_C do *not* give a new $P = \pm P^{conj}$. We must instead choose KP_C, when P_C contains the $I(ie_0)$ and $J(ie_0)$ mass parts. We find $-K(ie_0)P_{C6} \equiv P_{D6} = -P_{D6}{}^{\wedge}$.

Since the paper to *Eur. J. Phys.* was first rejected twice and the editor decided any new offering would need to be considered as a new paper, I included the following on the new idea of extending Dirac algebra in my October 1996 resubmission under the title *"Throwing sawdust into space: it will slow and stop so why is nature Lorentz covariant?"*

The hypercomplex number system for the real world seems to be the Dirac system or even an enlargement of this system. The C_ℓ algebra is $R\otimes Q^2$, the *D* algebra is $C\otimes Q^2$, and I venture to say that sub-proton physics will utilize $Q\otimes Q^2$, with 64 basis elements. Here, Q^2 means 2×2 matrices with quaternion elements. We see a natural pattern of evolving complexity here. The deeper we go into nature the more complicated things get. The basic equations should be written in this hypercomplex language, at least initially, when they are being guessed at. Look at this system as an extension of the complex number system: $a(e_0) + b(-ie_1)$, where there is one conjugation called complex conjugation, $(\ldots)^*$. This changes the sign of the second term. We have a rule for multiplying (e_0) and (ie_1), and $(-ie_1)(-ie_1) = -(e_0)$. Hamilton extended this system, after 10 years of struggle, to $a(e_0) + b(-ie_1) + c(-ie_2) + d(-ie_3)$, the quaternions, so he could deal realistically with the three dimensions of physical space, (x, y, z). The detailed multiplication rules for (e_μ) or (if_μ)

need not bother us here. See Appendix. The complex conjugation in complex numbers then needed extension as well in quaternions, and what worked best was to invent $(\ldots)^{\dagger}$, now called hermitian conjugation. By definition, it changed the sign of any i. This $\{(e_0), (-ie_k), k = 1,2,3\}$ sub-system is his original quaternion system, Q. It is a field but it is not large enough to do physics. We need zero divisors in relativistic physics, equivalent to indefinite metrics in matrices, and so we have to double the quaternions by inventing four new elements: $\{(ie_0), (e_1), (e_2), (e_3)\}$, with the rule $(e_1)(e_2) = (ie_3)$, *etc.* Now there is also another conjugation that is equally useful, $(\ldots)^\wedge$. It is called the quaternion conjugation and it is defined as follows:

$$(e_0)^\wedge = (e_0),\ (e_k)^\wedge = -(e_k),\ (ie_o)^\wedge = (ie_0),\ (ie_k)^\wedge = -(ie_k)$$

$$[i(e_0)]^\wedge = -i(e_0),\ [i(e_k)]^\wedge = i(e_k),\ [i(ie_0)]^\wedge = -i(ie_0)$$

$$[i(ie_k)]^\wedge = i(ie_k),\ k = 1,2,3$$

We see that $(\ldots)^\wedge$ changes the i *outside* to -i and e_k inside to $-e_k$, regardless of there being an i inside or not, and this is very important. (The [...] in equation 1 is not a special symbol here.) We will have things like $A = 3(e_0) + 2(ie_3) + 7i(ie_2)$ and $A^\wedge = 3(e_0) - 2(ie_3) - 7i(-ie_2) = 3(e_0) - 2(ie_3) + 7i(ie_2)$. We could use $7I(ie_2)$ instead to avoid the temptation to replace $i(ie_2)$ by $-(e_2)$, which is not correct. The $Q\otimes Q^2$ number system is in fact equivalent to extending this Dirac algebra by having the i's outside be replaced by {I,J,K} where these are the quaternion elements with $II = JJ = KK = -1$ and $IJ = K = -JI$, *etc.* This would double the Dirac algebra in a rather natural way. It gives 6-vectors instead of 5-vectors that are natural to Dirac algebra: $P_C = P^\mu(-if_\mu) + P^4 I(ie_0) + P^5 J(ie_0)$. This becomes ordinary Dirac if P^5 is negligible, P^4 is mc, and I is called i, but J and K are left out of the algebra. This would then be the QED approximation of nature. In Clifford algebra language, Dirac is an $R_{2,3}$ system and this extension is an $R_{3,3}$ system with metric $(+ - - - + +)$. We see a second possible mass appearing here for sub-proton particles and such.

So far, no one seems to be excited by the possibility here that there are several parts to mass at the subproton level. We don't see this at our level because the Klein-Gordon equation finds these all blend into one mass term. I am shocked that reviewers are not jumping up and down. Heaven knows, we are in desperate need of new ideas to continue to make any progress in subproton physics. This may not be the key, but it could be and that is enough to make us very excited about it. The math throws this at us as a possibility. We also see that the new mass parts come in two varieties. In P there is room for two very different mass parts and in M there is room for 6 more mass pieces. Since we know nothing about mass really, we should be excited when someone finds anything about it that may be exotic and yet still fit within the basic covariance requirements. Of course many new ideas will fail to fit reality and this may be just another one. Considering the importance of such a thing if it is real for our world, reviewers and editors must be shamed for their lack of support for its being published and circulated widely. This could clearly be the greatest advance in physics since the original Dirac equation was proposed. Time will tell if multi-mass is in our world or only a possibility that got left out in the actual design we have.

The associate editor at *Phys. Rev.* did show an interest in the question of what two extra parameters might mean in extending the Lorentz group. This is the kind of interplay I thought should follow from publication of this radical perspective in one of the establishments main journals. I still have hopes that the paper might make it into *Phys. Rev. D* if expanded so that the radical notation can be appreciated by traditionally trained theorists. This notation barrier is a severe one. Experts don't want to have to go back to school, and again that is why I have aimed this book at younger readers in the field.

Index

S